BIOMIMETICS IN ARCHITECTURE
ARCHITECTURE OF LIFE AND BUILDINGS

PETRA GRUBER

SpringerWienNewYork

Petra Gruber
Institute for History of Architecture and Arts, Building Research and Preservation
Vienna Institute of Technology
Vienna, Austria

Printed with financial support of Bundesministerium für Wissenschaft und Forschung, Vienna, Austria

This work is subject to copyright.
All rights are reserved, whether the whole or part of the material is concerned, specifically those of translation, reprinting, re-use of illustrations, broadcasting, reproduction by photocopying machines or similar means, and storage in data banks.

Product Liability: The publisher can give no guarantee for all the information contained in this book. This does also refer to information about drug dosage and application thereof. In every individual case the respective user must check its accuracy by consulting other pharmaceutical literature.

The use of registered names, trademarks, etc. in this publication does not imply, even in the absence of a specific statement, that such names are exempt from the relevant protective laws and regulations and therefore free for general use.

© 2011 Springer-Verlag/Wien
Printed in Germany

SpringerWienNewYork is part of
Springer Science+Business Media
springer.at

Cover Illustrations: Petra Gruber
Copy editing: Petra Gruber, Jo Lakeland
Image editing: Roman Bönsch
Typesetting: Petra Gruber
Printing: Strauss GmbH, Mörlenbach, Germany

Printed on acid-free and chlorine-free bleached paper
SPIN: 12594559

With 466 Figures

Library of Congress Control Number: 2010938940

ISBN 978-3-7091-0331-9 SpringerWienNewYork

CONTENT

1	**Introduction**	**7**
2	**Background**	**9**
	2.1 Architecture	9
	2.2 Bionics [Bionik] Biomimetics	13
	2.3 Transfer and methods	41
3	**Classical approaches to investigate overlaps between biology and architecture**	**50**
	3.1 Relationship between nature and architecture	51
	3.2 "Natural construction"	54
	3.3 Nature's design principles	96
	3.4 Parallels, differences and synergies between design in nature and in architecture	108
	3.5 Biomimetics in construction and architecture	109
4	**New approaches and application of biology's life criteria on architecture**	**110**
	4.1 Life, biology	110
	4.2 Architectural interpretation of life criteria	124
	4.3 Comments and hitherto unexplored fields	191
	4.4 A living architecture	194
5	**Case studies**	**196**
	5.1 Adaptation and evolution of traditional architecture on Nias Island	196
	5.2 Transformation Architecture	243
	5.3 Lunar Exploration Architecture	247
	5.4 Biomimetic Design Proposals	254
6	**Discussion**	**262**
	6.1 Transfer strategies and methods	262
	6.2 Suggestions	263
7	**Appendix**	**264**
	7.1 Literature	264
	7.2 Figures and Photography	270

1 INTRODUCTION

The aim of the project biomimetics in architecture - architecture of life and buildings - is innovation in architecture. The purpose of investigating the areas common to architecture and biology is not to draw borders or make further distinctions, or even to declare architecture a living organism, but to clarify what is currently happening in the overlapping fields. The accumulation of knowledge of individual examples is less important than the investigation of the methodology of translating knowledge gained from nature into technical solutions. The objective is to employ biomimetics as a tool in architectural design. The fields in architecture where this is applicable and necessary are diverse. Innovation will help to solve current problems in architecture and environment, and new fields of architecture and design will be explored, e.g. space design. The strategic comparison with biological paradigms will help identify areas for innovation. Best of all, biomimetics in architecture will help develop a culture of active environmental design.

This book entitled "Biomimetics in Architecture - Architecture of Life and Buildings" (Architekturbionik - die Architektur von Bauwerken und Lebewesen) gives a broad overview of overlapping areas in the fields of biology and architecture, investigating the field of what is called biomimetics in architecture (Architekturbionik).

A comprehensive comparative study of these overlapping areas has not yet been carried out. Numerous people have already delivered contributions to the connections between architecture and biology. Many such approaches have provided successful architectural developments.

Werner Nachtigall has compiled a vast collection of examples and made a heroic attempt to order the field of biomimetics as a whole. In his works "Baubionik"[1] and "Vorbild Natur, Bionik-Design für funktionelles Gestalten"[2] he concentrates on issues around design and building. Frei Otto and his group have tried to give architects and engineers a view on what he called "Natürliche Konstuktionen"[3] and developed an experimental approach to natural design. Otto Patzelt has compared growing and building in "Wachsen und Bauen"[4].

The Russian Juri S. Lebedew in the 1960s wrote the only comprehensive work done so far on "Architekturbionik"[5]. Recent developments in biomimetics in Germany and the UK occasionally touch architecture, but no comprehensive effort is being made.

The new approach carried out here transfers the biological characteristics of life onto the built environment and thus architecture.

In order to make the topic accessible to architects the basics of life sciences are presented, which cannot be omitted when dealing with nature and natural role models for design. An overview of the present state of research in the relatively young scientific field of biomimetics shows the potential of the approach.

Methods used for this investigation are diverse. Literature research, conference organisation and participation, and expert interviews were carried out.

Data for the case study about traditional architecture in Nias was collected in two field trips, with extensive architectural documentation and narrative interviews. The processing of the data was done with architectural and engineering tools. Students design projects in different intensities delivered examples for the biomimetic approach in architecture. Another case study in space architecture based on literature research and design studies in workshops was also carried out.

The basic assumption of the research is that the study of the overlapping fields of biology and architecture will show innovative potential for architectural solutions.

The important questions are: Can the combination of the biological characteristics of life and the built environment offer new solutions for more appropriate, more sustainable architectural designs? Can the new approach - searching for life's criteria in architecture - provide a new view of architectural achievements and make visible innovative potential that has not yet been exploited?

---

1 Nachtigall, W.: Bau-Bionik, 2005
2 Nachtigall, W.: Vorbild Natur, 1997
3 Otto, F. et al.: Natürliche Konstruktionen, 1985
4 Patzelt, O.: Wachsen und Bauen, 1974
5 Lebedew, J.S.: Architektur und Bionik, 1983

Why is it important to deal with architecture and biology?

- **Growth of cultural landscape**

At the beginning of 2007, 6.6 billon people are living on earth.[6] Almost all of them live in cultural landscapes. We have managed to transform the natural environment to fit our needs in many respects. About half of the world's population lives in densely populated urban areas. Built environment has replaced the former natural environment as man's "normal" surrounding.

Therefore the design of the built environment is becoming more and more important. Qualities that in former times could be found in nature have to be introduced into the artificial, cultural, or social, environment, in order to maintain quality of life and biodiversity.[7]

Architecture is mainly concerned with the built part of our environment, but must also refer to spatial planning on a larger scale.

- **Environmental concern**

Together with the growth of the world's population, the rapid development in technology and economy creates an enormous impact on the environment as a whole. The building industry's, and thus architecture's, share in the developments which are listed below are considerable:
- Natural land loss and irreversible destruction of biodiversity
- Exploitation of raw material
- Extensive use of energy
- Production of waste
- Emissions into soil, water and air

Architecture has to adapt to environmental changes. If pollution continues, architecture will in the future have to provide shelter from a potentially hazardous future environment.

The investigation of biology and natural processes makes architects aware of the ongoing processes and the influence they can achieve.

- **Technology**

Technological progress provides different means of planning and building, and has opened up opportunities, which allow a more generous interpretation of architecture in terms of functionality and mediation between humans and their environment. Current developments show an increase of life's criteria being implemented into architectural projects.

There seems to be a transition in architecture from providing unsophisticated shelter to a smart third skin for humans.

- **Sick building syndrome - sick environment syndrome**

The discovery of the "Sick Building Syndrome" has led to intensive research and development in the sector of building automation and technology integration.[8] The "Sick Environment Syndrome" has not yet been defined as a cause of illness, but their own concern will hopefully lead humans to pay more attention to ecology in the future.

- **Other future environments**

Both the expansion of civilisation on Earth and space technology have introduced new environments already dealt with by architecture. Until recently only functionality and materials were considered when designing for extremely hostile environments.

Once long duration missions are undertaken human factors have to be considered and these need to be answered by architectural means The different nature of space environments requires innovative architectural approaches, which will then influence architecture on earth.

- **Innovation**

The only way to solve some of our building-related problems is through innovative solutions. Role models taken from nature, which have developed over many years can enhance innovation.

- **Criteria of life**

Life has been introduced into architecture discussion, but life itself evades precise definition. Criteria of life attempt to provide a definition of life that covers the whole of contemporary understanding of life in its many manifestations. Offering all possible starting points for comparison and transfer, the paradigm of life's criteria is perfect for investigation of overlaps between biology and architecture.

- **Transfer of biological criteria of life to architecture - spin off**

The transfer of biological criteria of life to architecture requires discussion of areas, where biology and architecture actually meet. Instead of staying in the centre of each discipline, the boundaries have to be explored - in order to find that these fields are not as distinct as they seem to be.

The examination of the overlapping fields will hopefully result in further mutual understanding, convergence of disciplines and common action.

---

6 Millennium Ecosystem Assessment: Ecosystems and Human Wellbeing, 2005, http://www.maweb.org/en/Condition.aspx#download [03/2007]

7 Turner, W.R. et al.: Global Urbainzation and the Separation of Humans from Nature, BioScience, 2004

8 Daniels, K.: Technologie des ökologischen Bauens, 1999, p.41

2 BACKGROUND

For this interdisciplinary investigation it is necessary to define and describe the fields of both architecture and biology to provide the essential background.

2.1 ARCHITECTURE

"[Architecture is] the art or practice of designing and constructing buildings...
The complex or carefully designed structure of something"[9]

The second definition taken from the New Oxford American Dictionary already characterises the "nature" of architecture. The discipline that puts material or immaterial things in order is called architecture in many fields, not only in the classic architecture of building. There is a basic architecture in the design of life and organisms, and even in information technology we use the word architecture when describing the basic layout of computer programs.

What is architecture? What is not architecture? About projects, constructions and structures

Architecture is interpreted here as a widespread profession engaged in the design of the built environment. It includes design on all levels of scale, from urban and regional planning to small building projects. It is not exclusively referred to as "proper" architecture, which is designed by architects, but as a general term standing for the material structure that defines space and enables interaction.[10] Architecture contains life. As Kaas Oosterhuis says, *"architecture becomes the discipline of building transactions"*[11]: it is about to move beyond containing activity, taking an active role, not only influencing but interacting with living systems.

Making architecture is about making projects. "Project" has a wider meaning than a common intention or plan to do something. In architecture, projects are building tasks, which may already have been completed, or are still in "project phase" existing on paper (or encoded on hard disks), waiting for realisation, or having been already abandoned. The term does not refer to a specific size or scale. As innovation is the focus in this discussion, we will also take into account unbuilt projects, if they are important to illustrate developments in architecture. Some unbuilt projects became very famous in architecture history as exemplary designs. Being unbuilt, and often described only roughly, these projects on one hand still provide space for vision, and on the other hand the basic idea is not yet spoiled and watered down by the needs of execution. Frederic Kiesler's "Endless house", for example, was never built, but served as a kind of asymptote for many other attempts at organic space. Buckminster Fuller's visualisation of a transparent dome over Manhattan is tempting for anyone dealing with lightness in design, but still impossible to realise. The exactitude of expression is difficult to maintain in the process of application and execution, but this is also one of the big challenges in architecture.

Most interesting in the context of biological paradigms for architecture are projects which show a strong interrelation between form, function and structure or construction, so load bearing is a key function and will therefore be focused on.

"Construction" and "structure" are commonly used for elements of architectural projects which have to fulfil tasks of load bearing. The differentiation between construction and structure is somehow connected to that of structure and material. Following Jim Gordon: *"Structures are made from materials and we shall talk about structures and materials; but in fact there is no clear-cut dividing line between a material and a structure."*[12] In common use, a construction is something which has to be put together, typically a large element. The term structure is used in a more abstract notion, when we are talking about load bearing for example, but can as well be used for any important ordering element (which could be abstract), or a discernable pattern, even surface patterning.[13]

---

9 New Oxford American Dictionary, program version 1.0.1, 2005

10 Hillier B.: In his essay "Specifically Architectural Theory: a Partial Account of the Ascent from Building as Cultural Transmission to Architecture as Theoretical Concretion" Hillier defines "architecture" in contrast to "building" as a design process requiring purposeful and thoughtful innovative emphasis, other than building by what he calls "culturally bound competences". Eventually Hillier concedes innovative potential to development of and in traditional architecture. However, in general this differentiation is difficult to maintain, as the involvement of people as creative potential in building processes always brings a chance for innovation, whether systematic intent or practical understanding is underlying the new solution. http://eprints.ucl.ac.uk/archive/00001027/01/hillier_1993-specifically_architectural.pdf [11/2007]

11 Oosterhuis, K.: Hyperbodies, 2003, p.6

12 Gordon, J.E.: Structures, or why Things don't Fall down, 1981, p.29

13 The German terms "Konstruktion" and "Struktur" are not clearly differentiated, either. "Konstruktion" is used for a large-scale load bearing system, e.g. the steel construction for the roof of a stadium, but speaking in an abstract sense, "Struktur" can also be used. In dealing with the works of Jim Gordon and Werner Nachtigall, there seems to be a better understanding in translating "Konstruktion" with "structure" and "Struktur" with "material".

2.1.1 Which architectures are important in this context?

When focusing on development and progress, we have to think about tradition and technology as well as innovation and experiment. The architectural examples which will be used to illustrate the theoretical framework cover the gap between traditional building typologies that have developed over a long time and new designs that contain innovation of some kind. Many projects have come to be classic examples for their time or the technology they represent.

So when considering an imaginary scale of innovation, the two extremes are interesting: the typologies, where innovation has almost come to an end in a long optimisation process, and the projects, advancing development and innovation. Case studies performed by the author or done under her guidance will be used to explain specific aspects of a biomimetic approach to architecture in detail.

2.1.2 Categories of architecture

Only a small part of all built environment is designed by architects. Unfortunately, architects tend to restrict themselves to "proper architecture", which confines their influence.

There are many ways of categorising architecture in order to get a general idea of this huge field. We could use the scale of the project, the function of the building, the quality of the building, the region where we find it, the tradition or singularity, the construction method, the material used, the style, the date of building, and so on. All these categories overlap, and it is difficult to count or even estimate the number in each category. In the course of this discussion, "scale" will often be used for categorisation. Difference in scale implies different boundary conditions and thus different planning and development processes. For this reason, the categories "urban design", "building", "process" and "material" will be used for a comparison with criteria of life, and will then be explained in more detail. Other categories will be applied when needed.

The most important aspect for categorisation when asking about innovation and progress is the innovativeness implemented in a project. Quantification is not possible within the frame of this book. The categories' qualities are described and examples are mentioned. Figure 1 (opposite) shows architectural categories referring to a specific scale of innovation.

"Architecture of provision" is the lowest possible stage and is not identified in the scheme. Shelters are made of whatever can be used and no formal planning process is used. Today, unplanned settlements and slum architecture exist predominantly in warmer climate zones. Out of necessity and pragmatism, traditional architecture typologies have developed in a long empirical optimisation process. Traditional building typologies differ according to environmental conditions: society, climate, landscape, resources, technological standard, historical development and other influences. These typologies form the base for further development. All architecture is based on tradition: historical building typologies which have evolved over a long time, and which slowly adapted to the changing environmental conditions of the time, although these conditions may no longer exist.

The so-called "one-off projects" are singular phenomena in architecture. As the term says, their existence is a singular stroke of luck. The circumstances under which the implementation of such a project is possible include a visionary mind, the success of the technology being implemented, the necessary resources (ideally unlimited), excellent engineers, the support of the client as well as the support of society in terms of acceptance (building laws, political decisions). One-off projects are also well known by the general public, and judged by history. Few buildings become extraordinarily famous, e.g. the Centre Pompidou in Paris (Richard Rogers and Renzo Piano), Lloyds in London (Rogers) and the Hong Kong and Shanghai Bank in Hong Kong (Norman Foster), the Cupola of the Reichstag in Berlin (Foster). Some of these also become public landmarks of design, symbolising countries or even continents, for example the Eiffel Tower or the Sydney Opera House. These buildings represent a high-tech approach of a global architectural style. The innovation realised in these projects makes them role models for many other exemplary and high-quality buildings. The inventions (architectural features which can not be patented or otherwise copyright protected) developed in the course of the generation of these projects inevitably spread and become general knowledge, to the resentment of one and the delight of the other designer. Many renowned architects experiment on this 1:1 level, using their building tasks to develop specific ideas and push industry ahead.

Between these extreme positions the large field of "standard architecture" exists. Standard architecture covers a wide range of quality from high to low standard. This mass building (housing, office buildings, industrial...) is to only a small extent architect designed, and does not usually deliver outstanding experimental innovations.

architecture

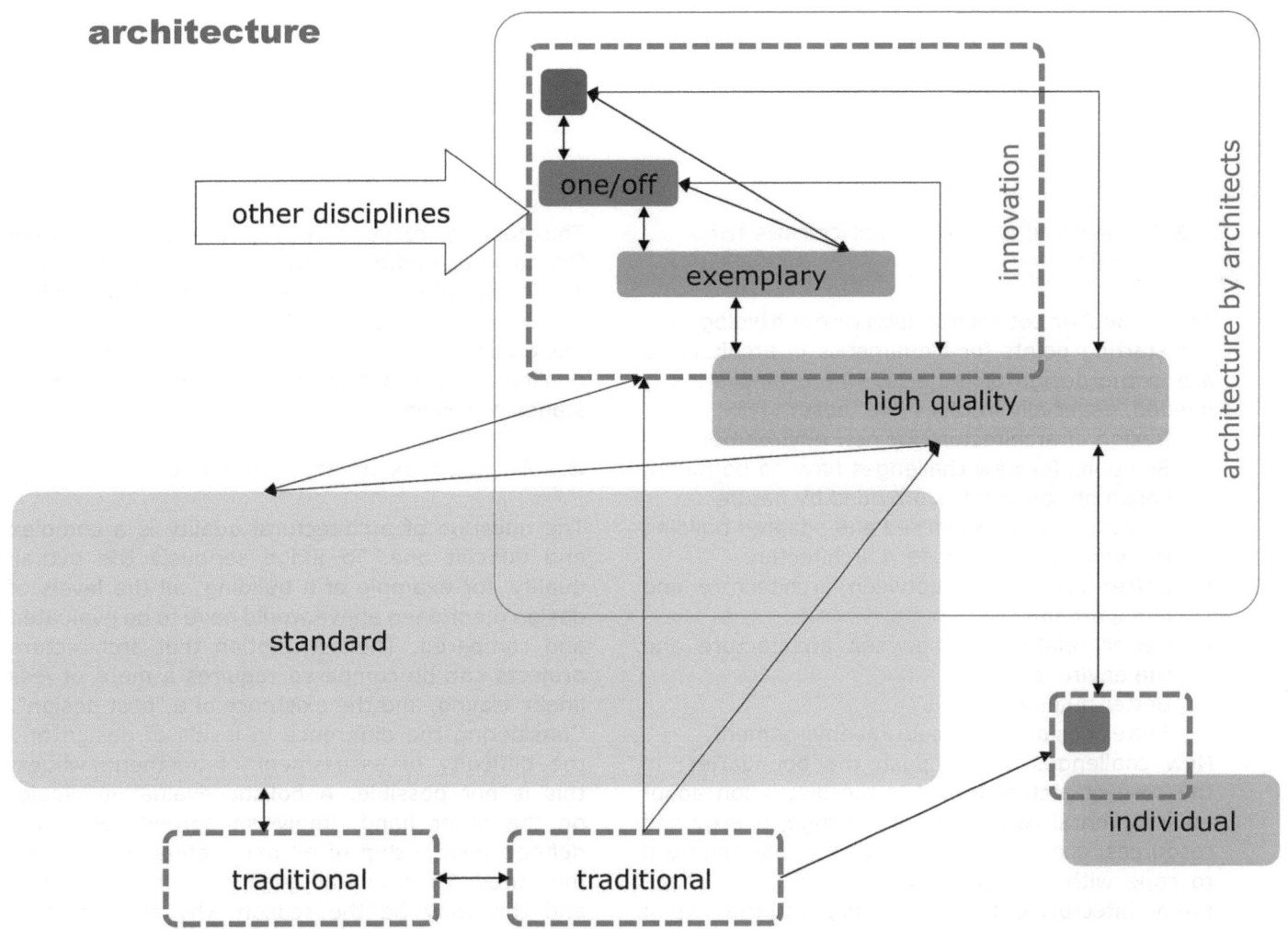

Fig.1 A rough categorisation of architecture, including the flow of information between fields. A frame in light grey indicates the sphere of activity of architects. The dashed frames indicate innovation.

Innovation achieved in this field concerns industrial economy, often conflicting with quality.
A small section of individual buildings show outstanding innovative potential - isolated visionary phenomena, designed by individual house owners or small companies.
Apart from that, innovation that raises quality usually occurs only in the upper upper right segment of Figure 1. The information flow between these categories of architecture is mutual.

Innovation from other disciplines and new technologies influences architectural development and is integrated into one-off or exemplary projects. Information then spreads slowly down to standard buildings (e.g. point fixing in glass, fingerprint access control - features which are now common even in single family housing). Innovative architecture is in many cases also influenced by traditional typologies - it is known that many famous architects have travelled extensively and studied traditional buildings typologies.
Occasionally, information of the individual section is taken up and spread into other fields. This is especially interesting, as these outsiders' achievements are often based on and integrated into a local environment.

- **Small field: experimental architectural research**

Experimenting with building tasks is the usual way for developing new solutions, but makes life hard for architects, companies and clients. Only small steps can be taken, and any innovation has to comply with specific building regulations, still provide security and functionality for the users, must have the same quality as the standard solution, and must not require extra resources. The planning costs for the design innovation are usually not calculated - the architect or designer and the building companies have to cope with this "loss" and all the dangers which a new solution may bring.
Strategic experimental research would require a platform free from the restricting 1:1 condition. Due to the lack of interest and funding, experimental design research is done by very few people and organisations. It is usually done at universities in the course of design programs, diploma and dissertation projects. Research bodies are funded if either the pressure to find new solutions is high enough to provide the necessary resources, or if economic success is expected in the near future.

2.1.3 Where biomimetics comes into play

The connection between architecture and biology and the starting points for biomimetics in architecture are in the design of projects, where innovation is needed, especially in cases like these:
- Design of architecture for new environments
- Solutions for new challenges have to be found, based on role models provided by nature
- Investigation of optimised and adapted building traditions informs modern architecture
- Better relationship between architecture and living organisms
- Better relationship between architecture and the environment
- Better quality of life
- Better design of the cultural environment

New challenges always push the boundaries: in the case of architecture it is the discussion about environmental issues (climate change, energy and resources, ecology and sustainability) and the need to cope with new environments which will shape the architecture of the 21st century with the help of nature's role models.

2.1.4 The many levels of a project

An architectural project is designed to fit many different and often conflicting requirements. The ultimate demand is to provide sheltered space for the user's activities. However, other than basic functional levels of design are not limited to: external and internal circulation, the relationship between privacy and public access, visibility and transparency, construction and material, colour, light and surfaces, supply systems and infrastructure, building physics and building services, detailing, intangible things like general idea, abstract concept, geometric order, aesthetic concept, style, significance of the building to its surroundings, functional and aesthetic relationships to the environment including flows of energy and matter, urban and regional planning issues, ecological issues...

The architect's task is the integration of all these levels in the final project. The value that the designer attributes to the different topics and the relative importance of the topics for the final project is part of the freedom of design and the individual approach to architecture in general. The significance of the topics considered and their relative importance for the given task is not always decided consciously, but chosen through personal intuition, and they may be ousted by the complexity of project development.

The factors responsible for the project's final "identity" can differ widely. And wide is the field of application of any novel idea - from highly detailed items to large urban structures.

The quality of the final project is also defined by the quality of investigation conducted in the important stages of design.

2.1.5 Quality in architecture

The question of architectural quality is a complex and difficult one. To judge seriously the overall quality, for example of a building, all the levels of design mentioned above would have to be evaluated and compared. The assumption that architecture projects can be compared requires a more or less linear scaling, and the existence of a "best design". Considering the difference in levels of design and the difficulty of assessment of aesthetic values, this is not possible. A holistic evaluation would, on the other hand, imply an agreed value and defined relationship of all parameters that can be measured. Thus, overall quality is not measurable, and this may be the reason why evaluation of architectural projects is not at all taken for granted. But specific qualities are very well measurable. For instance, energy processing of buildings is a serious current issue, as is the amount of embodied energy in building materials. Measuring is not limited to the world of matter and energy, but is also possible for activity and intangible parameters. The space syntax theory of Bill Hillier, for example, measures the value of "integration", which expresses a kind of urban importance of a space.

Summing up, it can be said that evaluation in architecture is not routine at all, and that even simple checks (e.g. the assessment of the internal circulation of a building) are not often carried out, so that consistent quality enhancement usually does not exist. Another reason for this is the individuality of architecture projects in contrast to product design in industry, where high quantities justifies the implementation of quality control.

For these reasons, it is generally the market size that decides on quality, and the criteria remain largely unexplored. The measured parameter is market price per m^2, which provides no information about resources and energy, nor about architectural quality.

Fig.2 Motorbike covered with fake fur, Vienna 2005.

2.1.6 Innovation and expectation

Expectation is a key to occupant satisfaction. The image of architecture in the user's mind settles his interpretation of space and potentials of activity. Obviously, unexpected events are not likely to be appreciated (e.g. malfunction, disorientation because of unclear spatial relationships), but slow change is. The capacity of humans to deal with new situations is amazing, but doing so costs time and energy. Relying on the well-known makes the day-to-day routine (and advancement in many other fields) possible, but palls at the same time. The happy medium has to be found - introduce the new and unexpected, but not too much at a time, so that people have a chance to change their minds. There are solutions in architectural design which have proved to work out very well. Here the introduction of novel ideas would lead to a decrease in quality. Nonetheless many architects continue to reinvent the wheel.

Satisfaction of users is crucial to the success of architecture as a discipline, which creates, forms and organises people's lives. The lack of connections between planners and users leads to continuing misinterpretations. On the other hand, integration of users into the planning process can be very difficult in terms of decision-making, time and therefore cost (regardless of the fact that the user's opinion does by no means guarantee satisfaction). As a matter of fact any field that can reasonably be influenced by the future user's wishes has to be defined clearly, as well as those fields where future users will have no say. The distinction relates to hierarchical levels in design, which will be explained later.

2.2 BIONICS [BIONIK] BIOMIMETICS

2.2.1 Terms, definitions and related fields

The terms bionics, bionik and biomimetics

According to Werner Nachtigall, the German-language term Bionik originally comes from the English word "bionics", which was coined by the US Air Force Major J.E. Steele at a conference entitled "Bionics Symposium: Living prototypes – the key to new technology" in 1960, supposedly as a combination of the words "biology" and "technics" or "electronics".

In German, the term "Bionik" has found a very expressive reinterpretation in the first and last syllables of the words Biologie [biology] and Technik [technology].[14]

- **The term bionics [Bionik]: a combination of two terms:**

biology, the science of life.
technology, the constructive creation of products, devices and processes by using the materials and forces of nature, taking into account the laws of nature.

---

14 Nachtigall, W.: Bionik, Grundlagen und Beispiele für Ingenieure und Naturwissenschaftler, 2002, p.5

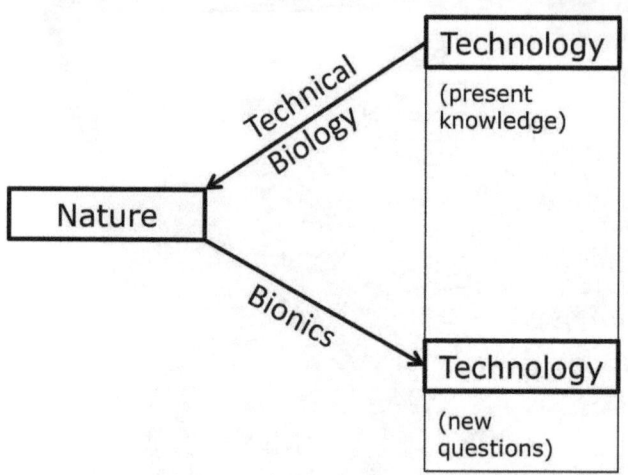

Fig.3 Diagram explaining the relationship between bionik (bionics) and technical biology, nature (left) and technology (right), Werner Nachtigall, translation by the author.

The term "bionics" nowadays has a narrow definition: robotics and replacement or enhancement of living matter, tissue, body parts and organs with mechanical versions. In the English-speaking world, the term "biomimetics" has appeared as equivalent to the German "Bionik" and is commonly used. Otto Schmidt coined this term in the 1950s.[15]

As the part "mimetic" suggests a mimicking of nature, the term is controversial. Recently, "Bioinspiration" has been used more often in the same context, but seems to be too general to prevail.

Another solution for the continuing terminology discussion is to also use the term "bionik" in English.

In the following, the three terms bionics, bionik and biomimetics will be used synonymously.

Interpretation

"Technical biology means 'understanding nature with the help of technology'. Bionics means 'learning from nature for the sake of technology'.
Neither perspective is trivial. They consciously transcend the disciplinary boundaries and do not involve the exclusion or rejection of knowledge. They hold the merging of biology and technology to be essential, but are conscious of their respective methodological peculiarities and of the limits on an integrated view. Technical biology is devoted to uncovering a small sector of the realm of the unknown. This is an independent civilizational and cultural task, like all basic scientific research. Bionics takes the results and renders them useful. This is a practical necessity of our times."[16]

Quoting Max Planck: "Application must be preceded by knowledge"; Werner Nachtigall presents the disciplines of Bionik and Technical Biology as antipodes:

"Technical biology... examines and describes constructions and processes of nature in light of the analyses and descriptive methods of physics and technology... is basic research, a task of civilization and culture...
Bionics... combs the thicket of nature's constructions and processes in search of inspiration for independent technological creation... is a practical necessity"[17]

Both disciplines complement each other in the cycle of continuing scientific technical development and progress.

The following definition of the term "bionik" was agreed in 1993 at a meeting of The Association of German Engineers (VDI), and extended by Werner Nachtigall in 1998:

"As a scientific discipline, bionics deals systematically with the technical execution and implementation of constructions, processes and developmental principles of biological systems. This also includes various forms of interaction between living and non-living elements and systems."[18]

Less complicated definitions are also used in Germany and the UK for communicating what the young discipline is about. On the website of the German Biokon network, Bionik is defined as

"Decoding of 'inventions of animate nature' and their innovative implementation in technology"[19]

and the Centre for Biomimetics at the University of Reading defines Biomimetics simply as *"the abstraction of good design from nature."*[20]

Bionik and biomimetics both are defined as activities, taking information from the field of biology (or nature in general) to technology.

---

15 Vincent, J.F.V. et al.: Biomimetics - its practice and theory, J.R.S., 2006, p.1

16 Nachtigall, W.: Bionik, Grundlagen und Beispiele für Ingenieure und Naturwissenschaftler, 2002, p.7

17 Nachtigall, W.: Bionik, Grundlagen und Beispiele für Ingenieure und Naturwissenschaftler, 2002, p.4

18 Nachtigall, W.: Bionik, Grundlagen und Beispiele für Ingenieure und Naturwissenschaftler, 2002, p.3

19 http://www.biokon.net/bionik/bionik.html [08/2007]

20 http://www.rdg.ac.uk/biomimetics/about.htm [08/2007]

Function, form and structure

Transfers from living nature make sense because natural models have come into being through the process of evolution, subjected to many and various conditions. They therefore represent extremely complex solutions, and their translation is not merely a matter of form.

Frei Otto wrote about natural and technical constructions:

"Constructions possess a shape or (geometrical) form and an inner structure. Form and structure come into being by way of a common developmental process, depending on physical and chemical laws or human creative power."[21]

Juri Lebedew stated, *"In nature, the principle of integration of function+form+structure is effective, and is adapted to the existence and interrelation with the environment."*[22]

The existing object always embodies both process and result. The integrity of form, structure and function in nature makes purely morphological translations worthless.

Optimisation

"It is said again and again that nature 'optimizes' everything, and bionics is on a constant search for specific optima. The concept of 'optimum' in nature and how it may resemble – or differ from – the same concept in business and technology is a subject of fundamental inquiry."

The statement of Werner Nachtigall introduces the discussion about the fundamental differences between otpimisation in nature and in technological environments.

"In both of these fields, optima can usually be arrived at via mathematical optimisation computations... In biology, the complexity of the influencing factors and optimisation criteria is usually too great to formulate a definite objective function."[23]

In contrast to biology, optimisation of technological structures can be carried out, nowadays also with help of design tools based on evolutionary strategy, as is presented in the example of the optimisation of a cantilevering framed truss.

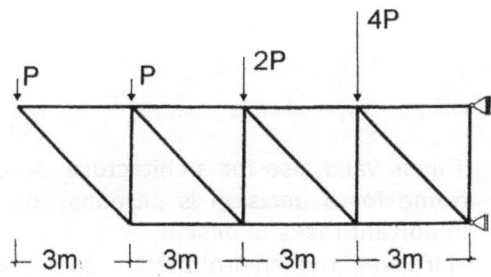

A

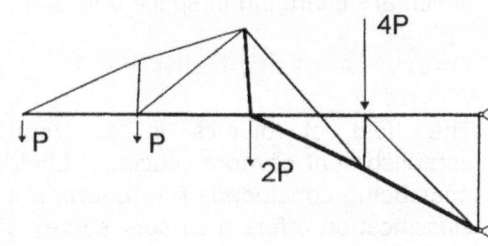

B

Fig.4 Optimisation of a framed structure: A Non-optimised framed structure – initial form, B Final form, optimised for defined conditions.

The expressions "nature designs" and "nature optimises" personalise nature, but nature possesses neither a creative mind nor the ability to perform creation, and nature has no interest.

"...Evolution is as value-free as the dance of Shiva, where all beauty and ugliness, creation and destruction are expressed or compressed into one complex symmetrical pathway...
The fascinating cases of adaptation that make nature appear so clever, so ingenious, may also be early steps toward pathology and overspecialisation."[24]

The natural processes that lead to this apparent perfection will be described later.

In fact time-variable environmental conditions, animate and inanimate, mainly trigger the optimisation development of organisms. The frame of reference, which is important for the existence and survival of specific organisms, is not known in detail. Statements about optimisation with regard to specific conditions are assumptions, which often enough prove to be right, but optimisation in an absolute sense does not exist.

In technology the framing for optimisation processes is also very specific and the outcome highly dependent on the starting design. Any mistake in the framing or a bad starting design renders the optimisation process worthless.

---

21 Otto, F. et al.: Natürliche Konstruktionen, 1985, p.24
22 Lebedew, J.S.: Architektur und Bionik, 1983, p.55
23 Nachtigall, W.: Bionik, Grundlagen und Beispiele für Ingenieure und Naturwissenschaftler, 2002, pp.257

24 Bateson, G.: Mind and Nature, 1979, p.163

This is valid also for architecture: setting the right frame for a decision is perhaps one of the most important tasks in design.

Moreover, the environmental conditions in nature are changing continuously. Therefore adaptation is an ongoing process, on ontogenetic as well as phylogenetic level. Optimisation in nature is transient, depending on dynamic circumstances, which are changing in space and time.

Overview of subfields

The field of bionics is so young that the establishment of more specific subfields is nowhere near being concluded. The following more detailed classification offers a cursory survey of the current state of research, and can by no means be regarded as complete.

All examples that can be found for bionics can be allocated to three broad, general subfields, which are according to Nachtigall further differentiated into subfields.

- **Structural bionics**
Nature's constructions, structures, materials
- **Procedural bionics**
Nature's procedures or processes
- **Informational bionics**
Principles of development, evolution and information transfer

Detailed subfields

- *"Structures bionics [Strukturbionik] (material bionics)*
Biological structural elements, materials and surfaces.
- *Device bionics*
Development of usable overall constructions.
- *Structural bionics [Konstruktionsbionik]*
Biological constructions, closely related to above structural and device bionics.
- *Anthropobionics (bionic robotics, bionic prosthetics)*
Issues of human/machine interaction, ergonomics.
- *Construction bionics [Baubionik]*
Light constructions occurring in nature, cable constructions, membranes and shells, transformable constructions, leaf overlays, use of surfaces, etc.
- *Climate bionics (energy bionics)*
Passive ventilation concepts, cooling and heating.
- *Sensory bionics*
Detection and processing of physical and chemical stimulation, location and orientation within an environment.
- *Locomotion bionics (bionic kinematics and dynamics)*
Walking, swimming and flying as primary forms of movement. Interaction with the surrounding medium.
- *Neurobionics*
Data analysis and information processing.
- *Evolutionary bionics*
Evolution techniques and evolution strategies, made useful for technology.
- *Process bionics*
Photosynthesis, hydrogen technology, recycling.
- *Organizational bionics*
Complex relationships of biological systems"[25]

The distinction between the subfields suggested by Nachtigall seems to be academic, but is still the most comprehensive effort to order the field. As different categories are used, many examples can be allocated to more than one subfield.[26] The subfields represent merely different emphases, but the differentiation implies that the field of bionics will be increasingly important. In some disciplines bionic approaches are already integrated in industry.

Some of the subfields are especially interesting for architecture: structural, climate, construction, locomotion and evolutionary bionics are promising fields.

For architecture as a whole, the expressions of "biomimetics in architecture" or "Architekturbionik" includes and transcends what Nachtigall calls "Construction bionics" [Baubionik].

---

25 http://www.uni-sb.de [2002], description and examples are found in Nachtigall, W.: Bionik, Grundlagen und Beispiele für Ingenieure und Naturwissenschaftler, 1998, pp.19

26 In the second edition of his book, Bionik, Grundlagen und Beispiele für Ingenieure und Naturwissenschaftler, Nachtigall changed this distinction of subfields, but the new version is not fully convincing either.

Definitions

Other terms occurring in conjunction with bionics:
- **Biomimicry**

Coined by a US group around Janine Benyus, is about using the genius of nature to develop innovation.[27] The holistic approach includes ecological design as well as interest in technological innovation.
- **Bioinspiration**

Is the most general expression for the design inspired by natural role models, including all levels of abstraction, also purely morphological interpretations.
- **Biomorphology**

Is the science of construction and of the organisation of living things and their components – organs, tissue and cells.
- **Structural morphology**

Refers to functional design in technology and functional anatomy in biology.
- **Micromorphology**

Examines and describes the form of microscopic objects and represents a treasure trove of functional forms.
- **Biomechanics**

Is the application of physical laws of mechanics to examinations of natural objects.
- **Biophysics**

Examines and describes biological objects with the terms and methods of physics.
- **Biotechnology**

Explores biological objects using technical methods. Recently the notion has shifted towards technologies using organisms for production purposes in biochemistry, e.g. enzymes, drugs and pharmaceuticals. Biotechnology is also related to genetically modified organisms.

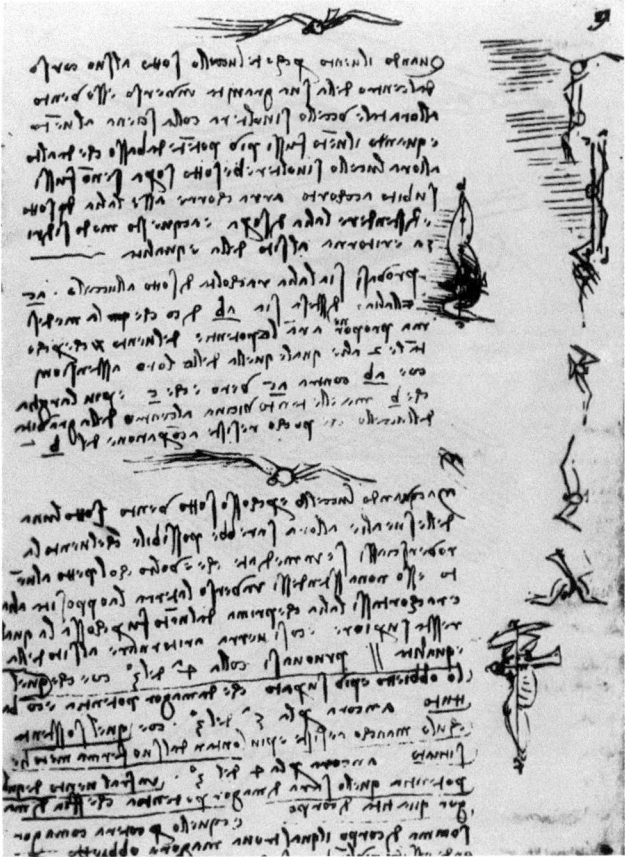

Fig. 5 Leonardo da Vinci's studies on birds flight.

2.2.2 Historical background and development of bionik

Researchers and scholars who have used biological role models for their work can be found very early in history. A good example is the historical development of human flight, a challenge which had occupied researchers and inventors for centuries, where a wide range of role models from nature were taken into consideration. Flight and plants as role models will serve as examples to illustrate the historical development of a discipline which is committed to learn from nature. Werner Nachtigall has contributed to the investigation of the historical background of bionics and to current developments in his extensive compilation "Bionik - Grundlagen und Beispiele für Ingenieure und Naturwissenschaftler".[28]

---

27 Benyus Janine M. : Biomimicry : Innovation Inspired by Nature, 1997

28 Nachtigall, W.: Bionik, Grundlagen und Beispiele für Ingenieure und Naturwissenschaftler, 2002, first edition 1998

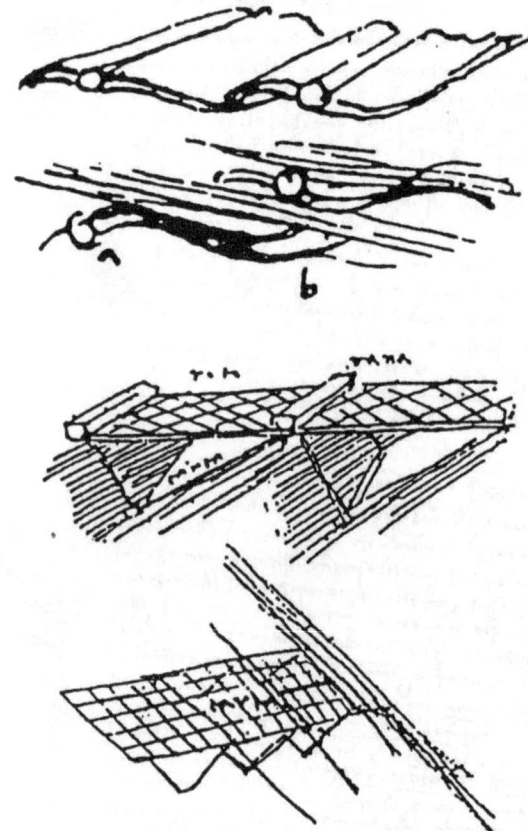

Fig.6 Leonardo da Vinci: sketches investigating the overlapping of the feathers and flow through the bird's and the technical wing (acc. to Giacometti 1936).

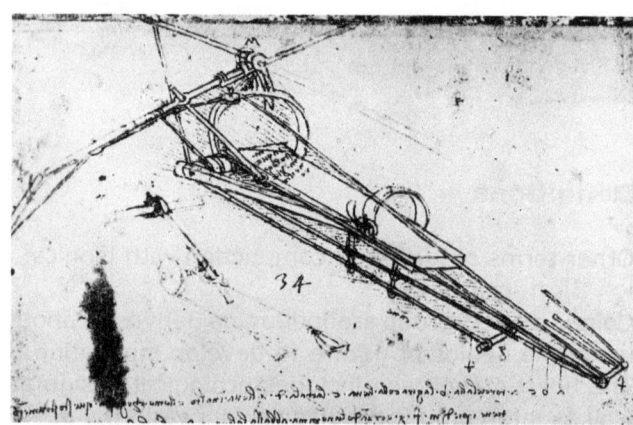

Fig.7 Design for a apparatus to fly, 1490.

Fig.8 Studies of water current, around 1508.

Flight

- **Leonardo da Vinci (1452-1519)**

In 1505 Leonardo da Vinci compiled a book on the flight of birds, "Sul vol degli uccelli". Werner Nachtigall describes one of his inventions in the following:

"The free pinions of birds wing close due to their peculiar bearing and asymmetry during the downward flap without gap; during the upward flap they open up creating airflow. (Following today's concept formation this is a basic observation of technical biology) As a consequence Leonardo suggested using flaps made of tracery of willow covered with linen for technical wings, which could close when moving down and open when moving up. (This is a typical bionic suggestion: the basic observation is not copied, but the principle of surface closure without gaps and surface opening with creation of airflow was abstracted in an appropriate way. The fact that this could never have worked is not relevant to the existing consideration. The proposal is embedded into the time's technical perception)." [29]

Leonardo also designed a kind of corset with buckled wings, following the flight of birds with flapping wings. The concept of technical flight was successful after the abandonment of this principle and the transition to volplane.

Leonardo has drawn numerous other ideas and observations from nature, which were not taken up during his lifetime, but have influenced countless inventors ever since.

The parachute designed by Leonardo da Vinci in 1483 was built and tested by Adrian Nichols in 2003. Nichols only used materials which were available in da Vinci's time. For the test he was taken up to a height of 3,000m by a modern hot air balloon, and slowly descended to earth with da Vinci's parachute. To avoid being hurt by the 85kg heavy construction, Nichols finally landed with a modern parachute.[30] Leonardo made many of his designs for military purposes. Research and invention for military purposes have been important sources of innovation until today.

---

29 Nachtigall, W.: Bionik, Grundlagen und Beispiele für Ingenieure und Naturwissenschaftler, 2002, p.8

30 Bürgin, T. et al.: HiTechNatur, 2000, p.91

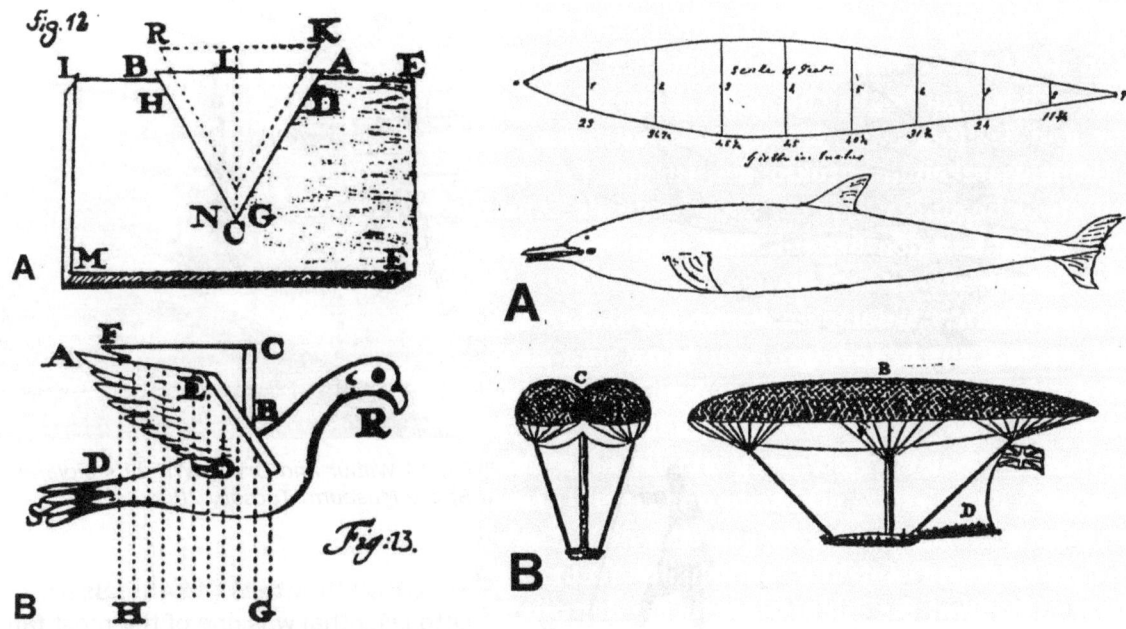

Fig.9 Alfonso Borelli: about the impact of wedges and the flapping wing.

Fig.10 Sir George Cayley's Studies on form and design of a balloon.

Fig.11 Sir George Cayley's sketch of tragopogon pratensis.

Other ideas about flight were developed by:
- **Alfonso Borelli (1608-1679)**

A professor of mathematics in Florence and Pisa, he explained the flight of birds by means of the physical impact of a wing as wedge-shaped displacement of air in "De motu animalum" (about the locomotion of animals).[31]

- **Sir George Cayley (1773-1857)**

The English nobleman and aeronautical explorer Sir George Cayley was concerned with streamlined shapes. He analysed the forms of a dolphin and a woodpecker by cutting their frozen bodies into slices. In 1816 he designed a balloon with allegedly very low air resistance and in 1829 he investigated the ability of the flying fruit of Tragopogon pratensis to sail in a very stable and calm balance. He modelled a kind of parachute on this fruit.[32]

Pioneers of manned flight

Manned flight remained a challenge until the end of the 19th century, when Otto Lilienthal, Igo Etrich and the Wright brothers made their contributions to progress.

---

31 Nachtigall, W.: Bionik, Grundlagen und Beispiele für Ingenieure und Naturwissenschaftler, 2002, p.40
32 Nachtigall, W.: Bionik, Grundlagen und Beispiele für Ingenieure und Naturwissenschaftler, 2002, p.40

Fig.12 Otto Lilienthal flying.

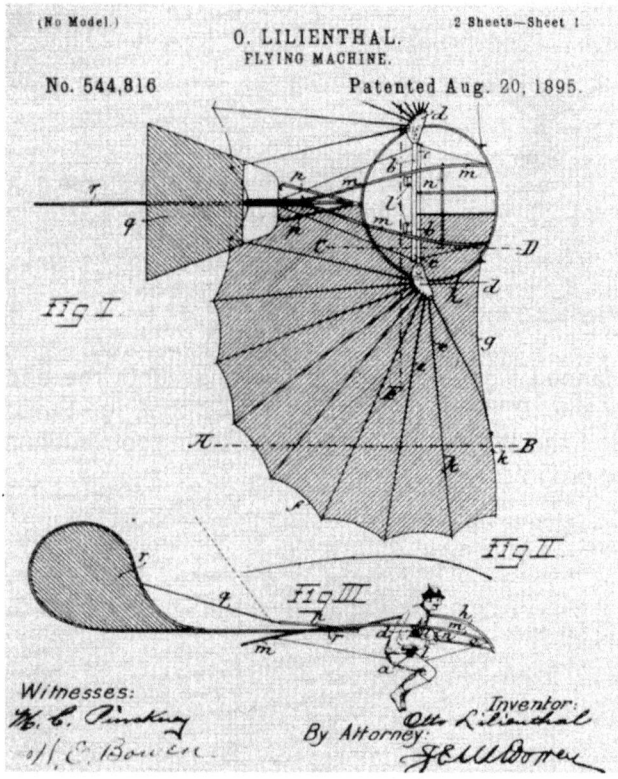

Fig.13 Patent of Lilienthal's glider, 1895.

Fig.14 Wilbur and Orville Wright's biplane in Pima Air and Space Museum, Tucson, 2004.

- **Otto Lilienthal (1848-1896), Germany**

Otto Lilienthal was one of the most famous pioneers in human flight. In particular he studied the flight of storks to develop his devices. He undertook over 3,000 flights with different apparatus and died in a crash in 1896.[33] His drawings show perfectly how a living creature can be described by means of engineering drawings. His research was serious and profound. Architects could make use of these kinds of drawings, as this is nature described in a technical language: classical technical biology.

- **Igo Etrich (1879-1967), Austria**

In 1907 Igo Etrich created an elegant monoplane called "dove" modelled on a flying plant seed, the tropical seed Zannonia Macrocarpa.[34] This device made history as "Etrichtaube", and a 1:1 model can be seen in the Technisches Museum in Vienna.

In spite of the use of a plant seed as role model, the final morphology of the plane followed the morphology of a gliding dove ("Taube" in German).

- **Wilbur und Orville Wright, USA**

Wilbur and Orville Wright succeeded in 1903 in the first motorised manned flight, at Kitty Hawk, North Carolina. Their achievement initiated the fast development of motorised flight. The biplane is on display in Tucson, Arizona, in the Pima Air and Space Museum.

The succeeding developments leading to today's commercial aircrafts are less bio-inspired. As soon as the successful functioning of the basic principle was more or less guaranteed, purely technological developments and amendments were in the foreground. Biomimetic approaches come into play again in an attempt to push boundaries; to integrate the more sophisticated phenomena of flight in nature, e.g. adaptiveness of wing tips and wing shapes, dynamic change of form according to flight situation etc.

33 Bürgin, T. et al.: HiTechNatur, 2000, p.81
34 Nachtigall, W.: Bionik, Grundlagen und Beispiele für Ingenieure und Naturwissenschaftler, 2002, p.233

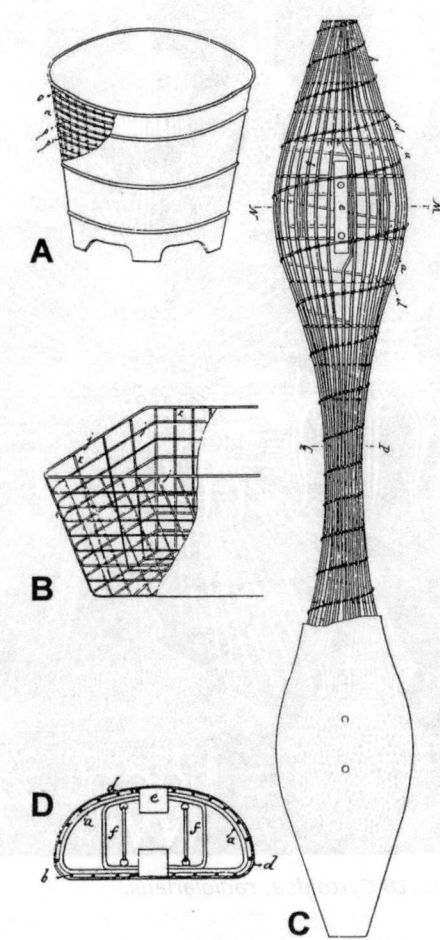

Fig.15 Drawings of Monier's patent specifications: containers for plants, railway sleeper.

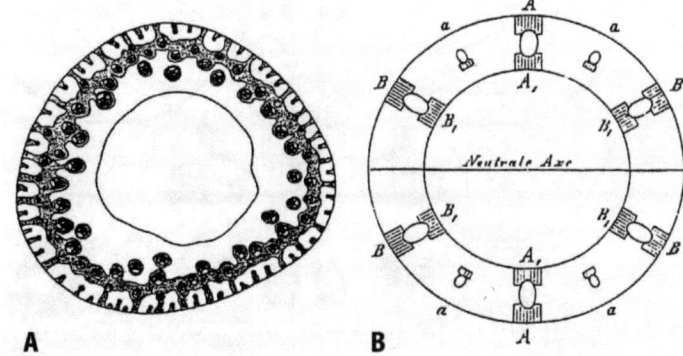

Fig.16 Cross-section of Cladium mariscus and structural interpretation by Schwendener.

- **NASA - vision of future airplane**

The visions of future airplanes that NASA has developed and published in 2000 show some adaptive features. The wingtips can change their form, the angle of the wings can be changed, and the surface of the wings can actively influence airflow. All these measures are common in bird flight.[35]

Plants as models

Plants have been used as role models ever since man began to use technology. For architecture, plants are especially important as they share some common problems with houses: most of them stay at one place and are dependent on local environmental conditions. Trees and houses are of a similar size, and subjected to similar influences of natural forces.

The famous comparison between a blade of grass and a tower will be explained later.

Plants as models and the biomechanics of plants had already been investigated in the 19th century by the Swiss botanist Simon Schwendener.

- **S. Schwendener**

"Without any doubt plants construct using the same principles as engineers, but their technology is much finer and more perfect." [36]

Schwendener found out that in corn stalks load bearing capacity and bending resistance is achieved with similar elements as in buildings.

- **J. Monier**

Joseph Monier, a gardener, made garden pots out of wire mesh and concrete. The sclerenchymatic fibre structure of decaying parts of opuntia and the problem of breaking garden pots inspired him to invent reinforced concrete. After the discovery of the principle, he made various types of products and, later, new elements for the building industry, e.g. railway sleepers, which were formed according to the forces they needed to withstand. Monier is considered the inventor of reinforced concrete. He patented his idea in 1867. Reinforced concrete combines the tensile strength of metal and the compression strength of concrete to withstand heavy loads.[37]

- **W. Rasdorsky**

Wladimir Rasdorsky interpreted the construction of plants as composite structure, strands of sclerenchyma corresponding to metal reinforcement and the parenchyma tissue to the concrete matrix (sclerenchyma cells are specialised in structural support; parenchyma cells are unspecialised, "typical" plant cells).

He was inspired by lectures about reinforced concrete construction in 1906/07. In 1929 he stated that *"there is a extensive analogy between the technical composite structures and the organs of plants concerning the whole construction principle..."*.[38]

Later authors affirmed these ideas.

35 http://www.hq.nasa.gov/office/aero/library/multimedia/videos/morph_small.mov [08/2007]

36 Schwendener, 1888 in Nachtigall, W.: Bionik, Grundlagen und Beispiele für Ingenieure und Naturwissenschaftler, 2002, p.41

37 Nachtigall, W.: Bionik, Grundlagen und Beispiele für Ingenieure und Naturwissenschaftler, 2002, pp.453

38 Rasdorsky in Nachtigall, W.: Bionik, Grundlagen und Beispiele für Ingenieure und Naturwissenschaftler, 2002, p.41

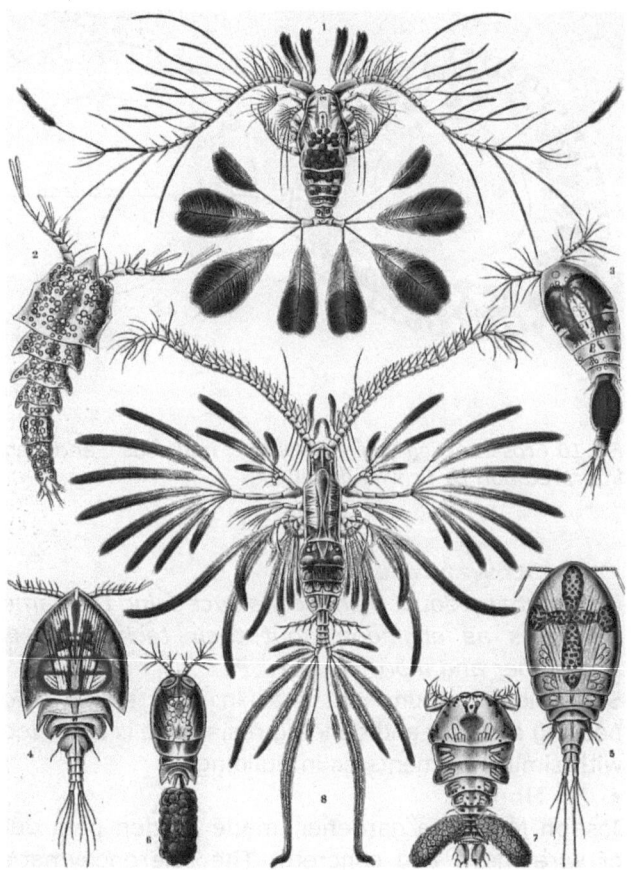

Fig.17 Ernst Haeckel's tables of marine organisms, here copepoda, small crustaceans.

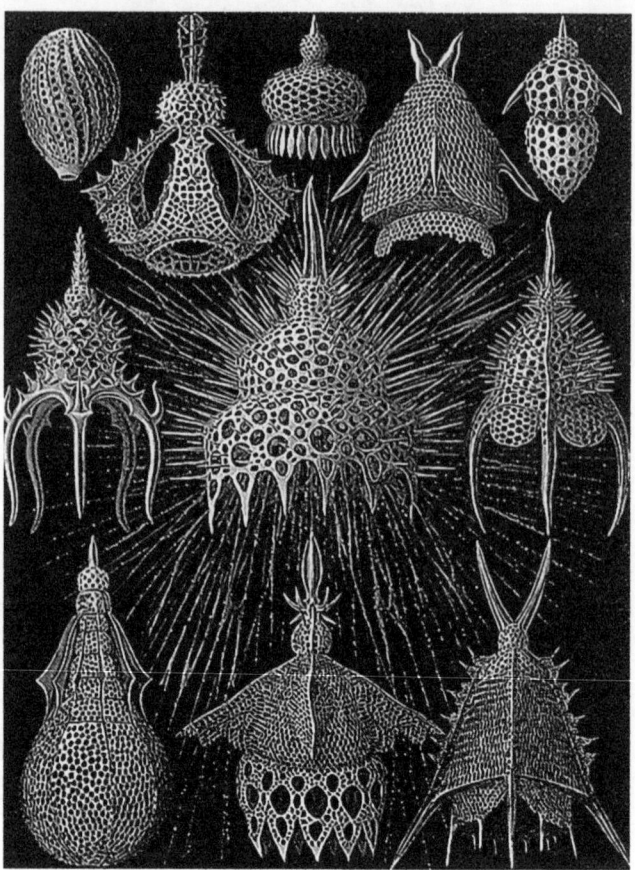

Fig.18 Cyrtoidea, radiolarians.

Other dimensions: Ernst Haeckel and Karl Blossfeldt

The influence that life sciences can have on other disciplines is highly dependent on the publication of universally understandable research output. Haeckel's and Blossfeldt's impact on the public and the broad effect of their work are, to a considerable part, due to their method of presentation.

Much of mankind's fascination with natural objects lies in the need to leave the human scale and immerse oneself into another world. The influence that life sciences can have on other disciplines is highly dependent on the publication of universally understandable research output.

- **Ernst Haeckel (1834-1919)**

Ernst Haeckel was both artist and scientist, and in his time ranked among the most famous biologists in the world. His book "Kunstformen der Natur" about protozoa and marine organisms reached a wide audience. The fascinating change of scale was made possible through improved microscopes.

During his numerous journeys and expeditions Haeckel produced countless sketches, which were later processed to become highly aesthetic engravings. The panels of marine organisms, protozoa so-called "first animals" and plankton have become enormously popular until today.

Haeckel was fascinated by the diversity of forms that were to be found in marine organisms, especially in the skeletons of radiolarians, which consist of silica and strontium sulphate, and form deposits on the sea floor several metres thick.

Frei Otto's Institut für leichte Flächentragwerke also worked in this field and published a specialised volume, IL 33 Radiolaria.

While aiming at a general morphology of organisms, Haeckel discovered the biogenetic law, which was published in the book "Generelle Morphologie der Organismen" in 1866. The biogenetic law, or theory of recapitulation, relates the embryonic development of an organism (ontogeny) to the evolution of the species (phylogeny). It says that in its embryological development an organism once again recapitulates the steps that the respective species has taken, once again. Haeckel's theory was very influential and controversial, but is no longer considered valid in its proposed form.

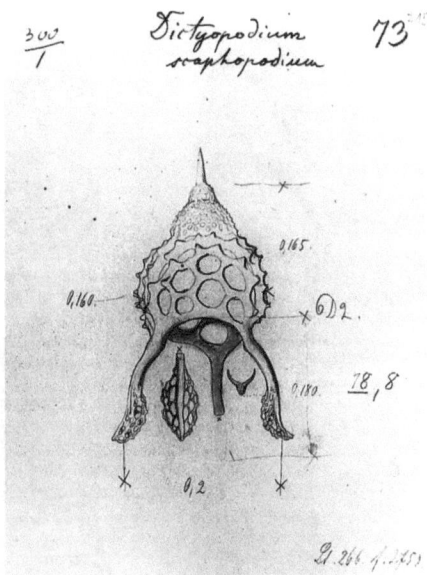

Fig.19 Design for a monumental entrance to the World exhibition in Paris 1900 by René Binet.

Fig.20 Ernst Haeckel, radiolaria.

*"Haeckel's idea was that an increase in complexity can be observed in his ordering of organic forms. The order is based on morphological characterisation of the organisms, but their tissues grow more differentiated with their ascending position in the row of evolution. The specific cells, which are symmetrically arranged, become integrated into these tissues. The row of forms, which can be described in embryologic development, is interpreted by Haeckel as a real historical succession of living organisms. According to Haeckel they form a series of singular creatures, the organisms not codified but with the ability to further differentiate their design... that the **specific forms in their development program recapitulate the succession of individual rules found: the individual development (ontogeny) replicates the evolution of the species (phylogeny).**"[39]*

Haeckel published his book on the morphology of organisms seven years after the publication of "On the origin of species" by Charles Darwin in 1859 that proposed the evolution theory and natural selection.

His aim was the general classification of all forms of life. For many reasons he did not succeed, but the influence of his research and his drawings on his contemporaries, including architects and designers was considerable. For example, the shapes of protozoa inspired the project of Rene Binét while working on his project for the world exhibition in Paris 1900[40], which is one of the rare examples of the form of a whole organism being translated into the form of a whole building (zoomorphism).

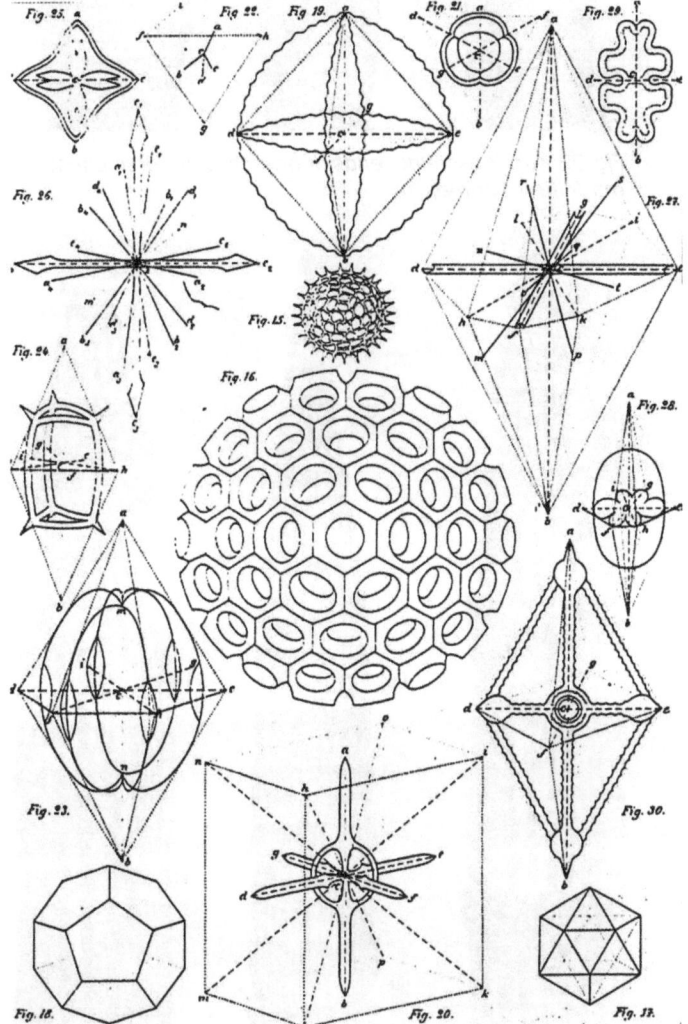

Fig.21 Ernst Haeckel, ideal basic forms, 1866.

39 Olaf Breidbach in Haeckel, E.: Kunstformen der Natur, reprint 1998, p.9
40 Irenäus Eibl-Eibelsfeld in Haeckel, E.: Kunstformen der Natur, reprint 1998, p.27

Background | Bionics [Bionik] Biomimetics

Fig.22 Karl Blossfeldt: ends of branches of horse chestnut, magnified 12times.

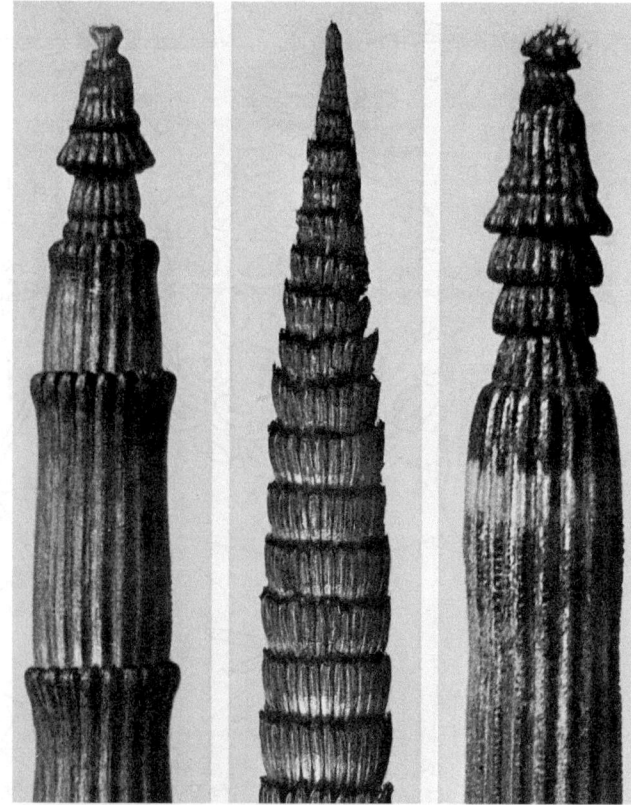

Fig.23 Different kinds of horsetail, magnified 4 to 18 times.

Fig.24 Pumpkin tendrils, magnified 4 times.

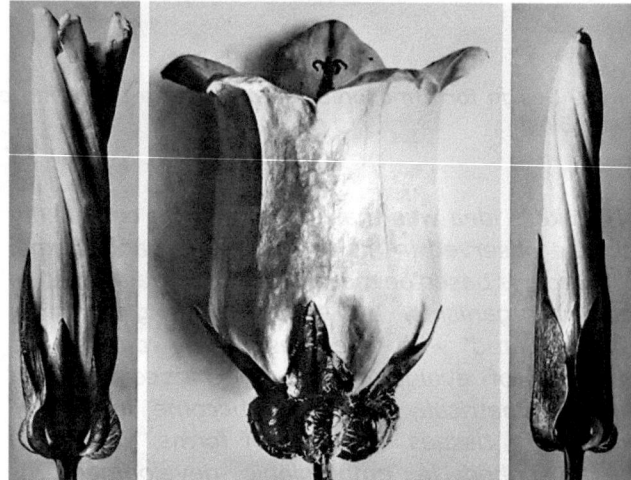

Fig.25 White hedge bindweed, bell flower.

- **Karl Blossfeldt (1865-1932)**

Karl Blossfeldt was a sculptor, university teacher and photographer. His photographic oeuvre is another very interesting example of the fascination that can be aroused by changing the scale of observation by a small degree. The difference in scale is not as significant as the difference in sensing would suggest.

Blossfeldt's interest in plants was of a documentary, artistic and didactic nature. He made the structure of plants, their organic design, the "highly artistic forms", which he considered "born out of necessity", visible and comparable in his strict photographic style. He published "Urformen der Kunst" und "Wundergarten der Natur" in the 1920s and with his work influenced contemporary artists, especially the Surrealists.

The ignorance of Blossfeldt's artistic work by the life sciences is just another example of how separated the disciplines are. However, for the general public and the artistic world his work is still important: a new edition of his book has been published.[41]

41 Blossfeldt, K. et al: Karl Blossfeldt, 1994

Fig.26 Front page of "Die Pflanze als Erfinder" of Raoul Francé, 1920.

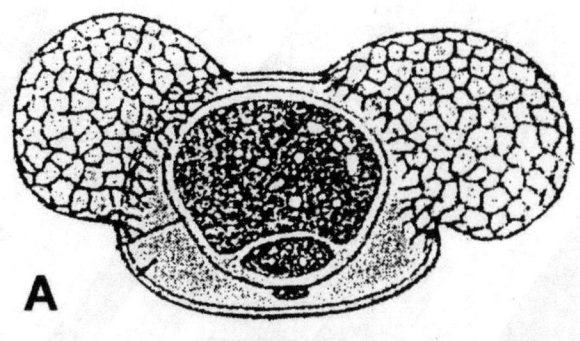

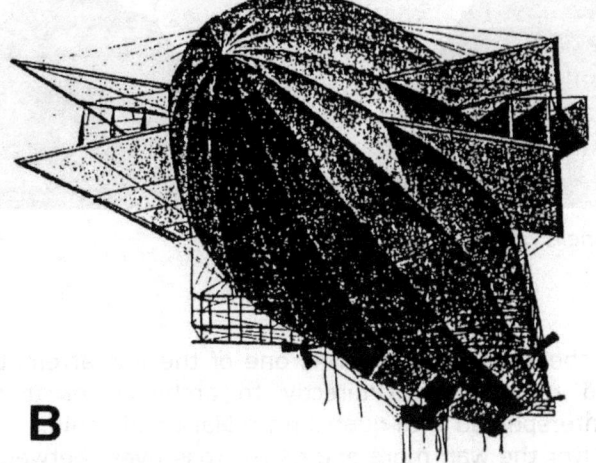

Fig.27 Raoul Francé, design for an air ship using pollen of a pine as role model, 1919.

First half of 20th century

Until the beginning of World War II many scientists and designers were concerned with applications, in particular locomotion in both water and air.
- **Raoul H. Francé**

Raoul Francé published a large number of articles and books, continuing with both research on and the development of structures, and mechanisms of plants, under the term "biomechanics". Although his numerous ideas for using natural role models in technology are often too direct and uncritical, he is a protagonist of a "biological technology" and his research is exemplary.[42]

"*Biotechnology is the crown of technology! Not the plants invent, not we ourselves invent, but the law of technical forms takes place in the dark and icy night of necessity.*"[43]

In spite of the importance of Francé's publications for "biotechnology" this seems to be a rather frustrated approach.

National socialism and communism

For different reasons, national socialism and communism paid attention to the idea of bionics. In the German Third Reich, with its over-emphasis on the natural and the grown, role models from nature were highly esteemed. Statements by Marx and Lenin justified bionic approaches in the Soviet Union.[44]
- **Alf Geissler**

Basing his work on Francé's findings, Geissler looked for role models from nature in many fields of technology, and developed analogies. His book "Biotechnik", published in 1939, contained ideological sections.[45]
- **Juri S. Lebedew**

The book "Architektur und Bionik" by Russian Juri Lebedew was published in Russia in 1977 and in the German Democratic Republic in 1983. Lebedew tried to give an overview of "Architekturbionik" and presented the architectural developments of the post-war period in this frame. His own interpretations of natural role models are partly astonishingly naive, and the designs seem to be much older than the publication itself.

---

42 Nachtigall, W.: Bionik, Grundlagen und Beispiele für Ingenieure und Naturwissenschaftler, 2002, pp.41

43 Raoul H. Francé in Blossfeldt, K. et al.: Karl Blossfeldt, 1994, p.34, original in Francé, R.H.: Die Pflanze als Erfinder, 1920

44 Nachtigall, W.: Bionik, Grundlagen und Beispiele für Ingenieure und Naturwissenschaftler, 2002, p.42

45 Nachtigall, W.: Bionik, Grundlagen und Beispiele für Ingenieure und Naturwissenschaftler, 2002, p.42

Fig.28 Shells of diatoms.

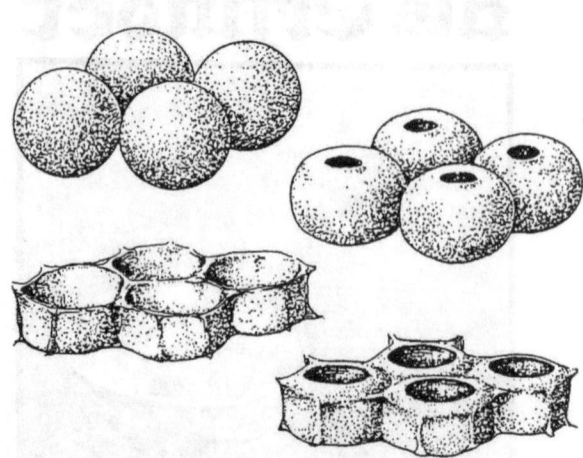

Fig.29 Johann Gerhard Helmcke, theory of form development in diatoms.

Lebedew's compilation is one of the few attempts to relate bionics directly to architecture. It is interspersed with quotes from Marx and Lenin.
After the war, more and more cross-overs between the disciplines have been made possible by progress in technology and the life sciences.[46]

Post-war development

The development after World War II is characterised by a transition to investigating form and function with respect to the development of form in nature. Only a few very influential researchers can be named here.

- **D´Arcy Thompson, (1860-1948)**

"On Growth and Form" was published in 1942. It discusses questions of how form in organisms develops, with radiolarians and some higher animals and plants serving as examples. It is still regarded as a bible for the development of form and structure of living organisms and many later works refer to Thompson's findings. Thompson discussed topics like magnitude, growth and scale, and investigated natural shapes in terms of mathematics and geometry.

His work is a still inspiring source which is taken forward through the discipline of mathematical biology, which investigates the spatiotemporal pattern formation in biology.

- **Heinrich Hertel**

Heinrich Hertel was an engineer and expert in aerodynamics who worked in the development of construction and actuation of airplanes. He established "Biologie und Technik" as a research discipline in Berlin, as one of the precursors of today's active research community in Germany.
He also experimented with fin actuation of boats, which did not succeed but which initiated the development of another technology for pumps. Hertel collaborated with the biologist Johann Gerhard Helmcke.[47]

- **Johann Gerhard Helmcke**

Johann Gerhard Helmcke primarily analysed the shells of diatoms. He published "Form und Funktion der Diatomeenschalen" in 1959. In 1964 the renowned Institut für Leichte Flächentragwerke was founded in Stuttgart, Helmcke being one of the important proponents of basic research.[48]
Helmcke found a plausible explanation for the form generation of diatoms: the cells synthesise droplets of a fatty liquid, which attach at the cell surface and flatten themselves. The cavities get filled with liquid silica, which hardens to a light and stable skeleton after the degradation of the droplets. Processes of self-organisation are supposedly responsible for this form generation. Helmcke's model experiments supported this hypothesis.

---

46 Nachtigall, W.: Bionik, Grundlagen und Beispiele für Ingenieure und Naturwissenschaftler, 2002, p.43

47 Nachtigall, W.: Bionik, Grundlagen und Beispiele für Ingenieure und Naturwissenschaftler, 2002, p.52

48 Otto, F. et al.: Natürliche Konstruktionen, 1985, p.8

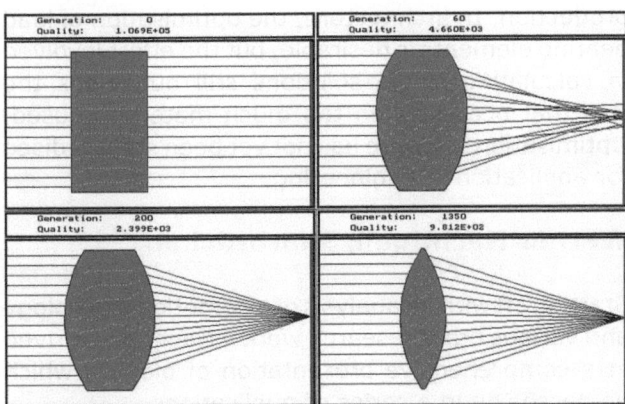

Fig.30 Evolutionary optimisation program, generation of a lens, Ingo Rechenberg.

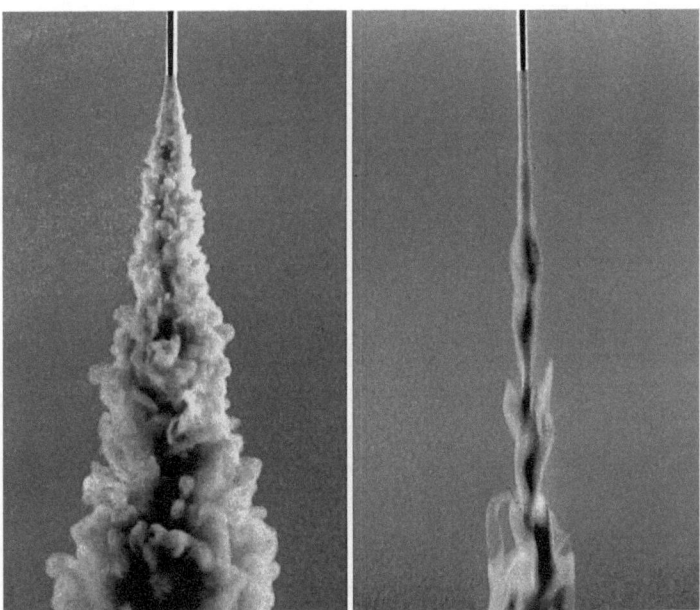

Fig.31 Flow experiments: reduction of turbulences in water mixed with fish slime (right), Rechenberg.

2.2.3 Recent views

There is a difference in the application or interpretation of bionics/biomimetics in different countries or cultural regions. Within the UK, the research institutions dealing with biomimetics are allocated in engineering faculties, but are active in many fields. In Germany there is a stronger connection to institutions of the life sciences, and meanwhile substantial collaboration with industrial partners. Japan and China have established biomimetic research in many fields, robotics, nanotechnology and medical technology being most important. Within the US, many institutions investigate robotic systems, material science and nanotechnology. Presenting the diversity of the approaches taken would go beyond the scope of this work.

The reason for the differences may just be one of historical development, but still the openness to other disciplines tells something about the field and about the urge to find innovative solutions. In the following, selected European researchers and institutions are described to give an idea of recent bionic research.

Ingo Rechenberg, Berlin

Ingo Rechenberg is a specialist in aerodynamics who discovered that mechanisms of evolution and natural selection can be applied to technical problems; especially to those difficult to solve by means of mathematics. In 1973 Rechenberg became head of the department for "Bionik und Evolutionsstrategie" at the University of Technology in Berlin, one of the few institutions where, at that time, research in bionics was being done.

Rechenberg published papers on his "Evolutionsstrategie" in the 1960s and the book "Evolutionsstrategie – Optimierung technischer Systeme nach Prinzipien der biologischen Evolution" in 1973[49].

An enlightening example, which is cited again and again, is the evolution of the human eye: with his ES (evolution strategy), simulating evolution and natural selection in nature, Rechenberg was able to demonstrate that the transformation from a rectangular-shaped transparent block to an optimal focussing lens took only 1350 generations. The optimisation criteria were the focussing of the simulated rays of light on a screen behind the lens and the reduction of the volume of the lens, which was equivalent to reducing its weight. After merely 60 generations the focussing was sufficient for imaging.[50] This very simulation opposes the theories of Creationists and Intelligent Design that claim that something as complex as an eye could only be shaped by a higher form of intelligence, and did not develop through evolution and natural selection. In fact, the basic principle of the eye has developed in several different manifestations in the course of evolution.

---

49 A more recent publication in Rechenberg's findings is Rechenberg, I.: Evolutionsstrategie '94, 1994.

50 http://www.bionik.tu-berlin.de/institut/s2anima.html [08/2007]

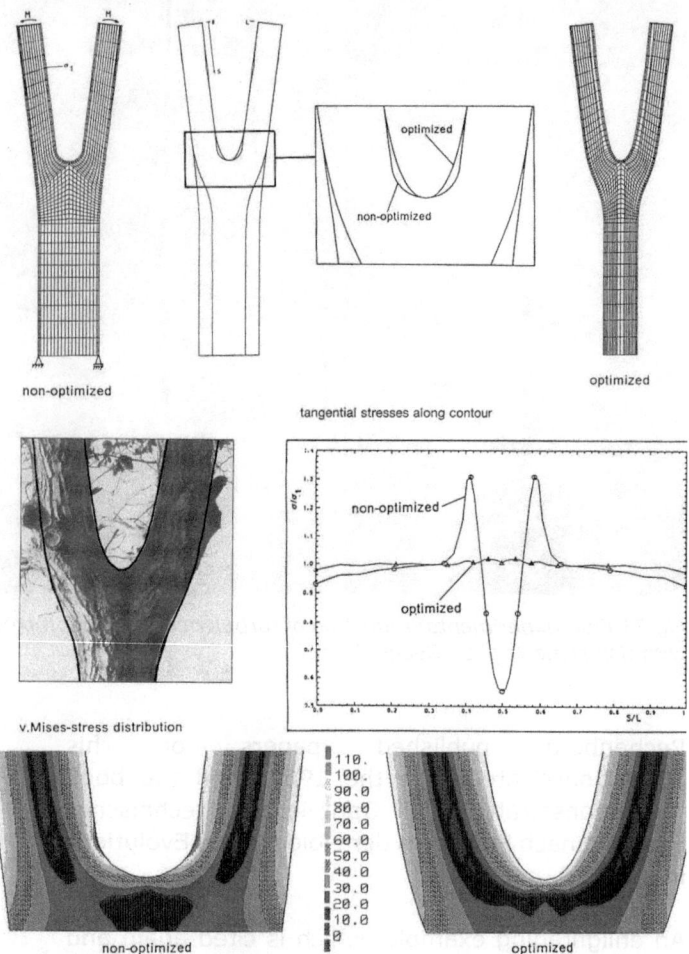

Fig.32 Comparison of stress in an optimised and a non-optimised fork, Claus Mattheck, FEM Uwe Vorberg.

This procedure is obviously affordable in mass production. In architecture, the optimisation of load bearing elements is desirable, but the effort involved in calculating single solutions still outweighs the loss that occurs when too much material is used. Optimisation software has not yet been standardised for application in engineering.

Werner Nachtigall, Saarbrücken

Starting off with an analysis of locomotion physiology and various other research works, Nachtigall arrived at a comprehensive presentation of bionics, which he passes on in a series of publications.
The book "Bionik - Grundlagen und Beispiele für Ingenieure und Naturwissenschaftler"[54] of 1998 is a standard work on bionics. In "Vorbild Natur"[55] he concentrates more on design. He is the cofounder of the Society for Technical Biology and Biomimetics (GTBB), which for a long time was the only forum for all interested persons and platform for interdisciplinary exchange in the German speaking world.

Rudolf Bannasch, Berlin

Rudolf Bannasch is member of the board of the German Biokon-network. He is a biologist, specialised in neurophysiology, who works in the fields of functional morphology of birds' flight apparatus, in fluid mechanics, underwater "flight" of penguins, comparative studies between swimming and flying organisms. A fellow of the Berlin's technical university, he founded his own company Evologics there in 1999. Within this framework he carries out research and development of diverse projects, e.g. autonomous submarine vehicles, new actuation systems, biomimetic propellers, underwater communication based on the role model delivered by the dolphin and application of the so-called "finray" technology, following the construction principle of fins.
With many others, Rudolf Bannsch and Thomas Speck belong to the "new generation" of bionics researchers in Germany, after the retirement of Ingo Rechenberg and Werner Nachtigall.

Claus Mattheck, Karlsruhe

Claus Mattheck invented the computer based optimisation procedure SKO (Soft Kill Operation) and CAO (Computer Aided Optimisation), which imitate adaptive growth of trees by technical means.[51] Natural trees grow according to the axiom of constant stress distribution and thus avoid peaks of tension at forks, but in technical elements this is not yet practised. During growth, trees add material specifically in tensioned parts. The computer model SKO works the other way: low stress initiates material cutback. CAO is a refined program that can do both: add and cut back material according to the stress distribution in the element.[52] The optimisation methods work well, if the basic layout of the structure is well designed. The Opel car company makes use of Mattheck's software in "topology optimisation" of constructive parts.[53]

51 Mattheck, C.: Design in der Natur, 1997

52 Nachtigall, W.: Bionik, Grundlagen und Beispiele für Ingenieure und Naturwissenschaftler, 2002, pp.375

53 Nachtigall, W.: Bionik, Grundlagen und Beispiele für Ingenieure und Naturwissenschaftler, 2002, pp.383

54 Nachtigall, W.: Bionik, Grundlagen und Beispiele für Ingenieure und Naturwissenschaftler, 2002, first edition 1998

55 Nachtigall, W.: Vorbild Natur, 1997

Fig.33 Swimming penguin, one of the subjects of investigation in bionics.

Thomas Speck, Freiburg

Thomas Speck is the director of the Botanical Garden Freiburg, Germany, runs the Plant Biomechanics Group Freiburg and, with his research group, is part of the Baden Württemberg Kompetenznetz.

Biological research in biomechanics and functional morphology is the basis for the further development of innovative materials and structures. His focus is on lightweight structures, gradient materials, composite fibrous materials and smart textiles - self-repairing and self-adaptive.

In developing bionic products, the group closely cooperates with other research institutions, industrial partners and designers.

George Jeronimidis, Reading

George Jeronimidis is the head of the Centre for Biomimetics in Reading, England. Research and development of his group include biomechanics and structural investigation of biological materials, e.g. bone and wood, composites materials, plant fibres, smart materials, biological sensors etc. George Jeronimidis also teaches at the AA in London, within the graduate programme on Emergent Technologies. Together with Julian Vincent, he is one of the key promoters of biomimetics in the English-speaking world. They initiated BIONIS (Biomimetics Network for Industrial Sustainability), an open network to anyone interested in biomimetics, integrating people from all over the world.

Julian Vincent, Bath

The biologist Julian Vincent was the head of the Centre for Biomimetic and Natural Technologies in Bath, England. Research there includes mechanical properties of biological materials and structures, the application of concepts from biology within engineering, texture of foods, mechanical design of plants, biology as a tool for innovation. Julian Vincent is now active in the company BioTRIZ and also collaborates with architects as a consultant in biomimetics.

Schools and Networks

Biomimetics is currently taught in Bremen, Germany as a full Bachelor and Masters study, led by Antonia Kesel. Courses in biomimetics are offered in Vienna, Freiburg, Berlin, Saarbrücken, Darmstadt, Ilmenau, Aachen, Reading and Bath. In 2009 a new master course in "Bionics in Energy Systems" was implemented at the University of Applied Sciences in Carinthia, Austria. On the whole bionics is becoming increasingly known to the public and influential in research and development, so networking activities have increased in the last years. The German GTBB was followed by the very active BIOKON network as a national platform, which is keen on contact with industry.[56] BIONIS in the UK takes a more open, international and scientific approach.[57] In many countries, Austria as well, national networks are being initiated to connect single research activities and increase contact and collaboration. BIOKON International[58] was established in 2009 and tries to integrate national efforts.

For many reasons architects are becoming increasingly interested in biomimetics, so at the time of writing new research groups are emerging, keeping in close contact with the classic research institutions in biomimetics.

56 http://www.biokon.net/ [11/2007]
57 http://www.extra.rdg.ac.uk/eng/BIONIS/ [11/2007]
58 http://www.biokon-international.com/ [10/2010]

Fig.34 Red pigment is washed away completely from a lotus leaf.

2.2.4 Fields of research and examples

In the following, some of the classic examples of bionics will be presented, according to the order suggested by Werner Nachtigall. The examples come from different disciplines and shall illustrate how the bionic approach is implemented.

Structures bionics [Strukturbionik]

The focus of structures bionics are
"Biological structural elements, materials and surfaces"[59]
(also packing, folding, patterns...). The scale of structures bionics is very small, many examples belonging to nanotechnology. The classic example of structures bionics is the Lotus effect.

Lotus effect

Meanwhile the "Lotus effect" has become famous as a term for self-cleaning effects. The same effect appears on surfaces of insect bodies and wings.

- **Biological basics of the lotus effect**

The surface of plants, especially the outer layer of the surfaces of plant leaves, is covered with fine wax excretions, which make the surface hydrophobic. This fine fractal structuring is also responsible for the weak adhesive forces of dirt particles, which can easily be removed with water. This effect is particularly distinct in the leaves of the lotus plant. The botanists Wilhelm Barthlott and Christoph Neinhuis were able to prove that the *"connection between the structure of the surface, reduced adhesion of particles and hydrophobic character is the key to the self-cleaning mechanism of many biological surfaces"*.[60] The pollution and the moistening was standardised for experimental testing of different plants. The removal of particles has diverse advantages for the plant: better efficiency for photosynthesis and protection from infection.

- **Physical basics of the lotus effect**

The wetting of surfaces is the basis for the phenomenon. The degree of surface tensions between water and air, water and solid surface, and solid surface and air, and the contact angle is explained by the Young-equation: a contact angle of 0° denotes complete wetting, a contact angle of 180° complete non-wetting.

---

[59] Nachtigall, W.: Bionik, Grundlagen und Beispiele für Ingenieure und Naturwissenschaftler, 2002, pp.57

[60] Nachtigall, W.: Bionik, Grundlagen und Beispiele für Ingenieure und Naturwissenschaftler, 2002, p.340

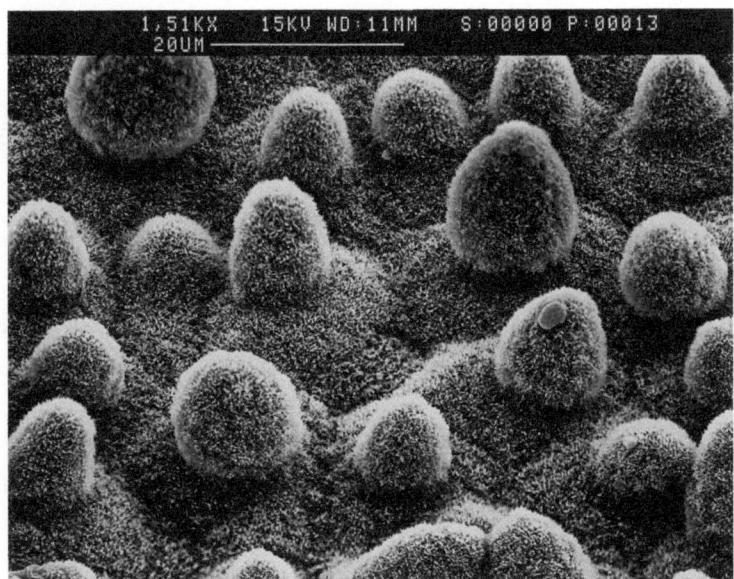

Fig.35 REM photograph of the lotus leaf surface.

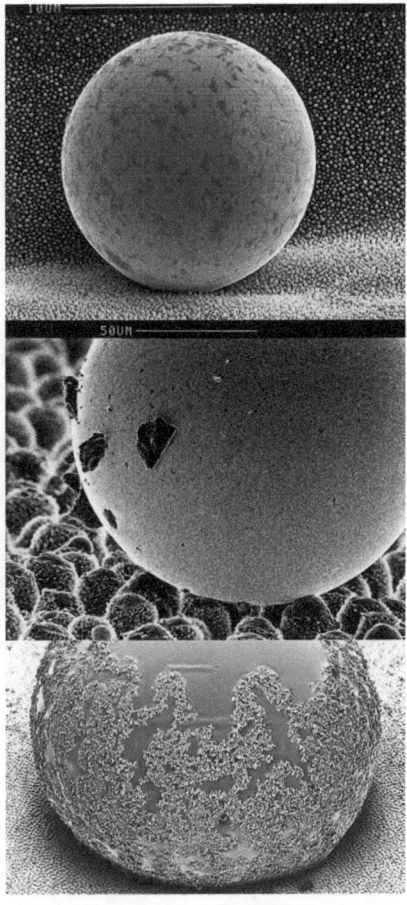

Fig.36 Mercury droplets on leaf surface.

Plant surfaces lie somewhere in between. On rough surfaces is the amount of interface between water and air increased, whereas the interface between water and solid surface tends towards a minimum. The water can gain little energy out of adhesion and takes the form of a sphere. The total energy of the system is lowest in this way. The contact angle of the droplet depends only on the surface tension of the liquid.[61]

- **Transfer of the lotus effect**

The Lotus effect was transferred to products that imitate the effect artificially. Lotusan is the name of paint with self-cleaning characteristics.

"Lotusan combines the ameliorated hydrophobic characteristic of the ispo silicon colours with the microstructure taken from the Lotus leaf. The contact surface for dirt and water is extremely reduced. Together with the high hydrophobic characteristics a super hydrophobic surface is created. Water rolls off and takes the loose particles with it. The façade stays dry and beautiful..."[62]

Other products that have been developed are ceramic surfaces (which repel even honey), self-cleaning foils and coatings. The self-cleaning effect depends on wetting. There is a noticeable difference in the grade of dirtiness, when the object has unwetted parts, as in the façades of buildings.

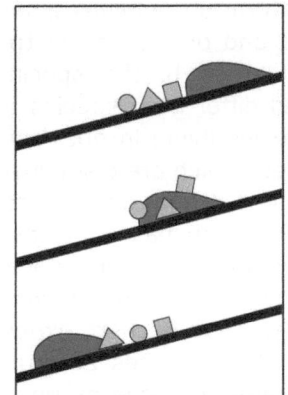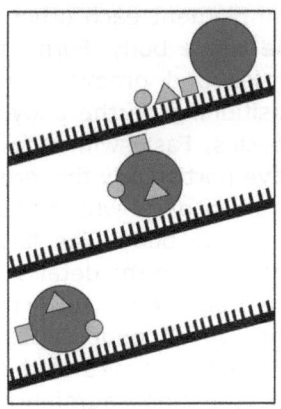

Fig.37 Effect of a smooth and hydrophilic surface in contrast to a structured and hydrophobic one.

---

61 Nachtigall, W.: Bionik, Grundlagen und Beispiele für Ingenieure und Naturwissenschaftler, 2002, p.342

62 Advertisment of ispo gmbh, Frankfurt, ispo.gmbh@ispo-online.de

Fig.38 Structure of sharkskin.

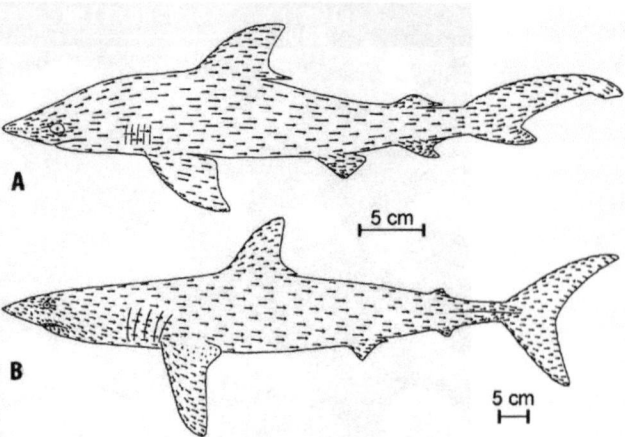

Fig.39 Form and pattern of groove in different shark species.

Shark skin

Sharks, with their long evolution history, display excellent adaptation to their environment. In the 1980s palaeontologist Ernst Reif discovered the structure responsible for the shark-effect. Particularly structured scales, so-called riblets, cover the skin of sharks. The grooves of the scales complement each other to form streamlines along the whole body. Form, size and orientation of the scales and grooves are adapted to the specific positioning on the body and differ from species to species. Fast swimming species living in open sea have particularly fine grooves, which are distinctive on body parts where the water stream disconnects from the body. The fine structure has a positive influence on the detailed dynamics of the interface: the drag is reduced. According to experiments with grooved foil conducted by Dietrich W. Bechert, expert on fluid dynamics in the DLR, Deutsches Zentrum für Luft- und Raumfahrt, drag is reduced by up to 10%. Experiments with mockup scales were carried out in oil channels because of similarities in fluid mechanics.

Applications exist in various products: 3M has produced foils to test on an Airbus for fuel reduction; Speedo and Strush company produced textiles for swimsuits; surface structuring of tubes makes pumps, heat exchangers and air conditioners more efficient.[63]
"Speedo will only focus on the management of existing forces... will never compromise sporting integrity." Speedo advertisement, 2000.

The aim of research was to improve the design of the traditional swim suit and to reduce its drag in the water. The structure of the "fastskin" suit imitates the surface of sharkskin: v-shaped grooves reduce friction and enable the swimmer to glide faster through the water.[64] The material used is super elastic, enhances the fit of the gear and compresses the muscles, which leads to more efficiency.
The cut of the suit was adapted by means of a body scanner to the swimming movements and muscle groups. The seams, too, are elastic and work like tendons together with the elasticity of the fabric. They create a tension within the suit, which enables full freedom of movement.[65]
Recent investigations have rejected the effectiveness of the fabric in terms of drag reduction, but *"it may support the muscles to some extent."*[66]

---

63 Nachtigall, W.: Bionik, Grundlagen und Beispiele für Ingenieure und Naturwissenschaftler, 2002, pp.204

64 Nachtigall, W.: Bionik, Grundlagen und Beispiele für Ingenieure und Naturwissenschaftler, 2002, p.210

65 http://www.speedo.com [2002]

66 Vincent, J.F.V. et al.: Biomimetics - its Practice and Theory, J.R.S., 2006, p.3

Devices bionics

"Development of usable overall constructions."[67]
Devices bionics focuses on the investigation, transfer and application of the vast variety of natural microsystems, taking role models primarily from insects and crustaceans.
Application fields are: microrobotics, microactuation, grabber systems, driving elements, kinematic chains, micro constructions, microtools, energy storage systems...[68]

Structural bionics

"Biological constructions, closely related to above structural and device bionics." [69]
Velcro is a comparatively old but famous biomimetic product that is used in many spheres of life. It is a real biomimetic product, following the principle of statistical hooking of burrs, patented in 1951 by the Belgian George de Mestrel, who took inspiration from the clingy plant seeds when freeing them from his dog's fur.

Fin

"Finray technology" is a biological construction discovered and applied by the research group of Rudolf Bannsch. They found that the tail fin of a fish reacts strangely to pressure. Pushing to one side, the fin does not bend to the other, as one would expect, but instead the whole fin vaults against the acting force. The cause of this is radial bracing within the fin. The scientists used the principle of "reversal of action" to design a revolutionary backrest: if pressed backwards - when somebody leans on it - the whole construction bends forward. This way it connects closely to the spine and enables support. In cooperation with experts on ergonomics they developed the chair design for industrial application.[70] The principle has already found its way into a wide range of products.

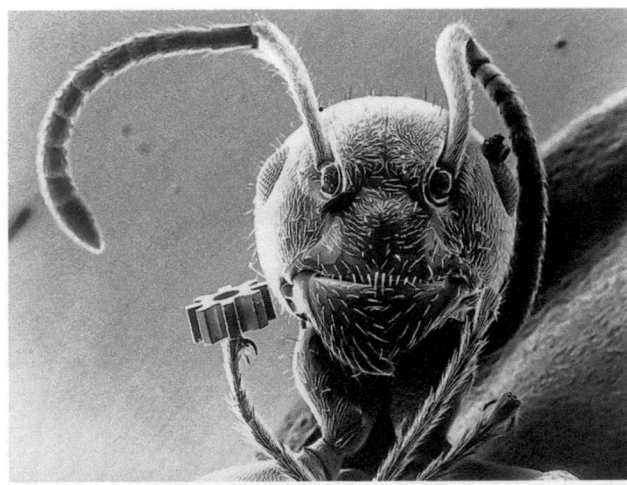

Fig.40 Ant with micro gear.

Fig.41 Velcro.

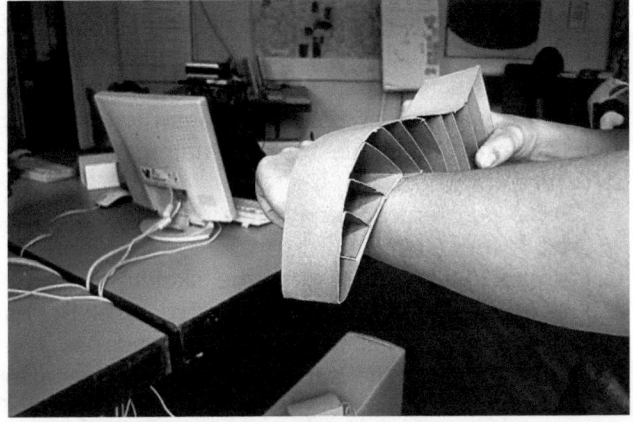

Fig.42 Student with a model of a fin, 2010.

---

67 Nachtigall, W.: Bionik, Grundlagen und Beispiele für Ingenieure und Naturwissenschaftler, 2002, pp.111
68 Nachtigall, W.: Bionik, Grundlagen und Beispiele für Ingenieure und Naturwissenschaftler, 2002, p.113
69 Nachtigall, W.: Bionik, Grundlagen und Beispiele für Ingenieure und Naturwissenschaftler, 2002, pp.111
70 http://www.kommwiss.fu-berlin.de/fileadmin/user_upload/wissjour/dim23.pdf [08/2007]

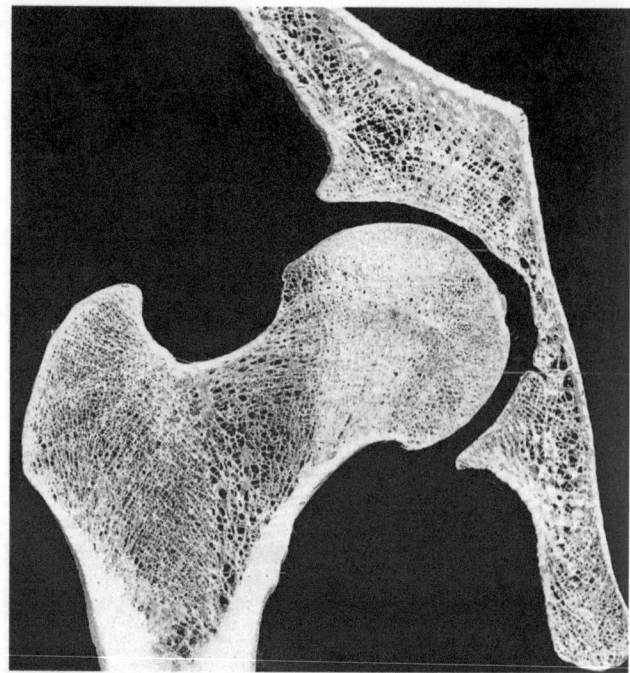

Fig.43 Section through the head region of the femur, the thighbone.

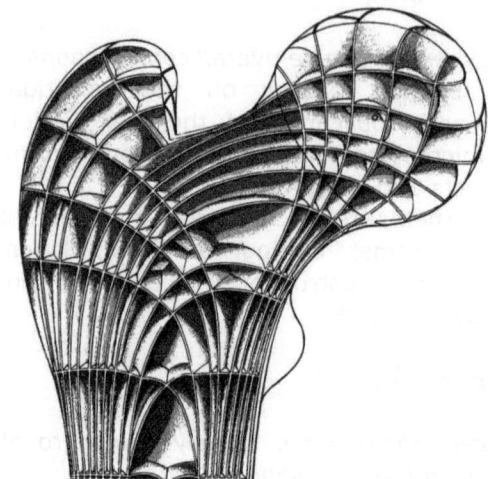

Fig.44 Reconstruction model of the spatial trajectory surfaces, showing direction of pure tension and pressure.

Bone

In its complexity, bone is a hitherto untranslated example of a biological construction, but some aspects have recently found translation, and research efforts are continuing.

Bone is also interesting as a role model for architecture, as it represents a natural lightweight construction which has some very dynamic characteristics helping to efficiently fulfil the task of the organism's primary construction. It is also a perfect example for a dynamic adaptable structure.

- **Bone as ideal construction, for example the human femur**

It is important to state that the skeleton is only a part of the whole load bearing structure. Bones work only in collaboration with tendons and muscles, forming the supporting apparatus, also called the musculoskeletal system. The object of the following observation is bone itself, without the other elements of the entire system.

- **Axiom of constant stress distribution**

As in trees, the axiom of constant stress distribution is also valid for bone. There are no predetermined breaking points and there is no waste of material. The principle of economic input of material is ubiquitous in biological constructions.

The lightness of construction of the mammal skeleton has priority over its strength.

The safety factor s - the ratio of load that would produce failure to the greatest load expected in use[71] - is well known in technology. For bones and tendons, the safety factor is between 2 and 6. With plants, the safety factor is between 2 and 4.[72]

- **Spongiosa**

The end region of our long bones is formed by a so-called "spongiosa" or cancellous bone, minute beams out of fine bone lamella. Their orientation follows tension and pressure trajectories, as can be seen in experiments in photoelasticity. The average direction of the resulting force is aligned exactly with the orientation of the pressure trajectories, both being crossed at right angles by tension trajectories.

Processes of addition and removal of material are active continuously, adapting the structure of spongiosa to the forces acting. In contrast to the adaptive growth of trees, the bones of vertebrates can disintegrate within their bodies. With old age or certain illnesses the dynamic process of adaptation is disturbed, so that the bones become brittle and fragile.

For reasons of energy management, nature renounces absolute strength, and accepts breakage when a certain load limit is exceeded. In case of breakage, the body creates an emergency splint out of fast growing connective tissue. After the fracture is overgrown, the bone is rebuilt to its normal state as well as is possible.

---

71 Vogel, S.: Cats' Paws and Catapults, 1998, p.243
72 Vogel, S.: Cats' Paws and Catapults, 1998, p.244

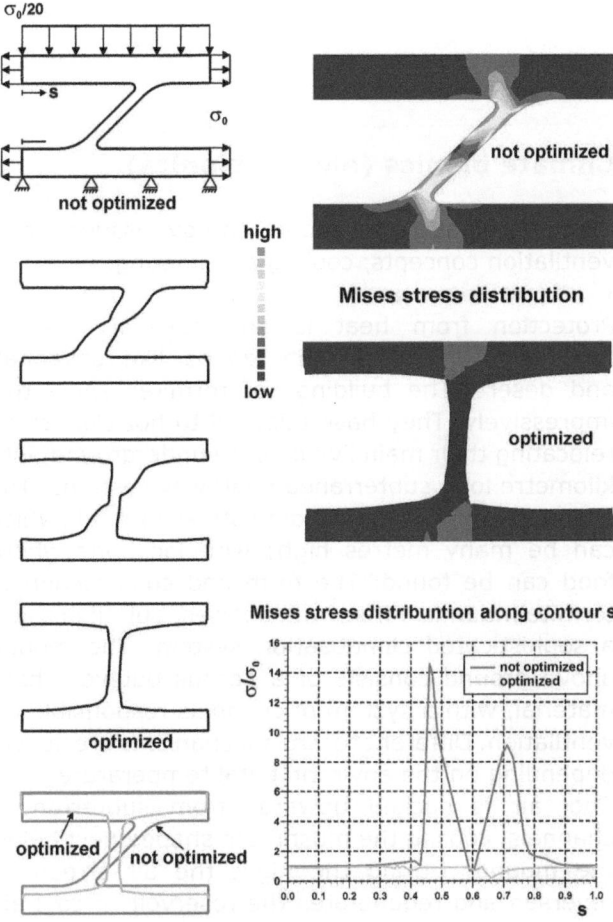

Fig.45 Principle of rearrangement of trabecular bone after change of stress, Claus Mattheck.

Anthropobionics

"Robotics, bionic implants"[74]
Anthropobionics investigates the whole issue of man-machine interaction and ergonomics.

Artificial bone material and prosthesis

Medical technology produces artificial bone material, which is used for damaged bone, where it is integrated through growth, in time disperses and is replaced by newly grown material. If whole parts are affected, only implants can help. Materials and processing aim at avoiding infection and rejection. A serious problem with implants is the change of stress distribution. These changes can lead the body to remove bone material in border areas, which loosens the connection of the implanted part. Scientists try to avoid this reaction by using porous material, in which natural material can grow. Elastic bars made of titanium with a plastic coating can mimic the elastic properties of bone.[75]

Mimicking the process of creation of bone material is the starting point of many different research groups. Application of the process is much more interesting than mimicking the static structure, but more difficult to investigate.

It is known that electrical voltage which is created by bending is responsible for the control of these processes. Piezoelectricity within the bone material creates a grid of excitation, which functions as guidance for the cells transporting the calcium substance. The process of bone growth and degradation is highly dynamic: within days and weeks the strength of human bone can change considerably, as has been learned from spaceflight.[73]

This kind of fast adaptation to environmental conditions is typical for structures in nature, and represents process-like optimisation, which is obviously needed for efficient survival.

Applications relating to the construction of bone are presented in the next chapter on what is called anthropobionics.

73 Nachtigall, W.; Blüchel, K.G. (Ed.): Das große Buch der Bionik, 2000, pp.244

74 Nachtigall, W.: Bionik, Grundlagen und Beispiele für Ingenieure und Naturwissenschaftler, 2002, pp.285

75 Bürgin, T. et al.: HiTechNatur, 2000, p.47

Fig.46 Termite mound.

Construction bionics [Baubionik]

"Light constructions occurring in nature, cable constructions, membranes and shells, transformable constructions, leaf overlays, use of surfaces, etc."[76]

Baubionik deals with all topics that relate to building. Large-scale natural constructions are interesting role models. Building physics is also integrated in this field, and animals' buildings may serve as role models. Construction bionics is the most promising field of bionics for architecture.

Climate bionics (energy bionics)

Climate bionics is about energy issues, e.g. ventilation concepts, cooling and heating.[77]

- **Termite mounds**

Protection from heat is an essential criterion for design in hot climate zones like savannah and desert. The buildings of termites show this impressively. They have adapted to hot climate by relocating their main living space underground, into kilometre long subterranean pathway systems. The channels connect the building above ground (which can be many metres high) with locations where food can be found. The form and construction of termite mounds differs with species, but all possess a sophisticated climatisation system. The mound above ground consists of a porous but very hard material, with a system of channels responsible for ventilation. Different control mechanisms are active, depending on the environmental temperature.

Cool air is sucked upwards from subterranean channels, cooling the mushroom shaped nest below the mound. During the night the air stream is reversed and regenerates the reservoir of cool air. Particular systems use deep channels to the ground water to gain additional cooling energy through evaporation. Compass termites even orient their asymmetrical flat shaped mound with the long axis from east to west, to avoid the hot summer sun.[78] The reason for all this effort is to control humidity and temperature, in order to provide a stable environment for the termites and their offspring in the nest, and for some species also for their symbiotic fungi. The termite mounds serve as a role model for an effective passive ventilation system for the control of the internal climate. Efforts have been made to translate this principle in architecture. Detailed examples will be given later. The material that termites generate out of sand and their saliva is porous but very hard. In suitable climate zones the material can be used for adobe buildings as an additive to clay, due to its prestructured nature.

76 Nachtigall, W.: Bionik, Grundlagen und Beispiele für Ingenieure und Naturwissenschaftler, 2002, pp.149

77 Nachtigall, W.: Bionik, Grundlagen und Beispiele für Ingenieure und Naturwissenschaftler, 2002, pp.154

78 Nachtigall, W.; Blüchel, K.G. (Ed.): Das große Buch der Bionik, 2000, pp.306

Robotics - sensory bionics

"Sensory bionics, neurobionics and locomotion bionics"
Bionics of sensors investigates detection and processing of physical and chemical stimulation, location and orientation within an environment. [79]
It is strongly related to the fields of locomotion bionics and neurobionics, which are about data analysis and information processing.

- **Robotics, artificial insects**

Walking with six legs is the most stable variation of legged animal locomotion. When three of the legs make the next step, the others form a stable triangle. Attempts to mimic this kind of locomotion artificially (classic locomotion-bionics) would be impossible without complex control circuits of sensing and regulation.

Research of the locomotion control of stick insects showed that exchange of information takes place along the legs of each side; the front leg informs the next one, the middle one the last one. Control is decentralised by six independent nodes, which are inter-connected with each other. Each leg has its own pace generator. The decentralised control unit approximates the principle of neural networks.

Lauron II (Laufender Roboter II), developed in 1996, served as platform to investigate concepts of locomotion control rather than as a vehicle for unstructured terrain (where it would have a big advantage over wheels). Lauron II is equipped with 150 sensors with different purposes controlling its locomotion.[80]

Results for locomotion of this kind find applications in the development of machines, vehicles and robots for regions that for some reason are difficult to access (sewage canals, forests, mined areas, Mars surface...). The two-legged human walk still remains a challenge for researchers.[81]

Fig.47 Stick insect.

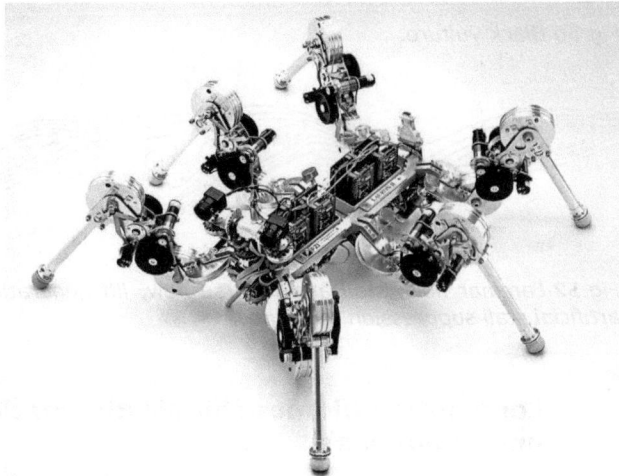

Fig.48 Lauron II.

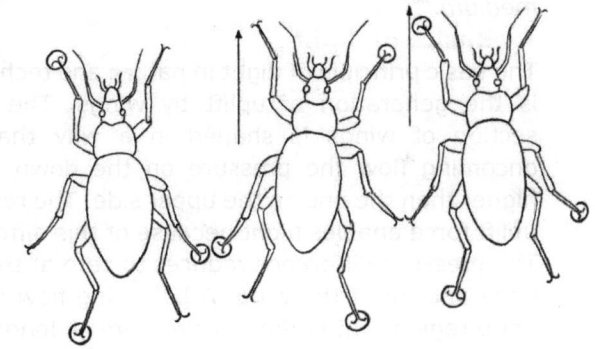

Fig.49 Motion pattern of six-legged walk.

[79] Nachtigall, W.: Bionik, Grundlagen und Beispiele für Ingenieure und Naturwissenschaftler, 2002, pp.175, pp.243
[80] Siemens Forum: Bionik, Zukunfts-Technik lernt von der Natur, 1999, p.36
[81] Examples from MIT leglab http://www.ai.mit.edu [08/2007]

Fig.50 Black vulture.

Fig.51 Seagull.

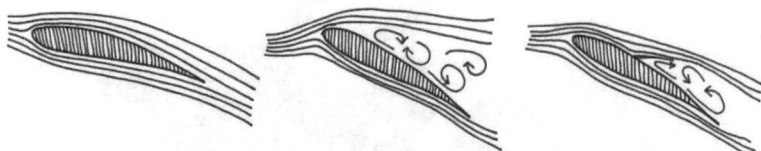

Fig.52 Laminar flow, stall and reverse flow, lift-generating flow with artificial stall suppression measures.

Locomotion bionics (bionic kinematics and dynamics)

"Walking, swimming and flying as primary forms of movement. Interaction with the surrounding medium."[82]

- **Basics of flight**

The basic principle of flight in nature and technology is the generation of uplift by wings. The cross-section of wings is shaped in a way that with oncoming flow the pressure on the down side is higher than the one on the upper side. The resultant uplift force enables flight because of this difference. The pressure difference reduces to zero at the back edge and tip of the wing. A balancing flow tries to unite regions with different pressure. A long vortex is created in addition to the main air stream, which flows off to the back, its drag wasting something between 30% and 50% of the total energy required for propulsion. Seabirds solve this problem with very long wings with narrow tips and small drag vortices. For other birds and airplanes a wide wingspread is cumbersome. They have solved the problem with split wingtips, creating many small vortices, with less drag. But the distance between the "fingers" must be large enough that the small vortices do not unite into a single big vortex, multiplying drag. For this reason, and because the dynamics during flight does not allow a static solution, the so-called mono-winglet is common as a technical solution for drag reduction at wingtip.

With Rechenberg's Evolution Strategy, a Mulit-winglet was optimised whose best glide ratio is 11% better than that of the rectangular version.[83]
Michael Stache and Rudolf Bannasch developed the so-called split-wing-loop further by using the principle of loop-shaped "blades" for propellers and wind turbines, to reduce turbulence and produce less drag and less noise.[84]

- **Kinds of flow**

"Inertia also affects how fluids flow. At small sizes, viscosity predominates, and flows are orderly, laminar affairs. Each bit of fluid does nearly the same thing as its neighbours. At larger sizes, inertia increasingly offsets viscosity, so the bits of fluid tend to keep doing whatever they have been, despite any different motion of their now more temporary associates; we call such flows, with their chaotic eddying, turbulent."[85]

- **Boundary layer**

Sufficient uplift for flight is created through profiled aerofoils, when oncoming flow takes a specific angle of approach and no separation between flow and aerofoils occurs. With growing angle of approach uplift increases to a maximum, but then drops abruptly due to laminar separation and reverse flow of the boundary layer on the upper side of the airofoils. With airplanes, stall happens during excessive flight manoeuvres and at insufficient flight velocity. Laminar flow around the aerofoil is then disturbed, and regeneration of the flow is unsuccessful in most cases.

---

82 Nachtigall, W.: Bionik, Grundlagen und Beispiele für Ingenieure und Naturwissenschaftler, 2002, pp.175

83 Siemens Forum: Bionik, Zukunfts-Technik lernt von der Natur, 1999, pp.28

84 http://www.evologics.de [08/2007]

85 Vogel, S.: Cats' Paws and Catapults, 1998, p.52

In risky manoeuvres birds also risk laminar separation and reverse flow in the boundary layer of the upper side of their wings. But in birds wings the highly elastic covert feathers of the pinions deploy automatically and prevent the dangerous reverse flow. This principle was already known in the 1940s, and is now being tested as a kind of brake for reverse flow by means of overlapping flaps with low aerial permeability. It is important that the reaction of the flaps is initiated automatically, in a passive mode, when separation begins. In an experiment with stripes of perforated foil the model withstood separation until an angle of approach of 40°, compared to 18° with a comparable model without flaps.[86]

Other measures to prevent stall are removing the small vortices from the upper side of the wings through active exhaustion, or covering the wings with microscopically small plates plates that will influence the boundary flow. These plates will be controlled by microscopic sensors and regulation elements connected to an efficient processing unit.[87]

Neurobionics

"Data analysis and information processing."[88]
The development of neural networks has been inspired by the findings in neurobiology and biocybernetics. Neural networks are used for data processing in the control of automated systems, for example locomotion control of robots.

Evolutionary bionics

"Evolution techniques and evolution strategies, made useful for technology."[89]
The Evolution Strategy developed by Rechenberg is used for optimisation tasks for a wide range of applications. Processes of mimicking evolutionary development are important when the solution has not yet been found by with numeric methods.
The connection between the concepts of technical design by humans and design in nature holds advantages.

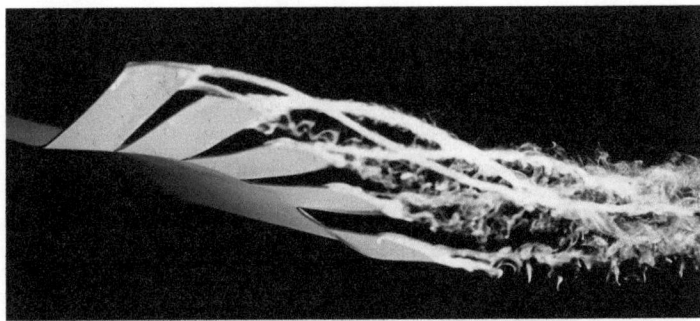
Fig.53 Drag vortices on a multi-winglet in a wind tunnel.

Fig.54 Multi-winglet on airplane.

Fig.55 "Berwian", biomimetic wind turbine.

---

86 Patone, G., Müller W. in Siemens Forum: Bionik, Zukunfts-Technik lernt von der Natur, 1999, pp.31

87 E. Covert from MIT in: Spektrum Spezial Nr.4 Schlüsseltechnologien im 21. Jahrhundert, 1995, p.59

88 Nachtigall, W.: Bionik, Grundlagen und Beispiele für Ingenieure und Naturwissenschaftler, 1998, p.24

89 Nachtigall, W.: Bionik, Grundlagen und Beispiele für Ingenieure und Naturwissenschaftler, 1998, p.25

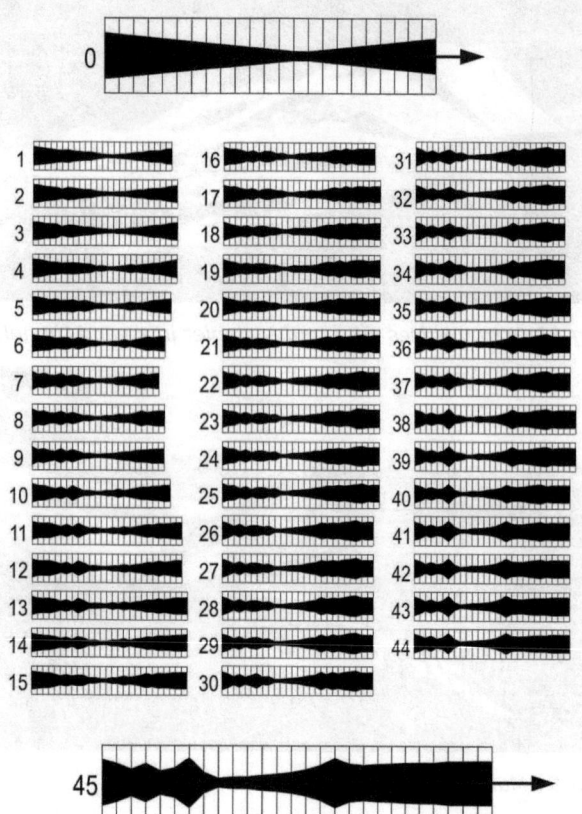

Fig.56 Optimisation of a two-phase supersonic nozzle by means of artificial evolution, by Ingo Rechenberg 1994.

Principles

The mimicking of evolution is done by simulating the individual processes of mutation, recombination and natural selection. The characteristic parameters of the technical system which has to be optimised are subject to mutation in form of changes. The resulting systems are combined and reproduced using nature-like strategies. The new generations of the system are tested, and their adaptation to the examined criteria noted. Two different methods of selection can be applied: in the so-called "+ selection" the better offspring deposes the parent generation while in the "- selection" only the best offspring is developed further to a new generation. The process will lead to an increased adaptation to its environment for the investigated characteristics, so that the system can be considered optimised in relation to the specific requirements.[90]

Efficiency optimisation of a two-phase supersonic jet

The problem of inefficient nozzles occurred during the research of generation of energy in satellites. At that time, in 1968, theories for calculating complex flow had not yet been developed, so there was no way of arriving at a better theoretical solution. Hans-Paul Schwefel and Ingo Rechenberg succeeded in increasing the efficiency of the nozzle by applying artificial evolution to the problem. After 45 generations an unforeseeable optimal geometric form emerged.[91] Meanwhile the theory of the form had become understood. The Evolution strategy is used primarily in mechanical engineering, but applications in architecture are emerging.

Procedural bionics

"Nature's procedures or processes"[92]
Biological processes are investigated, imitated or applied in technology. Photosynthesis, the development of the artificial leaf, hydrogen technology, recycling processes, desalination are examples which relate to energy and ecology issues.[93]

- **Organic solar cells**

Polymer based solar cells were investigated and developed worldwide by many research and development institutions. The technology to print thin films of organic solar cells was investigated by Serdar N. Sariciftci at the Kepler University in Linz.[94] Recently the Konarka company announced an efficiency enhancement of up to 6.5%. Compared to common silicon solar cells, the new technology can be produced incredibly cheaply, and the light, flexible and translucent foils can be integrated into all kinds of products as well as architectural elements.[95]

Organisational bionics

"Complex relationships of biological systems"[96]
Integration into ecosystems, networks, self-organisation, strategies for recycling, packing etc. is subject of organisational bionics. Nachtigall suggests the integration of bionics into a so-called "bio-strategy", to ensure sustainability and survival.

---

90 Nachtigall, W.: Bionik, Grundlagen und Beispiele für Ingenieure und Naturwissenschaftler, 2002, pp.175, pp.362

91 Nachtigall, W.: Bionik, Grundlagen und Beispiele für Ingenieure und Naturwissenschaftler, 2002, p.14

92 Nachtigall, W.: Bionik, Grundlagen und Beispiele für Ingenieure und Naturwissenschaftler, 1998, p.25

93 Nachtigall, W.: Bionik, Grundlagen und Beispiele für Ingenieure und Naturwissenschaftler, 2002, pp.311

94 http://www.ipc.uni-linz.ac.at [2001]

95 http://www.konarkatech.com [08/2007]

96 Nachtigall, W.: Bionik, Grundlagen und Beispiele für Ingenieure und Naturwissenschaftler, 2002, pp.391

Fig.57 Luigi Colani Model, Design museum exhibition 2007.

2.3 TRANSFER AND METHODS

How does this transfer of ideas work? To apply ideas from nature to technology, and especially architecture, strategically, an important precondition has to be fulfilled.

2.3.1 Interdisciplinarity

Interdisciplinary work is essential for the investigation and the design of natural constructions.
Basic biological research must form the foundation for biomimetics. Interdisciplinary working methods are required on both sides; the discipline delivering research and the discipline dealing with application. To this end, the linguistic barriers between the disciplines must be overcome. For this reason: the study of biomimetics includes both life sciences and engineering.
Working methods have to ensure suitable communication. The quickest results in the cooperation of different disciplines can be obtained through intensive workshop-like situations, where differences in approach, methods and understanding are detected immediately and resulting problems are resolved fast. Mediated approaches are more difficult to handle. In an ideal situation, experts of any special field that appears can be called in.
The generalist approach that architects inevitably have to take when designing buildings qualifies them to work in biomimetics. Their usual working practice involves contact and cooperation with highly specialised consultants and professionals. Their specific approaches, ideas and the requirements of their respective profession have to be integrated into one single project.

Fig.58 Long Island Duck.

Architects have to grasp these facts fast, while not neglecting the other dimensions of the project. On the other hand, it is necessary for architects to present their ideas lucidly to a more or less untrained audience, which is not an easy task at all. In general, good cooperation and communication depends on the ability of both sides to put themselves in the other's position. A basic understanding of the other disciplines is therefore a precondition for work in biomimetics.
Architects usually lack a basic grounding in life sciences. As, for good reason, they are used to being very pragmatic about information input from specialists, they tend to cut short information input from life sciences too early and do not fully understand the whole phenomenon that initially fascinated them. Consequently, the formal approach is often taken to be the first and most obvious option, and not the one consciously selected among other options.

Fig.59 Diatom shell.

Fig.60 Two-dimensional polyhedron network, soap foam, Frei Otto.

2.3.2 Inspiration

Quite commonly, architects, designers and artists take inspiration from nature. Luigi Colani's exhibition in the London Design Museum in spring 2007 is a good example: it is called "translating nature" and juxtaposes the work of Ernst Haeckel with the famous designer's own projects, most of them elegantly shaped technical devices, vehicles, trains and airplanes.

"Whenever we talk about biodesign we should simply bear in mind just how amazingly superior a spider's web is to any load-bearing structure man has made – and then derive from this insight that we should look to the superiority of nature for the solutions. If we want to tackle a new task in the studio, then it's best to go outside first and look at what millennia-old answers there may already be to the problem."[97]

The nature of the kind of transformation from biology to design is not visible, which is typical of this kind of biomorphism, interested in the creation of nature-like form. In the case of Colani's design, the outcome is evidently very successful, but even more is possible, looking at nature as a role model. On the other hand, biomorphism as the sole objective can lead to projects which take sole form as the only reference. In some cases, mimicking the form of organisms in architecture (zoomorphism) leads to funny designs, like the Long Island "Duck" built in 1930-1931 and originally used for selling for Long Island Ducklings.[98]
Going beyond the mere translation of form is the challenge that biomimetics in architecture [Architekturbionik] is about.

2.3.3 Analogy

"...A comparison between two things, typically on the basis of their structure and for the purpose of explanation or clarification"[99]

Analogies serve as a starting point for bionic translation. Common features connect elements of nature or technology.
Nachtigall stresses the importance of analogies and the research of analogies, as a pre-scientific stage of investigation, which is the starting point for studies. The term analogy was already mentioned by Rasdorsky, and used in bionic connotation by Helmcke in the 1960s.[100]
Analogy means similarity, correlation, and equivalence in terms of function or behaviour.
Juri Lebedew stated that *"...the product of a long-time exertion of similar functions in different organisms is a concrete structural similarity."*[101]
Similar functions require similar structures, and the research in this field can deliver new insights. According to Steven Vogel it is uncertain that analogies between nature and technology can deliver information regarding development processes as well, and that it is important to discern mechanisms from development history.[102]

---

97 http://www.colani.ch [06/2007]
98 Aldersey-Williams, H.: Zoomorphic, 2003, p.16
99 New Oxford American Dictionary, program version 1.0.1, 2005
100 Nachtigall, W.: Bionik, Grundlagen und Beispiele für Ingenieure und Naturwissenschaftler, 2002, p.9, 41
101 Lebedew, J.S.: Architektur und Bionik, 1983, p.27
102 Vogel, S.: Cats' Paws and Catapults, 1998, p.289

- **Analogies between nature and technology**

Analogies between the different fields are supposed to bring innovation. Frei Otto said about analogy:

"Objects can be similar are equal in form, gestalt, construction, structure and material. They may have acquired this analogy through identical, similar or completely different development processes. The development processes play a key role in research of analogies. Typical technical and artificial products differ from creations in animate nature by a basically different development process. However, the process of selection often is very similar... crude, and artificially drawn analogies are called 'trivial analogies'."[103]

- **Werner Nachtigall's interpretation of analogy research**

Nachtigall states that the nature of qualitative analogy research is impartial, open-minded comparison. If examination shows that the comparison makes sense, further questions can be raised and more detailed investigations e.g. on formal and functional features carried out. In the transfer phase, the traps that await those applying similarity laws have to be taken into account. Nachtigall presents numerous examples of insect micromorphology and relates functional mechanisms to technological examples in a visual comparison:[104]

- **Zipper**

The pleidae, a water bug species, connect their elytras with a zipper system out of chitin. The chitinous structures extend and lock, to close the connection.[105]

- **Joints**

Antennae of insects can move in all directions. Many insects have a ball joint as attachment.[106]

- **Saws**

Wasps have a particular kind of pad saw, whose parts can move against each other and saw quickly into plant material.

Yang Gao has designed a drill for application in space using the ovipositor drill of the wood wasp as a role model. The tool is intended to collect planetary samples.[107]

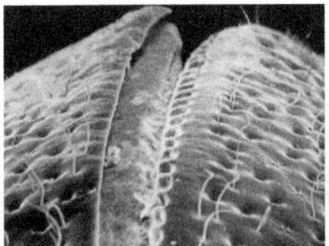

Fig.61 Zipper system in water bugs and technical zip.

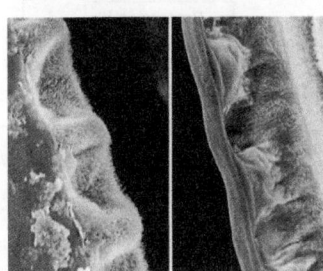

Fig.62 Chitinous structures lock into each other.

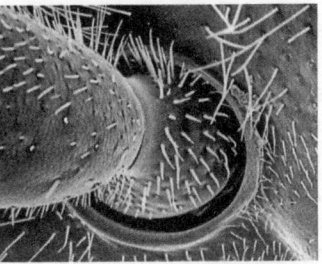
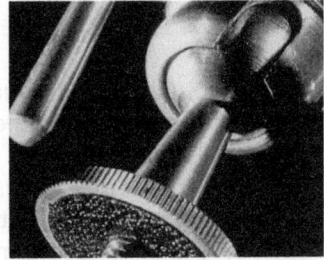

Fig.63 Antenna joint of bee beetle, and ball joint of tripod.

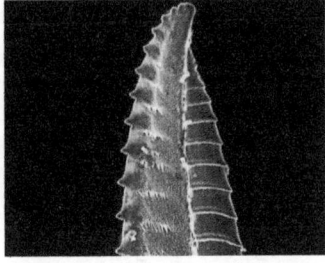
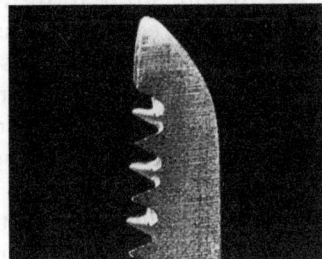

Fig.64 Padsaw of a wasp and technical saw on a pocket knife.

---

103 Helmcke, J.G. et al.: IL33 Radiolaria, 1990, p.158

104 Nachtigall, W.; Blüchel, K.G. (Ed.): Das große Buch der Bionik, 2000, p.214f.

105 Nachtigall, W.; Blüchel, K.G. (Ed.): Das große Buch der Bionik, 2000, p.220

106 Nachtigall, W.; Blüchel, K.G. (Ed.): Das große Buch der Bionik, 2000, p.221

107 Gao, Y. et al.: Bioinspired Drill for Planetary Sampling, 2007

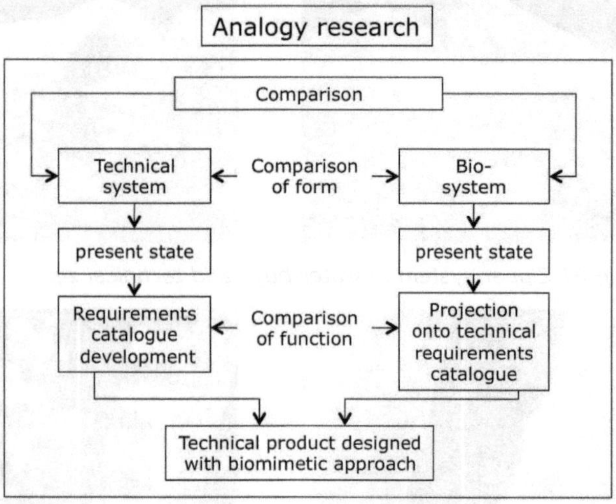

Fig.65 Concept of analogy research according to Nachtigall, translated by the author.

- **Frei Otto's interpretation of analogy research**

Frei Otto distinguishes between analytic and synthetic approaches to analogy research. The analytic approach evaluates similarities and development processes, the synthetic approach initiates experimental self-organisation processes in order to compare the outcome with the role models from nature.[108] Analogue physical models demonstrate form-generating forces, which depend on material and construction. The famous research work of Frei Otto's group on soap bubbles and minimal surfaces considerably increased knowledge of the characteristics of pneumatic structures in nature and technology.

The following statement by Paolo Portoghesi illustrates the power of analogy in using the whole capability of human mind:

"Analogy intervenes as an exploratory and unifying process capable of disclosing the general perspectives and harmonic or regulatory relations which the logic of identity alone permits neither to be perceived nor identified... Contrary to the logic of identity, which is a prevalently conscious act present in all abstract thought processes, the logic of analogy is characterised by its solid anarchism, its unconscious thematic organisation as well as the affective and emotional charge it is capable of projecting onto all objects encountered in life."[109]

2.3.4 Similarity and scale

Size matters, and the scale of things is important. Steven Vogel compares the sizes of organisms with the sizes of man-made technical devices. The range of sizes of both categories differ considerably. Whereas organisms range from 100 nanometres to 100 metres, mechanical technical devices use a range from millimetres to kilometres. This has changed through technological progress in micro and nanotechnology. Architecture uses the scale from millimetres to 100 metres and more.

The relationship between length, surface and volume is not linear, and all physical processes are affected by this phenomenon.[110] When resizing phenomena, which is often necessary in bionic translation, one has to take the effect of these differences into account.

The size of organisms is subject to constraints, which allow their existence only in a specific range of size. The size of cells, for example, is limited to 1 to 100 micrometres by the relation between surface and volume for the exchange of matter and supply.

Grass as a model for buildings

The comparison between plants and reinforced concrete has been well known since Simon Schwendener's work (1888), and the classical grass as a model for buildings was investigated again by Nachtigall and Lebedew, and is exemplary for translations in construction and structure.

- **Lebedew, 1977**

Lebedew compares grass with a chimney, as an elastic structure. He discerns several characteristics:
- Cavities for reduction of mass
- Differentiation of cell structure according to function
- Change of form in longitudinal and cross section
- Damping and elasticity mechanisms
- The static system of the blade is a cantilevered beam with rigid restraint, the chimney is a structure based on dead weight.
- Lebedew also describes the function of the nodes as damping system and stabilisation elements.[111]

The influence of scale on forces depends on the investigated parameter:
Impact and resistance, in this case wind load and section modulus against bending. For other impacts and parameters other rules of similarity are valid.

---

108 Helmcke, J.G. et al.: IL33 Radiolaria, 1990, p.158
109 Portoghesi, P.: Nature and Architecture, 2000, p.15

110 Vogel, S.: Cats' Paws and Catapults, 1998, pp.40
111 Lebedew, J.S.: Architektur und Bionik, 1983, p.98

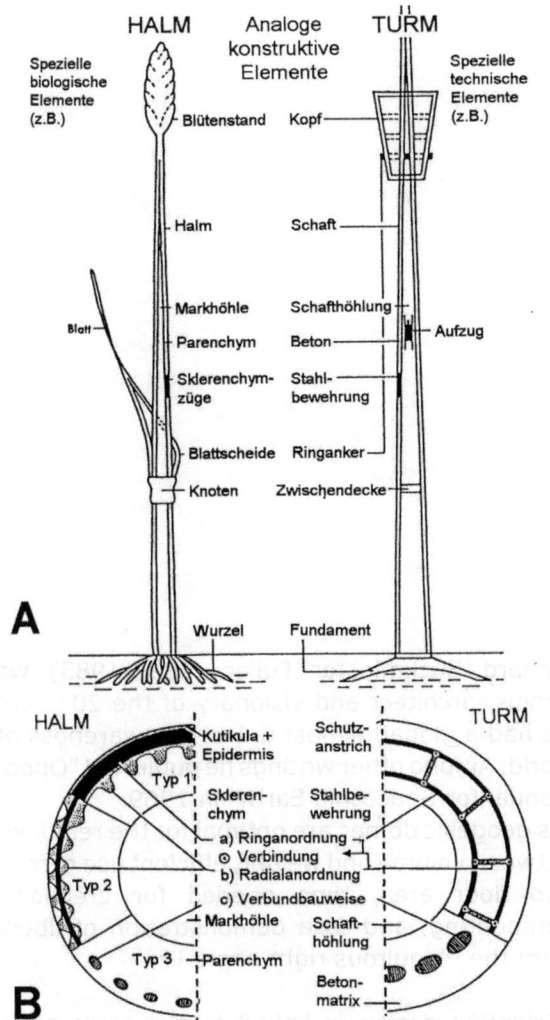

Fig.66 Comparison grass blade - television tower by Wisser and Nachtigall.

- **Nachtigall and Wisser, 1986**

Nachtigall and Wisser continued with a variation of the same comparison. They investigated a structure-functional comparison between a blade of grass and a television tower. Rules of similarity prevent direct causal comparability, but the kind of implementation, inclusion and embedding of reinforcing tension elements are interesting for building technology.

Biomechanic research on trees as building construction show that dead load increases with mass (and therefore with volume, or height³) but resistivity only with cross sectional area, $A=\pi(d/2)^2$. In spite of the same geometry higher trees are affected by increasing discrepancy between dead load and resistivity. Structural similarity of very high constructions require increasing plumpness, which increases in the ratio of $d \approx h^{1,5}$. This law is also valid for the grass blade example.[112]

With transfer of this example to another scale, the impact of dead load has to be taken into account.

112 Nachtigall, W.: Biomechanik, 2000, pp.376

The slenderness ratio of a grass blade can not be obtained when scaling up.

All the comparisons neglect the fact that grass and many other natural constructions rely on other principles, which make their construction easier in some ways, but more difficult in others: biological structures usually allow large deflections, which is not acceptable in architecture. The flexibility and softness of plant parts protects them from large forces.

Rules of similarity

"Again, since the weight of a fruit increases as the cube of its linear dimensions, while the strength of the stalk increases as the square, it follows that the stalk must needs grow out of apparent due proportion to the fruit: or, alternatively, that tall trees should not bear large fruit on slender branches, and that melons and pumpkins must lie upon the ground."[113]

The characteristics of constructions can be optimised in other scales if rules of similarity are respected. The relation between surface, volume and specific weight implies that a system doubled in length is around eight times as heavy.[114]

Big and small systems

Big systems are subject to gravity and inertia (and thus the impact of their own dead weight). The smaller the system, the more dead weight loses importance, and other forces and criteria come into play: surface tension, viscosity, diffusion (physical phenomena below the scale of cm).[115]

Phenomena from nature where molecular forces are most important cannot be scaled up. A famous example is the locomotion of the water spider on water surfaces.[116] Yet this phenomenon enables architects to build physical models in smaller scales, e.g. 1:50 or 1:100.

113 Thompson, D.W.: On growth and form, 1992, pp.26
114 Vogel, S.: Cats' Paws and Catapults, 1998, p.41
115 Vogel, S.: Cats' Paws and Catapults, 1998, pp.51
116 Vogel, S.: Cats' Paws and Catapults, 1998, p.49

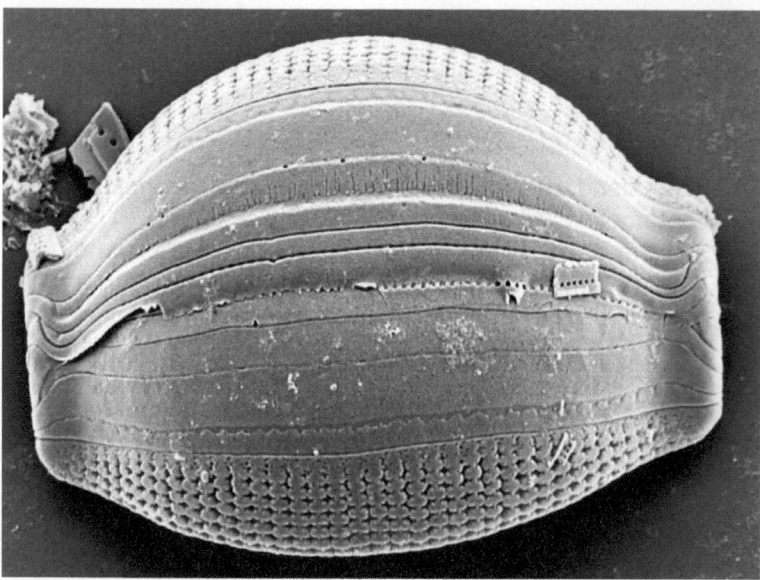
Fig.67 Diatom shell.

2.3.5 Convergence

We speak of convergence, if organisms and/or technical devices have developed similar characteristics. Converging developments exist in all fields.

- **Convergence in nature**

Attempting to classify all living organisms, biologists in the 19th and at the beginning of the 20th century made the mistake of relying on characteristics in phenotype. In evolution convergent development exists and creates analogy. Species of different branches of evolution can be very similar in appearance if the ecological forces were similar and nature had developed analogue adaptation. This phenomenon occurs frequently and leads to a variety of similar biological systems.[117] Clamping and coupling mechanisms, for instance, exist in a variety of solutions.

- **Convergence between nature and architecture**

Frei Otto stated "...like in 1962 our American friend Richard Buckminster Fuller was almost overwhelmed by the new stereoscopic photographs of the shells of diatoms... He saw the newest pictures from five- to fifty fold magnification, which he could not have seen before... Had he known the shells of diatoms before, all world would have said that he had copied [his shells] from living nature."[118]

Convergent development also happens between nature and technology, as similar problems seem to generate similar solutions.

Frei Otto referred to Fuller's geodesic domes, his most famous invention.

Richard Buckminster Fuller (1895-1983) was a genius, architect and visionary of the 20th century. He had a global, almost universal awareness of the world. Among other writings he published "Operating Manual for Spaceship Earth" in 1969.

His geodesic domes are optimal for the relationships between volume and weight, efficient use of material and floor area, time needed for erection and demounting, and as a demonstration of liberation from the ubiquitous right angle.[119]

"Scientific design is linked to the stars far more directly than to earth. Star-gazing? Admittedly. But it is essential to accentuate the real source of energy and change in contrast to the emphasis that has always been placed on keeping man 'down to earth'."[120]

Fuller patented his geodesic domes in 1954. The geometry of the domes is derived from the basic geometry of the icosahedron, a volume with 20 equal faces, a Platonic body. The edges are projected onto an inscribed sphere, generating sections of great circles, which are connected to a regular trigonometric pattern.

---

117 Nachtigall, W.: Vorbild Natur: Bionik-Design für funktionelles Gestalten, 1997, p.20

118 Otto, F. et al.: Natürliche Konstruktionen, 1985, p.8

119 Bürgin, T. et al.: HiTechNatur, 2000, p.17

120 R.B. Fuller in Krausse, J. et al. (Ed.): Your private sky, R. Buckminster Fuller 1999, p.4, original in Fuller, R.B.: Nine chains to the Moon, 1938, p.67

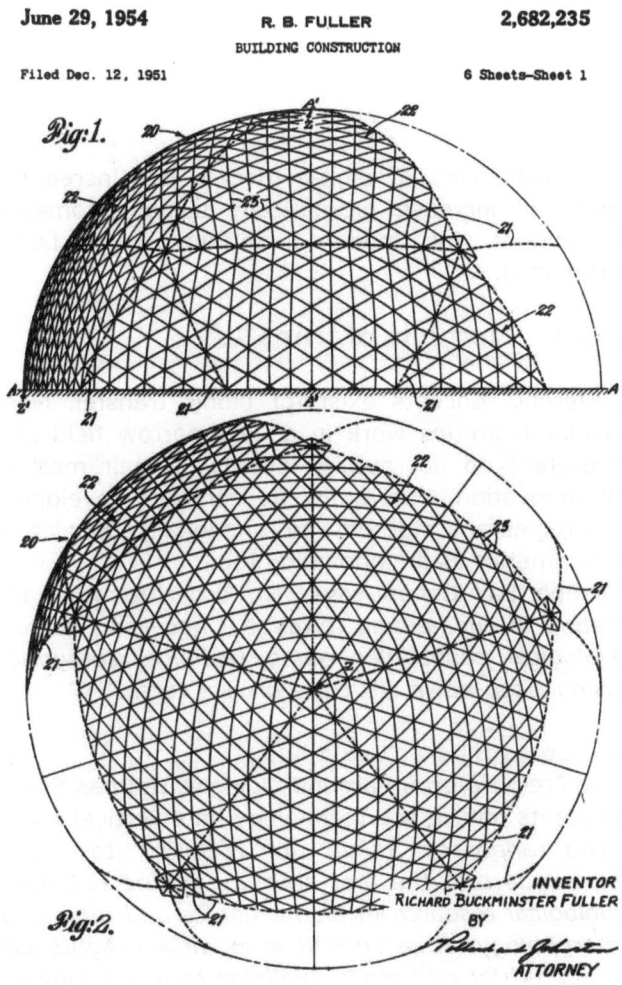

Fig.68 Excerpt of patent specification of Fuller's domes. geometric definition.

Fig.69 Dome in Montreal, world exhibition 1967, Fuller.

Fig.70 Baksei Chamkron in Angkor, 10th century.

- **Convergence in architecture**

Convergent developments can also be registered in architecture. In most cases, the typologies with converging characteristics have appeared at different times in history relating to a respective cultural stage of development, and convergence is observed in the presence of the artefacts. Erich Lehner discusses the astonishing similarities between the "Baksei chamkron" temple in Angkor, Cambodia, and the "Templo II" in Guatemala, being an early example of convergence in architectonic evolution.[121]

Lehner explains time independent parallel developments of architectonic characteristics with the general principles governing architectonic developments, e.g. construction principles and correlation between spatial arrangement and social importance.

---

121 Lehner, E.: Wege der architektonischen Evolution, 1998 pp.505

2.3.6 Reflow of information

Reflow of information from technological development to a new understanding of nature is common, as the following examples illustrate.

The generation of carbon compounds called "fullerene" in the 1990s was inspired by Fuller's geodesic domes. In the paper that first published the discovery of a C_{60} molecule in 1985 the molecules were called "Buckminsterfullerene".[122]

Pathologist Donald Ingber was inspired by Fuller's work to find a new interpretation of the cytoskeleton of cells.

"We introduced the concept that living cells stabilize their internal cytoskeleton, and control their shape and mechanics, using an architectural system first described by Buckminster Fuller, known as 'tensegrity'."[123]

Recently this method of information flow is being called "reverse biomimetics" in analogy to the expression "reverse engineering".

2.3.7 Models and abstraction

Abstraction is the key to transfer of ideas from one discipline to another. Models are abstractions from nature. In research, biologists use models to investigate phenomena. Experimental imitation seems to be the key to understanding in technical biology. In bionic translation, the phase of abstraction plays a key role.

In architecture, working with models as abstraction is common; both physical and digital models are made for design and presentation purposes. Models and plans used for design are:
- regional planning 2D large scale
- urban planning 2D, 3D data on city scale still rare
- 2D planning of building, design, plans for planning permission, implementation plans
- 3D models
- building appearance, including material, light
- load bearing behaviour, influence of forces
- building physics simulation
- models for realisation, interface to computer aided manufacturing (CAM)
- models for operation of buildings, control
- facility management

There is no integration of the different models yet; even the different stages of 2D planning are completely unconnected.

The availability of digital models has increased, and will increase dramatically using automated techniques for model generation (e.g. Laser scanning).

2.3.8 Methods of transfer

Different concepts exist for bionic transfer. Most research groups work in a very narrow field and therefore do not publish details of their method of innovation, but a few do consider developing strategies and methods, to functionalise bionics or biomimetics as an innovation tool.

Yoseph Bar Cohen says, *"In order to approach nature in engineering terms it is necessary to sort biological capabilities along technical categories using a top-down structure or vice versa."*[124]

- **Bottom up or top down**

The Freiburg group around biologist Thomas Speck presents nature and technology on a vertical scale.

"The approach taken in Freiburg to doing bionics is that the first step is carrying out basic biological research in biomechanics and functional morphology. In a second step, new insights are prepared for and made available to technology for further processing (bottom up). Additionally, we also follow an alternative strategy, e.g. searching for possible biological model solutions for specific technical problems (top down). From our point of view, the latter approach enables the development of bionically inspired products in shorter time, whereas the former approach has the potential to yield greater steps in innovation."[125]

- **Stochastic investigation**

Working with large databases including all sorts of phenomena from nature is one method of systematically accessing information delivered by life sciences for questions occurring in technological disciplines. The European Space Agency, for instance, uses a "biomimetics database" to look for suitable role models from nature. The ordering system for the database is the "technology tree", ordering natural role models as follows:
- Structures and materials
- Mechanisms and processes
- Behaviour and control
- Sensors and communication
- Generational biomimicry[126]

---

[122] Kroto, H.W. et al.: C60: Buckminsterfullerene, Nature 318, pp.162, 1985

[123] http://www.childrenshospital.org/research/ingber [08/2007]

[124] Bar-Cohen, Y. (Ed.): Biomimetics, 2006, p.496

[125] http://www.biologie.uni-freiburg.de/data/bio2/botanischer_garten/PDF-Dateien/Freiburg%20-%20profile%20-%20English.pdf [08/2007]

[126] Ayre, M.; ESA: Biomimicry - a review, 2004, p.7

The differentiation of role models according to the ESA technology tree seems to be easier and more suitable for the diverse natural phenomena which are used as role models, than Nachtigall's ordering system.

Another database providing substantial information on nature's phenomena is provided by the Biomimicry Institute, based in the US. It is called "Ask Nature"[127] and is open to the public.

The Centre for Biomimetics and Natural Technology in Bath is working on the further development of TRIZ (the "theory of inventive problem solving" designed by Genrich Altshuller) towards the so-called BioTRIZ. TRIZ was developed from the statistical analysis of a vast collection of inventions, and it serves as a tool for the transfer of various inventions and solutions from one field of engineering to another. BioTRIZ includes natural role models in the system, to expand the capabilities onto the transfer from biology to engineering.
Using TRIZ implies having an obvious engineering problem. In architecture, problems in design often affect many levels of the project, and often they are difficult to define. As a consequence, variations of TRIZ seem to be too specific a tool to deal with the complexity of architectural design, but may be suitable to solve specific design tasks and questions.

- **Purposeful, systematically, or at random?**

In the investigation of the connections between technology and nature, it is interesting to know if transfer was carried out deliberately, or was just a convergent development.
For the discipline of bionics or biomimetics, knowledge about the intention is a matter of self-affirmation, as the systematic approach differentiates the discipline of biomimetics from singular efforts which have always happened in history.
In architecture this is a controversial issue. The development of an architectural project is not easily understandable, as it follows an intuitive creation process. Inspiration from nature may not be obvious in the final design, and the retrospective interpretation of design is very common. In spite of that, the connections between architecture and biology can be investigated, and known purposeful translations examined and presented. Their successful implementation of aspects from nature justifies the approach, even if the process of creation is unknown in many other cases.

127 www.asknature.org

Background | Transfer and methods

Fig.71 Humble administrator's garden in Suzhou, China, 16th century.

3 CLASSICAL APPROACHES TO INVESTIGATE OVERLAPS BETWEEN BIOLOGY AND ARCHITECTURE

Architects and builders have always drawn inspiration from nature. Countless analogies can be found in the architecture of all ages. The examination and application of nature's materials, the symbolic or structural transfer of natural form, the interrelation of the edifice with the environment, all these aspects have been considered by builders at all times. Few people, however, have accepted the challenge to investigate the overlaps between architecture and nature.

Paolo Portoghesi has compiled a vast collection of analogies in "Nature and architecture", using a grid of archetypal elements of architecture with examples from all ages.[128] It is not a mere collection of formal elements. The system of using archetypes (the wall, the tower, the bridge...) ensures parallels in functionality and structure. He refers to famous artists and philosophers dating back to classical Greece, and includes many examples of vernacular architecture as well.

The examination of "Natural Constructions" carried out during the 1960s in Germany by Frei Otto and his group takes a more technical approach. The structural functioning of natural role models is the most important feature investigated.

The so-called "Sonderforschungsbereich 230 Biologie und Bauen", which was the frame for many efforts, partly laid the base for the active bionics community in Germany today.

Strategic investigation of the overlaps of architecture and biology has also been carried out by the Russian Juri Lebedew, culminating in the only publication on "Architekturbionik" in 1971, which is a comprehensive collection of natural constructions of his time.[129] In 2003, Werner Nachtigall published his book "Baubionik" that focuses on construction, orders the field in terms of technology, and offers a collection of role models.[130] "Vorbild Natur" discusses general design principles, focusing on product development.[131] The Design and Nature conference series by the Wessex Institute of Technology provides a forum for a diversity of approaches to this generic topic[132]. Some of the numerous other publications dealing with single aspects will be mentioned in the chapters following.

128 Portoghesi, P.: Nature and Architecture, 2000, p.10

129 Lebedew, J.S.: Architektur und Bionik, 1983
130 Nachtigall, W.: Baubionik, 2005
131 Nachtigall, W.: Vorbild Natur, 1997
132 Brebbia, C.A. (Ed.): Design in Nature series, launched in 2002

3.1 RELATIONSHIP BETWEEN NATURE AND ARCHITECTURE

3.1.1 Inside or outside

"Being an integral part of nature ourselves, we shall never be able to talk about it from the outside but only from the inside, uncertain whether to consider something created and produced by man as being 'outside' nature."[133]

Paolo Portoghesi's statement touches the general dilemma in the relation of nature and technology. This philosophical question will not be discussed within the frame of this project. We can definitely say that architecture is a phenomenon of life, and thus is an integrated part of nature, but imitates nature in many respects.

Bien and Wilke, members of Frei Otto's group express a dialectic perspective on the "constructions of nature":

"Not until the principle of imitating nature is understood as an integrated part of a cycle, or better: part of a hermeneutic-poietic spiral, whose complementing section is the constructivist conditionality of every knowledge- or activity seeking observation of nature, one can attribute a relative right to the postulate of imitation."[134]

"...that the opposing interpretations of the formula 'nature's constructions' [nature constructed as representation of god, or man constructs the reality of nature] could at the same time be held and correlated to each other and that both incorporated theses are at the same time right and valid... We grasp nature by constructing it according to the role model delivered by our inventions. We explain nature by interpreting it according to the composition of our inventions. In this sense nature is a product of our culture."[135]

3.1.2 Positions

The attitude towards man-made and natural things is best expressed in the relation of prevailing values. Significant cultural differences exist between "Western" and "Eastern" attitudes. Human ignorance prevents us from valuing phenomena that emerge without human interference, e.g. old natural environments, not to mention biodiversity as a whole. Structures resulting from self-organisation may have a long history, and their future influence and importance cannot be predicted. Eastern cultures, and many traditional societies around the world, value things differently. In China and Japan things that cannot be manufactured, for example crack patterns in ceramics, the growth of moss on stone, stone itself, are highly esteemed. On the other hand a strong tradition has evolved to cultivate these processes in terms of setting an environmental condition for self-organisation. In some respects, Japanese design has domesticated self-organisation.

Western culture - in general - values more what can be manufactured and produced, and does not value what grows by itself, which "merely" takes resources and time. The *"rapprochement of Western and Eastern civilisations"*[136] is inevitably taking place in the course of increasing globalisation, and movements to protect the environment have already changed the public's point of view.

However, in Western culture nature and technology are still considered to be in opposition. But *"the complementary nature of opposites, the ability to see two aspects of every phenomenon and every object without contradicting unitary synthesis... have all been repeatedly taken into account in the Western world, particularly at the beginning of this [20th] century."*[137]

Movements in architecture history have taken their own approaches to nature, and found their specific position towards nature, and architects continue to define a relation to nature.

"Despite the damage that the Modern Movement undoubtedly did to our sense of connection with nature, its greatest practitioners, men such as Alvar Aalto, Wright and Le Corbusier, all strove in their way to lay emphasis on this connection.[138]

Technology is one of the main driving forces for development in architecture. At the turn of the 19th century, the proponents of the futurist movement were enthusiastic about the incipient technological development, and designed urban utopias far away from the formal natural interpretations, as can be found in Jugendstil. Vienna Secessionists, "stile floreale", and Art nouveau movements around 1900 excessively explored the use of steel and glass.[139]

---

[133] Portoghesi, P.: Nature and Architecture, 2000, p.9
[134] Bien G. in Teichmann, K. et al.: Prozess und Form Natürlicher Konstruktionen, 1996, p.21
[135] Bien G. in Teichmann, K. et al.: Prozess und Form Natürlicher Konstruktionen, 1996, p.23

[136] Portoghesi, P.: Nature and Architecture, 2000, p.31
[137] Portoghesi, P.: Nature and Architecture, 2000, p.31
[138] Aldersey-Williams, H.: Zoomorphic, 2003, p.18
[139] Aldersey-Williams, H.: Zoomorphic, 2003, p.16

Fig.72 "The Cloud", 1968-72, © COOP HIMMELB(L)AU / Erwin Reichmann

The genre of Organicism with Imre Makovecz as one of the representatives[140] was influenced by Rudolf Steiner's anthroposophy movement and tried to translate biological compartments to technical functional elements. Art Nouveau and Organicism further influenced Bauhaus architects, and Frank Lloyd Wright "...who took 'organic' architecture to new heights".[141]

Functionalism, however, rejected the use of organic forms. *"For some time even modern architecture was convinced that it could turn simple conservative functions into the cornerstones of existence and consider every other necessity of life as a luxury, as something superfluous."*[142]
"The revival... of organic architecture in the 1950s was made possible by Second World War developments in concrete construction... Pier Luigi Nervi, Gio Ponti and Oscar Niemeyer exploited concrete's structural potential, while others pushed in a more biomorphic direction. The triumphant projects... are Eero Saarinen's aquiline TWA Terminal at New York's Kennedy Airport and Jørn Utzon's polysemous Opera House."[143], not to forget the asymptote of design of organic space: Frederic Kiesler's so-called "Endless house".
Frei Otto's group used an experimental approach aiming at the understanding of natural structures and processes, and finally making use of the physical laws that had been established, to design new structures, e.g. work with minimum energy surfaces (soap film models) brought about the development of membrane constructions. A similar strategy had been used by Antoni Gaudi much earlier, and is still used by Heinz Isler, who over decades has refined methods of experimental formfinding using hanging models made of various materials to build his amazingly ephemeral shell structures. In contrast to purely formal approaches, mimicking is the key to understanding and using functional and constructive aspects.
Also in the 1960s, the Metabolism movement became prominent in architecture, particularly in Japan. Kisho Kurokawa's Nagakin Capsule Tower in Tokyo shows the main principles: packing and repetition of industrially fabricated units, inspired also by new technologies and spaceflight.

Visionary architects like Buckminster Fuller developed a new kind of light construction. Organicism projects were also designed and experimented with by Archigram in the UK, Haus-Rucker-Co Austria, and Ant Farm US in the 1960s and 1970s. Again, the development of these new kinds of architecture relates to the availability of new materials; flexible fabric-based tensile or pneumatic structures. [144]
Since the energy crisis in the 1970s, the ecological design movement is concerned by energy efficiency and the influence of building onto the environment.

140 Kuhlmann, D.: Metamorphosen des Organizismus, 1998
141 Aldersey-Williams, H.: Zoomorphic, 2003, p.16
142 Portmann, A. in Portoghesi, P.: Nature and Architecture, 2000, p.38
143 Aldersey-Williams, H.: Zoomorphic, 2003, p.17
144 Aldersey-Williams, H.: Zoomorphic, 2003, p.17

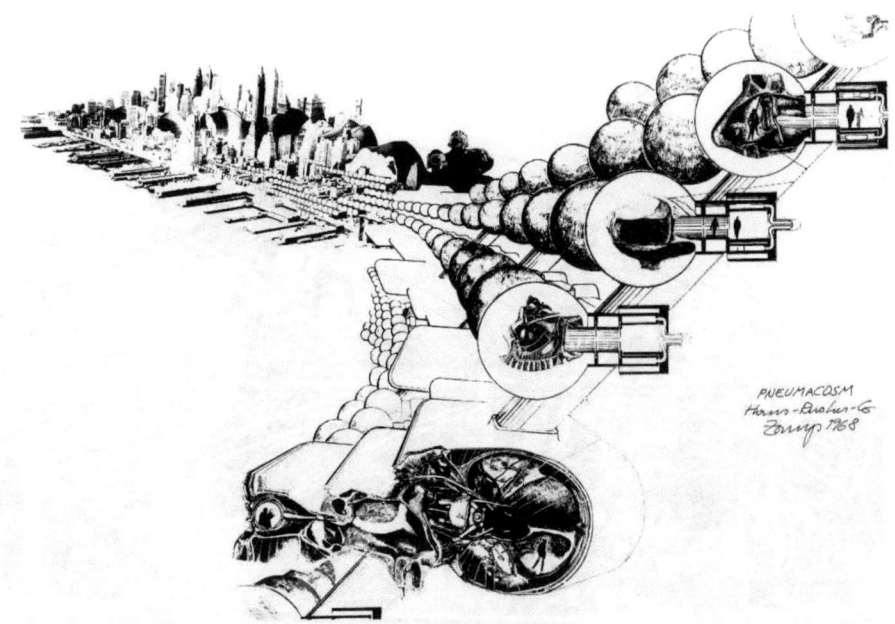

Fig.73 "Pneumakosm" - Erweiterung Manhattan, Haus-Rucker-Co, 1967.

"The cloud is an organism for living. The structure is mobile, the space can be modified. The building materials are air and dynamics."[145]

More recent developments in architecture include deconstructivism to overcome traditional concepts of order and gravity (e.g. Coop Himmelb(l)au), spiritual approaches trying to enhance human connection with the universe by introducing cosmic ordering systems (Charles Jencks), and most interesting in the context of biomimetics - the introduction of processes into architecture through e.g. morphogenetic and evolutionary design principles (e.g. FOA, Greg Lynn).

Within the last decades the use of biology's paradigm in architecture has increased, as Aldersey-Williams states *"...it is now becoming evident that a new strand of biomorphism is emerging where the meaning derives not from specific representation but from a more general allusion to biological processes... [Biomorphism] signifies the commitment to the natural environment... is apt to be considered more environmentally responsible (or responsive)..."[146]*

The next step taken is attributing architecture an active role in environmental design. Ken Yeang has developed a strategic approach for high-density buildings[147]. An integrated strategy for designing the built environment is still being worked at under many titles: "sustainable" and "ecological design", "bio-architecture" and "green architecture" to name but a few.

Fig.74 "Balloon for Two", Haus-Rucker-Co, Vienna, 1967.

145 Coop Himmelb(l)au, 1968, http://www.coop-himmelblau.at [12/2007]
146 Aldersey-Williams, H.: Zoomorphic, 2003, pp.19
147 Yeang, Ken: The Green Skyscraper 1999

Classical approaches | Relationship between nature and architecture

Fig.75 University of Minnesota Geodesic dome in the process of being covered with a plastic skin by students in Aspen Colorado, June 1953.

3.2 "NATURAL CONSTRUCTION"

Interpretation of the expression "natural construction" according to Frei Otto:
"Talking about 'natural constructions', we do not mean just any object out of the infinite diversity, but the typical. We mean those constructions that show with particular clarity the physical, biological and technical processes, generating the objects... Even if technology is a tool of the natural object man... we interpret it as a product of man and therefore nevertheless as a part of nature." [148]
"Natural construction" will be used here deliberately in the widest possible way, to include all aspects which are usable for architectural innovation.
- Constructions and structures made by animals
- Constructions and structures of nature - materials, processes, technologies
- Constructions and structures found in nature: non-living and living structures
- Constructions and structures made by humans, which are "natural" in some respect

Considering the separation between nature and technology in common understanding, whereby technology is understood as "everything made by man", the term "natural architecture" may seem paradox. We can understand the appearance of architectonic phenomena in nature or the application of natural phenomena in architecture, and integrate both views in the investigation.

The term "green architecture" has come to be used for projects that integrate aspects from nature in any way, e.g. through the selection of material, construction, applied processes or just their function. Direct and obvious applications according to bionics are rare. It is mostly single aspects and "bionic" elements that can be found. In order to reach high quality, the architectural design has to develop to self-reliance and thus leave the natural role model within a process of transformation. Considering this and the methodic interpretation of bionics and biomimetics, a "bionic architecture" or "biomimetic architecture" can not be postulated. But "bionic design", or "biomimetic design" can be accepted due to its notion of activity.

Bionics and biomimetics in the context of architecture do not imply sole biomorphism. The mere transfer of form is insufficient to reach the inherent qualities of natural constructions.

The following principles should be pursued to reach the qualities of "natural constructions": rigorously applied light construction, active and passive use of energy, and the investigation and use of natural processes and development principles. [149]

In the following, "natural constructions" will be presented.

---

148 Otto, F. et al.: Natürliche Konstruktionen, 1985, p.7

149 Nachtigall, W.: Vorbild Natur, 1997, p.85

Fig. 76 Spiderweb.

Fig. 77 Insect building.

3.2.1 Buildings of animals

"With buildings of animals all kinds of constructions are found: caves, beam structures (many birds' nests), membrane and rope-constructions (webs of spiders and caterpillars), shells... folded structures (honeycomb structures of bees), vaults (above ground ant-hills), massive constructions (puttied, cast and high strength termite mounds)."[150]

The technologies of animals' buildings are to a high degree genetically determined; evolutionary developments explain the growing complexity of animal buildings with phylogenetic age. Evolutionary very old species, e.g. insects, have developed highly complex construction systems, whereas the youngest large group, the vertebrates, make few and almost primitive constructions. On the other hand, the use of tools is almost exclusively reserved to the human species.

"The limited use of tools contrasts sharply with the way animals use materials from the environment to construct domiciles: caddisfly larval tubes; worm tubes; hermit crab shells; nests of fish, reptiles, birds, and mammals; beaver dams; and so forth."[151]

The buildings of insects, especially of spiders and termites, have been investigated thoroughly by Frei Otto's group of the Institut für Leichte Flächentragwerke, and have served as role models for membrane and tensile constructions, and for passive ventilation systems. Frei Otto also states that animals usually do not form the environment according to their needs, and mentions the beaver as exception of this rule.

"It seems that direct intervention in the ecology of a landscape is reserved for higher animals (in exceptions) and humans."[152]

The processed materials, e.g. silk, are extremely interesting role models. As they do not differ in their complexity and subtlety from living tissue, the same characteristics are valid and will be described later.

3.2.2 Traditional architecture

Including traditional architecture in "natural constructions" means that traditional building technologies can serve as source of innovation in the same way as biological role models.[153]

We talk about architecture and building techniques being traditional, when long established and regionally discernable techniques and typologies exist. Traditionally, locally available resources are used for building, and typologies evolve with time, more through trial and error than through abstract modelling of some sort. The influencing factors for the design are the existing environment and the social and cultural situations. Environmental issues include the availability of space, building material energy, climate and ecology among others. Social and cultural situations include knowledge and available technology, needs of society and culture, symbolism, rules, etc. Within the existing typologies, diverse influences from a long time span have been applied and stored, and much is still readable today, even if the developmental process and the environmental conditions of former times are not known. But the end product, the surviving typology, is here to be investigated.

Knowledge to apply this kind of architecture is usually passed on by local tradition. The term "vernacular" architecture is mainly used for residential buildings excluding temples and palaces. Yet, as constructions for bigger building tasks deserve investigation as well "traditional architecture" is the term used here.

---

150 Otto, F. et al.: Natürliche Konstruktionen, 1985, p.20
151 Vogel, S.: Cats' Paws and Catapults, 1998, p.309
152 Thywissen, C. in Otto, F. et al.: Natürliche Konstruktionen, 1985, p.21

153 Gruber, P.: Bionik in der Architektur, in Rieger-Jandl A. (Ed.): "Architektur transdisziplinär", in press

Fig.78 Karo Batak architecture in North Sumatra (Barusjahe), Indonesia, 2003.

In the context of bionics and biomimetics, the "evolution" and "adaptation" of traditional architectures is most interesting. The inherent qualities and the integrated information of traditional typologies can deliver priceless hints and solutions, when investigated and interpreted properly. The importance of traditional architecture as a source of innovation is inadequately identified and used. The case study on the adaptation of the traditional architecture on the Indonesian Island of Nias will present an approach to identify architectural qualities for further application.

Most old architectural typologies which have survived until today are not adapted to the requirements of a modern civilised life. For this reason, a "back to the roots" approach would certainly not be appropriate. The qualities and information found in tradition have to be applied to an independent, new solution. The transformation process, which inevitably has to take place when using principles coming from tradition, is similar to the one used with biological role models. Abstraction and separation from the role models has to be found in order to develop a contemporary and modern solution. The independence of the new design is of utmost importance.

Many architects and researchers have sensed the importance of vernacular architecture, and have focussed on this field.[154]

Best known, because of impressive photographic documentation, is Bernard Rudofsky's work on anonymous architecture, published first in "Architektur ohne Architekten" in 1965[155]. Rudofsky has made architecture traditions from around the world known to a wide public.

Amos Rapoport contributed to the interpretation of these structures with cross-cultural and comparative studies of environment-behaviour relations, published in his first book "House Form and Culture" in 1969. There is often criticism that our global architecture lacks the adaptation to regional features because of the ignorance of architects towards the needs of users, as Rapoport argues.[156] A comprehensive collection of vernacular architecture can be found in the works of Paul Oliver, who published the "Encyclopaedia of Vernacular Architecture of the World" in 1997[157].

154 See for instance the publication discussing adaptibility of architecture by Otto, F. et al.: Anpassungsfähig bauen, IL 14, 1975, with numerous contributions also referring to traditional architectures.

155 Rudofsky, B.: Architektur ohne Architekten, 1993
156 Amos Rapoport in Schaur E.: Il 41, Intelligent Bauen, 1995, p.237
157 Oliver, P.: Dwellings, 2003

Fig.79 Model of the "Endless House", Kiesler, 1959.

3.2.3 Architecture with aspects from nature

There is also a long tradition of using aspects from nature in architecture. Meanwhile not only aspects from, but also aspects for, nature are taken into account considering sustainability and ecology. In the following, a range of aspects will be presented, together with classic or very recent exemplary projects.

Spatial concepts

The translation of organic space into architecture
As representative for many efforts on the design of organic space and as an iconic project which still influences organic design today, Frederic Kiesler's "endless house" is selected here.
"The endless house is called the 'endless', because all ends meet and meet continuously. It is endless like the human body - there is no beginning and end to it."[158]

- **Friedrich Kiesler (1890-1965), the "endless house"**

Friedrich Kiesler was an architect, stage designer, painter, sculptor and visionary. He emigrated to the US in 1926. After decades of examining architecture, arts, theatre and perception he designed the first model of the "endless house" in 1950, which after alteration and extension in 1960 is exhibited in the Museum of Modern Art in New York. Kiesler felt like a pioneer for a new society and like a prophet for a new era. According to Kiesler, the "primary house", [Urhaus], was the egg, and light was considered an integral part of architecture. Kiesler demanded the creation of a new "biotechnology" [Biotechnik] in harmony with man.

"The 'endless house' is not amorphous, not a free-for-all form. On the contrary, its construction has strict boundaries according to the scale of our living. Its shape and form are defined by inherent life forces, not by building code standards or the vagaries of décor fads... The endless house is not a machine for living. It is rather a living organism with a very sensitive nervous system..."[159]

Kiesler demanded flexibility of space, formulated in his "elastic space" concept. The technology of reinforced concrete allowed the construction of shell structures, thus the metamorphosis of compressive forces into continuous stress. His later project, the Shrine of the Book in Jerusalem, was opened in 1965, and Kiesler died in the same year.

- **Translation into natural space: egg, nest, cave**

The organic forms express the need for safety and security. Many examples of organic form can be found in the architecture of the 1960s and 1970s.

---

158 Kiesler, F.: Inside the Endless House, 1966, in Museen der Stadt Wien (Ed.): Friedrich Kiesler, 1997, p.136

159 Kiesler, F.: Notes on Architecture as Sculpture, 1966, in Museen der Stadt Wien (Ed.): Friedrich Kiesler, 1997, p.140

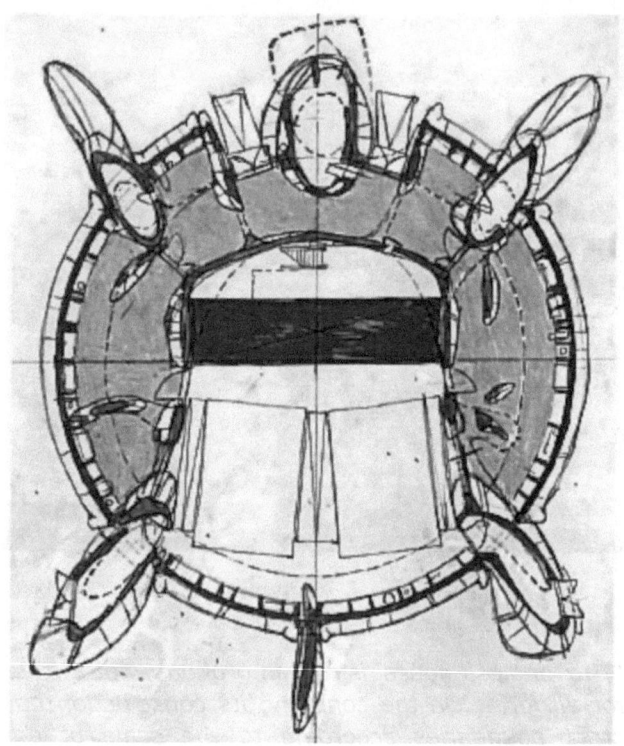

Fig.80 Floor plan of Turtle Portable Puppet Theatre, Michael Sorkin, 1995.

Fig.81 Jellyfish hotel for Tianjin, China, Michael Sorkin, 2010.

Formal analogies: zoomorphism and anthropomorphism, organizism, morphing, "animate form"

"...for nature, appearances are no less important than their corresponding functions and that, for all their splendid independence, living organisms are made to interact with both organic and inorganic environmental reality and that this interaction is one of the reasons that determines their forms."[160]

Form is an important issue in architectural design. In the context of bionics, sole formal translation is not appreciated or valued highly. The connection of form to function, construction, material, development process, and also matters of interaction with the environment, are regarded as the important issue. Admittedly, the interrelation of form and function of role models from nature is a more promising field for innovation and progress, but in architecture formal translation can have authority in its own right if a purpose is achieved by it.

Zoomorphic, Anthropomorphic and Biomorphic

Zoomorphism takes animal morphology as the role model for architecture projects. Animal representation occurs in three-dimensional imitations of whole or parts of animals, or two-dimensional mappings transferred into houses.

"The symbolism attributed to animal bodies by numerous civilisations made it possible for the architect to communicate ideas and confirm collective values. The traits of the animals and each part of their bodies... have sometimes been transferred to buildings for magical purposes. An example of this possible transference is the defensive action of the turtle, the liberating flight of the bird and the enveloping shape of the snail."[161]

Historical examples of zoomorphism include the Renaissance monster theme park in Bomarzo, Italy, from the 16th century, and the visionary French architect Jean-Jacques Lequeu with his animal-shaped designs, the so-called "architecture parlante", although this was never executed.

Later zoomorphic designs include "Lucy the elephant", a seaside attraction in New Jersey, designed by James Lafferty in 1883 and the Big Duck in Riverhead, New York, built in 1931. Robert Venturi and Denise Scott Brown used this example to coin the term "duck" in architecture for a building, where every parameter is subordinate to form, which predominates function.[162] Hugh Aldersey-Williams published a collection of zoomorphic buildings and even managed to draw lineages according to zoological classifications.

---

160 Portoghesi, P.: Nature and Architecture, 2000, p.38

161 Portoghesi, P.: Nature and Architecture, 2000, p.28
162 Venturi, R. et al: Learning from Las Vegas, 1977

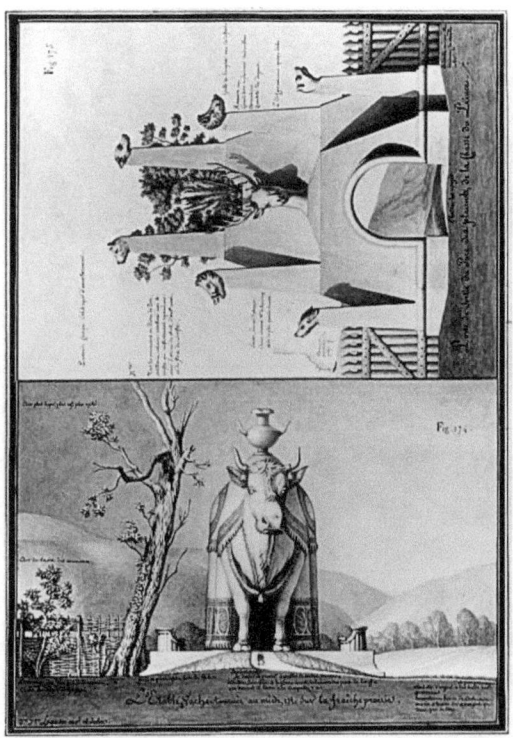

Fig.82 Zoomorphic design by Jean-Jaques Lequeu 1757-1826.

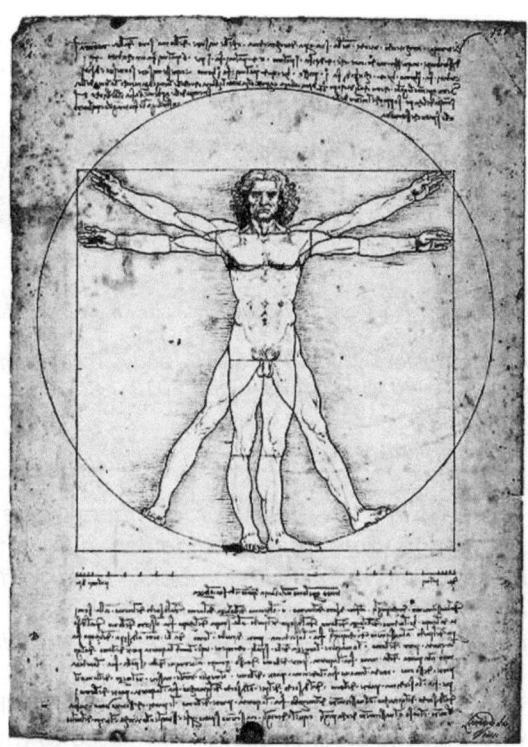

Fig.83 Leonardo da Vinci, Vitruvian man.

Aldersey-Williams discerns several zoomorphic designs: animal as symbol, animal by function - static or dynamic, and animal by accident. In this way, he can implement projects into his work which are perhaps bionic or bio-inspired in some aspect, but labelling it zoomorphism is, in some cases, rather far-fetched.

Anthropomorphism, the use of human form, is a very old concept, and thus occurs in many traditional building cultures - mostly in floor plans, defining the layout and design of settlements and buildings. Günther Feuerstein compiled a wide range of so-called biomorphic projects. In his book "Biomorphic Architecture" he discusses both human and animal forms in architecture.[163]

Geometry of nature

Since the beginning of scientific investigation nature's geometry has been the object of discovery. Laws of proportion and symmetry were already investigated and described in ancient Greece. Bilateral symmetry in creatures is due to the development of a symmetry loss from a spherically symmetrical fertilised egg. *"Air breathing animals must fight against the force of gravity, any up-down symmetry... is lost... independent locomotion... destroys symmetry along the front-back axis."*[164]

In architecture bilateral symmetry is due to gravity, rules of perception and culture. *"Perhaps, as Greg Lynn suggests, it is merely 'the cheapest form of beauty', nature's default setting."*[165]

Theories of harmony play an important role in all applied arts. Theories of proportion in architecture aim at toolkits of measurements for the creation of harmonic and beautiful designs. It is believed possible to achieve particularly well-proportioned and aesthetic spaces and arrangements by taking the measurements of nature. The integration of human measurements promises good adaptation to the human body and therefore good usability.

A canon of measurements already existed in the high culture of Egypt. Leonardo da Vinci studied the works on proportion by Vitruv (1st century B.C.) intensively, and arrived at the famous representation of the human body within a circle and a square.

163 Feuerstein, G.: Biomorphic Architecture, 2002

164 Aldersey-Williams, H.: Zoomorphic, 2003, p.13
165 Aldersey-Williams, H.: Zoomorphic, 2003, p.14

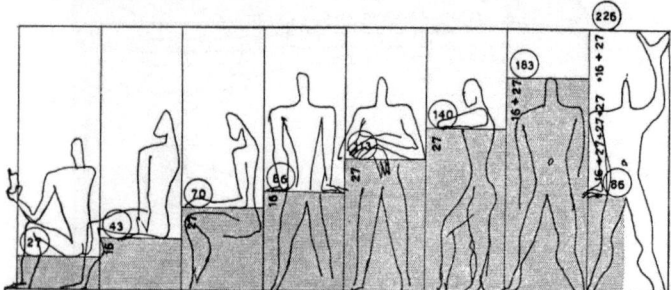

Fig.84 Le Corbusier, "Modulor" with body formations, 1952 "Der Modulor" 1980 © FLC/VBK, Wien 2010

- **Golden section**

"Experts in the theory of numbers have known for a long time that the 'golden number' is the most irrational of all numbers. It is 'difficult to approximate' by rational numbers, and if one quantifies this difficulty, it is the worst of all."[166]

The golden section is a ratio dividing a line in a way that the smaller section is related to the bigger section in the same ratio as the bigger section is to the whole line. The golden section has for centuries been used in architecture as a tool for the design of particular harmonic proportions, and has been considered a divine proportion. The golden section can easily be constructed with a drawing compass. The difference between two consecutive numbers of the Fibonacci-series approximates the golden number $\phi = 0{,}618034...$ The limiting value is $\phi = (\sqrt{5} - 1) / 2$. It can be approximated by 5/8. The angle between branches of plants that try to gain as much light as possible can be calculated by using a most irrational angle between the branches, $360 (1 - \phi)$ degrees.[167]

- **Le Corbusier (1887-1965)**

Le Corbusier developed the so-called "Modulor", a system of measurements based on the relation of the golden section and the proportions of the human body of the average size of 1.83m and 1.75m. He applied the system in his works, especially in the Unités d'habitation. The system continues to have an effect until the present day as an effort to establish a human measure in architecture.

The measurements used in traditional architectures were almost always derived from the human body. For aesthetic, magical or practical reasons the use of fingers, hands and limbs as measurement units was common in all societies before developments in science led to the introduction of a common metric system in the 19th century.

Fig.85 Interior space, spiral ramp.

Spirals in architecture

The special geometry of the spiral is extremely often used in architecture. The connection of these forms to growth will be discussed later on in this book.

- **Frank Lloyd Wright (1869-1959)**

Wright represented organic architecture and postulated the view of a building as an organism. Some designs are reminiscent of skeletons of marine organisms or spiders' webs. In the New York Guggenheim Museum, built 1946-1959, he applied the spiral to the main circulation space of the exhibition hall. This was a controversial concept until today, but the resulting space is impressive. The prominent site bordering the Central Park enhances the impression. Wright also used the spiral as a circulation ramp in smaller projects, like the Morris house in San Francisco, which was built 1948-1949.

---

166 Stewart, I.: Die Zahlen der Natur, 2001, p.169
167 Stewart, I.: Die Zahlen der Natur, 2001, p.166

Fig.87 Building phases.

Fig.86 Truss wall house, Ushida+Findlay 1993, site in suburban area.

Chaos theory and architecture

Formal approaches are also common in today's architecture, but the formality is no longer restricted to simple geometries; computational methods and new building technologies have made the control of complex geometries possible. The discoveries in physics have led to the investigation of more-dimensional space and the use of fractal geometry.

- **Eisaku Ushida und Kathryn Findlay, Truss Wall House, 1993**

"...by setting up a coordinate system, a time-space matrix, with one axis for the life-span of the architecture and another for the spatial layout of its surroundings... the landscape is the matrix of time-space."[168]

The architecture of Eisaku Ushida and Kathryn Findlay is based on the contemporary theory of fractals and chaos. The spiral as a fractal form produces self-similar fractal forms, however it is split. This geometric advantage is also exploited in construction.

The "Truss wall house" of Ushida and Findlay is included to represent the many projects that could be shown here. Despite its obvious intuitive spatial qualities, the statement of the architect suggests a mathematical modelling for the design.

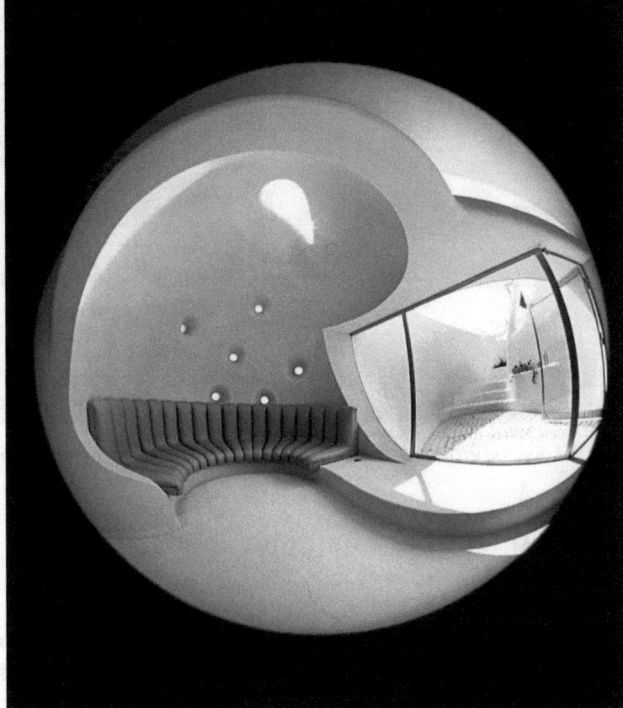

Fig.88 Interior space.

Fig.89 Isometric view.

168 Sato, A. (Ed.): Eisaku Ushida + Kathryn Findlay, 1996

Fig.90 Sydney opera house, Jørn Utzon 1972, photograph 2005.

Functional analogies: constructions, lightness

Functional transfer of phenomena from nature to architecture means predominantly the function of load bearing. Other functions such as protection, supply, storage, are considered less frequently, but important in other fields of bionics.

The principle of light construction requires minimising material and effort in construction through structuring and design determined by the flow of forces according to the laws of physics.

The integration of form, function and the process of creation in biological constructions is what architects and engineers admire, and try to pursue. Form and construction is a vast field, in research as well as in application, and is usually ordered in construction types and systems.

Fig.91 Exhibition model.

Finding a construction for a form

- **Sydney Opera House, 1972**

The Danish architect Jørn Utzon won the competition for a new opera house in Sydney in 1957 with a design reminiscent of an arrangement of mussel shells. Engineer Ove Arup and Peter Rice were involved in the execution of the project. Although the opera was not finished according to Utzon's original design, the project can be regarded as a triumph over practical, political and bureaucratic difficulties.

Fig.92 Sydney opera house in today's skyline.

Fig.93 Roof geometry and tiling.

- **Geometry**

The building was to be started soon after the competition. This is why the planners were under considerable time pressure, which resulted in many problems. In the competition's design the curves were drawn by hand, but later the design had to be expressed accurately, through geometry. A range of variations was thought about, drawn and dismissed. Circular segments, ellipses, parabolas were all tried to define the geometry of the shells.

- **Spheres**

The breakthrough was made by the interpretation of the shells as segments of a sphere. The single pieces are all parts of the same sphere, have the same curvature and were thus designable and buildable in the same way (more congruent parts in those times meant less energy and effort and thus cost).[169] After many difficulties and the dismissal of the architect the opera house was built, rather differently regarding the basement levels. In 2003, after several decades Utzon was awarded the most prestigious Pritzker prize.

- **Implementation**

The primary structure consists of identical ribs of reinforced concrete, produced on site. All ribs are great circles of a sphere (the centre of the circles and the sphere are the same - rotation of the circles would create the surface of the sphere).

The construction was covered with a set of equally sized arrow-shaped prefabricated tiles. It accentuates the curvature of the shells through colour, arrangement of the tiles and detailing of the gaps. At different distances, different features take effect in a very subtle design. The inside spaces are totally different, rather small-scale and with no overall concept. Meanwhile the basement floors, which face the harbour, have been renovated and their functions changed (new theatre space, facilities for visitors).

The project was highly successful in its symbolic significance for Sydney and the whole continent. After all those years, and in spite of the lack of an overall concept in the internal features it is still a landmark of Australia.

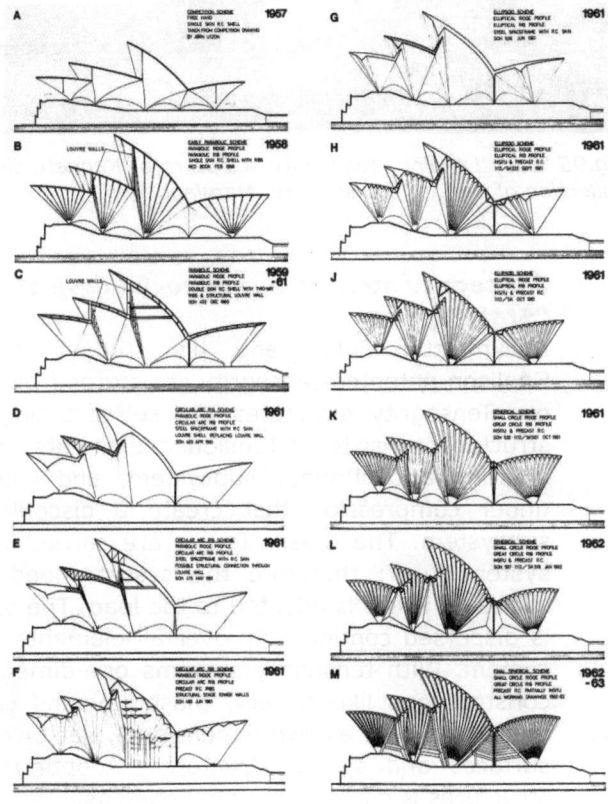
Fig.94 Development of geometry and structure of the shells of the Opera house.

169 Rice, P.: Peter Rice 1994, p.62

Classical approaches | "Natural construction"

Fig.95 "Easy Landing" tensegrity sculpture of Kenneth Snelson, 1977, stainless steel, 9x26x20m, collection of the city of Baltimore, Maryland.

Tensegrity structures as special case of "Stabtragwerk"

Buckminster Fuller and his student Kenneth Snellson patented tensegrity structures from 1959 on. Tensegrity structures are self-stressing. The structure consists of tensioned elements (ropes) creating a continuous subsystem, and elements under compression that create a discontinuous subsystem. The tensile forces are carried by the system itself; therefore the system needs pre-stressing which is adapted to the load. The tension is dispersed continuously over all elements of the system. With tensegrity systems one-dimensional constructions like towers, masts, beams can be created as well as two-dimensional, flat or curved surfaces and spatial structures. Adopting these structures is difficult for many reasons:
- Collapse of a single element leads to total collapse of the structure
- Pre-stressing forces are very high
- Assembly is quite complicated
- Large and complex design space is needed

The great advantage of the structures is their efficiency; their remarkable mechanical capacity in relation to the amount of material needed. This mechnical capacity and thus energy related aspect makes the existence of tensegrity structures in nature comprehensible. The construction of vertebrates and the cytoskeleton of cells seem to follow the same construction principle.

The fact that in spite of their good mechanical characteristics tensegrity structures in architecture are rather rarely applied, suggests that matters of security together with spatial and procedural control are more important in buildings.

The spokes of a wheel is a special case of tensegrity structure. The wheel absorbs the tensile stress; this construction is that used for stadium roofs.

Fig.96 Tree inside the Crystal Palace during the first Great Exhibition of 1851.

Girder constructions [Rippenkonstruktionen]

The easiest method to save material and mass is to distribute loads of the elements onto rib structures of any kind. Connecting thin plates to ribs oriented in the direction of the main stresses is a very economical construction. The support and transfer of loads can happen through a hierarchy of beams. John Paxton's Crystal Palace is said to have used the Giant Water Lily as a role model. According to photographs it looks more like a folding structure than a structure with ribs, and works as a roof construction in a completely different way to the floating leaf. Although the Crystal Palace may be an example of retrospective interpretation, leaves in nature are very often structured with ribs.

Fig.98 9th of October Bridge in Valencia, 1986-89.

Fig.97 Palazetto dello sport in Rome, Nervi and Vitellozzi 1957

Curved surfaces with girders

- **Pier Luigi Nervi (1891-1979)**

Pier Luigi Nervi designed many buildings using this kind of structuring in the 1960s, often in organic forms - following the flow of forces. He became famous for his concrete constructions, halls and stadiums with ribs, crossing or net-like prefabricated concrete elements. The ribbed ceiling of the Gatti wool factory was built in 1953. *"The arrangement of the ribs corresponds to the isostatics of the main points inside a system subject to stress."*[170] One of his masterpieces is the Palazetto dello sport in Rome, whose cupola of fine network transforms into V-shaped supports.

Between beam and shell constructions

- **Santiago Calatrava**

Following the civil engineering tradition of Nervi, Candela and others, Calatrava's main works are bridges, buildings for traffic, which structures combined construction with distinct architectonic forms. His focus is on the elegant and impressive balancing of masses and forces.

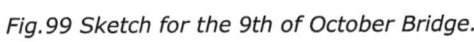

Fig.99 Sketch for the 9th of October Bridge.

The dynamic form often overrides functional and constructive requirements. Movement and locomotion as an interesting aspect of Calatrava's work is mostly expressed through form.

170 Portoghesi, P.: Nature and Architecture, 2000, p.159

Fig.100 Gateway Arch, for St.Louis, Eero Saarinen, 1961.

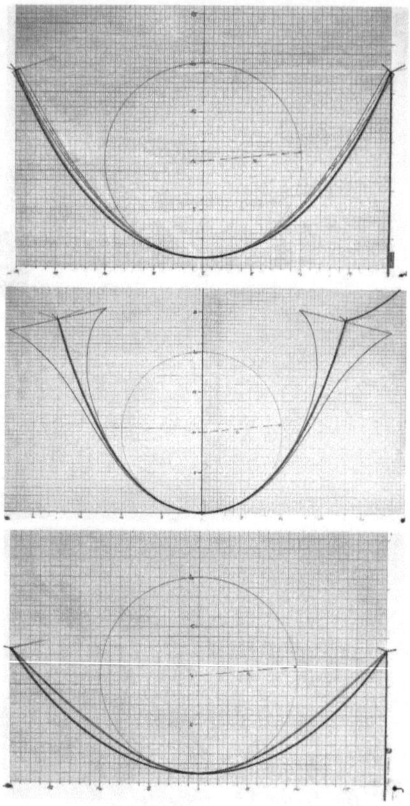

Fig.101 Catenary curves, records of Frei Otto's experiments, 1962.

Arcs and vaults

The form of elements should be adapted to the functional requirement, in the case of construction to the flow of forces. The use of parabolic curves for the construction of arcs and vaults was already known in ancient times, discovered by trial and error. By Roman times this knowledge had been lost again, and circular and spherical forms prevailed. Antoni Gaudi, Frei Otto and Heinz Isler made use of catenaries to develop the shape in experimental models. In the 1960s, numerous examples of organic architecture made use of free forms of curves and shells. The free forms of today's architecture are as difficult to control, and computational methods help with formfinding.

- **Eero Saarinen 1910-1961**

The famous TWA Terminal at JFK Airport New York, built 1956-1962, has a dynamic and bird-like form. Eero Saarinen said

"...a building in which the architecture itself would express the drama and specialness and excitement of travel... a place of movement and transition... The shapes were deliberately chosen in order to emphasize an upward-soaring quality of line. We wanted an uplift."[171]

The size and functionality of the interior space no longer meets today's requirements, and only societies for architectural conservation prevented demolition. The expressionist organic form of the building does not promote extension or flexibility of use.

Eero Saarinen also designed the famous St. Louis gateway arch. This project was built between 1961 and 1966, and has also become cultural heritage. Its steel construction follows a catenary curve.

---

[171] http://www.galinsky.com/buildings/twa/ [08/2007]

Fig.102 Rib construction of the Casa Mila, Antoni Gaudí, 1905.

Fig.103 Restaurant in Xochimilco, Felix Candela 1958, recent photograph.

Fig.104 Interior space.

Fig.105 Shape of the inner concrete surface, showing traces of the formwork.

Shells

Shell constructions are very efficient when the forces can flow in the direction of the material. This makes extremely thin constructions possible.

- **Felix Candela (1910-1997)**

Felix Candela used a shell-like form for the construction of a project in Mexico in 1957-1958; the restaurant in Xochimilco. In contrast to a natural shell, where the curves are focussed on a single centre, Candela's project is a cross vault consisting of four connecting hyperbolic paraboloids. Only parts of its form and the constructive principle of load bearing of shells are analogous to the shell in nature.

Fig.106 Centre Pompidou in Metz, France, Shigeru Ban architects, 2010.

Fig.107 Pattern of its wooden gridshell.

Grid shells

The next step in the minimization of material is to break up the surface of the shell. In the extreme, this is happening with grid shells. Examples of grid shells in architecture are found in vernacular architecture, and also in Frei Otto's oeuvre. A recent grid shell project, which also integrated the issue of recycling, is the Japanese Pavilion for the Expo 2000 in Hanover, and the Centre Pompidou in Metz, France in 2010.

- **Shigeru Ban**

Shigeru Ban has realised a number of constructions made of paper tubes. The structure of the Expo-hall was developed in collaboration with Frei Otto and Buro Happold civil engineering. The construction combines a series of laminated wood beams with a grid structure of paper tubes, which are simply connected by bands of polyester. The cladding consists of a special fire and water resistant paper. The roof of the Centre Pompidou in Metz is based on a hexagonal pattern of laminated wooden units and a fibre glass and teflon membrane, covering a surface area of about 8000m².

Rope constructions, tent and pneumatic constructions

The ultimate minimisation of material is done with rope, tent and pneumatic constructions.

- **Frei Otto**

Frei Otto is a pioneer in the field of hanging roof structures. From 1964 onwards he taught in Stuttgart, and was head of the Institut für leichte Flächentragwerke, which carried out fundamental research in the following years. His most famous project is the covering of the Olympia park in Munich, developed and executed together with the Behnisch office in 1972. Membrane buildings belong to the lightest constructions buildable at present. But to calculate the dead load correctly in order to compare it with conventional constructions, all foundations and bearings needed for pre-stressing have to be included. The materials, constructions and technologies for processing have developed fast within recent decades.

As membranes only transmit tensile forces, the form of these constructions is closely interrelated to the flow of forces. The development of form, formfinding, is an essential process in contrast to rigid systems. These so-called form active systems have to be prestressed in order to control all loading conditions which can occur in the construction. Of particular importance in mechanical prestressing is minimal surface: the surface with the smallest area between any form of closed polyline. At every point of such a surface the sum of the radii of curvature is zero. A minimal surface is the optimal form for load bearing behaviour under the precondition of equal tension forces in all directions. Depending on the geometry of the bordering curves, flat or saddle formed surfaces - anticlastic curvatures - are created.

Fig.108 Exhibition model of the Olympia Park in Munich, 1972.

Fig.109 Four point membrane, hyperbolic paraboloid, Frei Otto.

Fig.110 Folding structure applied on building façade, Future Systems, Oxford Street, London, 2009.

Folded structures

- **Rigid folded structures**

Folding thin plates is another method to gain structural performance without unduly increasing weight. Isotropic thin materials, which are not elastic, cannot be bent along a folding edge (more general: along the generating lines of a cylinder or a cone). The principle of folding is ubiquitous in nature: leafs and insect wings, for instance, are folded for reasons of structural performance. The same principle is applied in architecture in folding structures. The largest application field for folding structures is in material structuring, e.g. paper, sheet metal.

- **Moveable folded structures**

Moveable folded structures can either deploy at the same location or be folded and moved to another location. Deployable structures that provide a flexible internal space are still rare in architecture. On the other hand, the folding and deployment of surfaces in architecture is very common: diverse systems exist for sun-protection, windows, doors and gates, separation wall systems, connection elements, damping elements, furniture work with folding and deploying.

Classical approaches | "Natural construction"

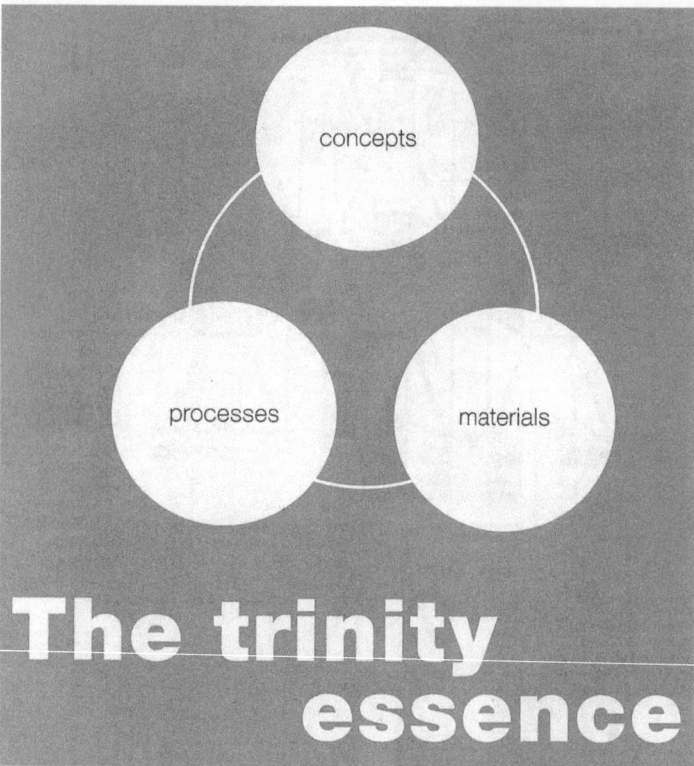

Fig.111 Illustration of relation between concept, process and material, Beukers und Van Hinte, 1998.

Materials and surfaces

"Every single functional object that we make evolves from a process that turns material into a functional shape, the inevitable trinity of technology. There is no shape without material and effort."[172]

Many of the materials and structures mentioned in this chapter are object of structures bionics, which investigates, describes and compares biological materials and structures and explores the applicability of specific, even unconventional materials and structures for specific purposes.
The differentiation between structure, material and surface is a matter of viewpoint. A surface is the border with another material. The structure of the surface also defines material characteristics (see Lotus effect, Shark skin), concerning interaction with the environment.
The architects committed to light construction stress the integrity between concept, process and material.
Nicola Stattmann presents a comprehensive compilation of new materials in her book "Handbuch Material Technologie" about new materials and technologies.[173]

Characteristic values of materials

Characteristic values of construction materials
- **Strength**

$\sigma_{rupture}$ = force [kN] / surface [cm²]
The strength of a material is also expressed as its rupture stress.
- **Specific weight, density**

γ = weight [kN] / volume [cm³]
Density is inversely proportional to the insulation characteristics of a construction material.
- **Stress and strain**

$\varepsilon = l / L$
ε measures the relative change of length of a material exposed to tension- or pressure forces, often expressed as a percentage.
- **Elasticity, Young's modulus**

E = stress/strain = σ / ε [kN/cm²]
Young's modulus is an important characteristic for the danger of bending and buckling of a material. The higher the modulus, the smaller the danger of breakdown. The value expresses the relationship between stress and strain.
- **Tensile strength**

$L_{break} = \sigma / \gamma$ [cm]
The breaking length is a real performance characteristic. It is independent of the scale of elements and tells how long a bar, rope, thread or similar element can be until it breaks under its own weight.
Building and construction materials are considered good quality if they have high strength, a high Young's modulus and a breaking strain of 20-30%. Biological materials do not occur with completely desirable characteristics or every characteristic optimal; their values are consistently average. The efficiency of biomaterials is due to the perfect cooperation of material, structure and form, and not to the selection of a perfect material (see the universal materials wood and chitin). Further important parameters for construction materials are thermal conductivity and ecological criteria.

---

172 Beukers, A. et al.: Lightness, 1998, p.23
173 Stattmann, N.: Handbuch Material-Technologie, 2000

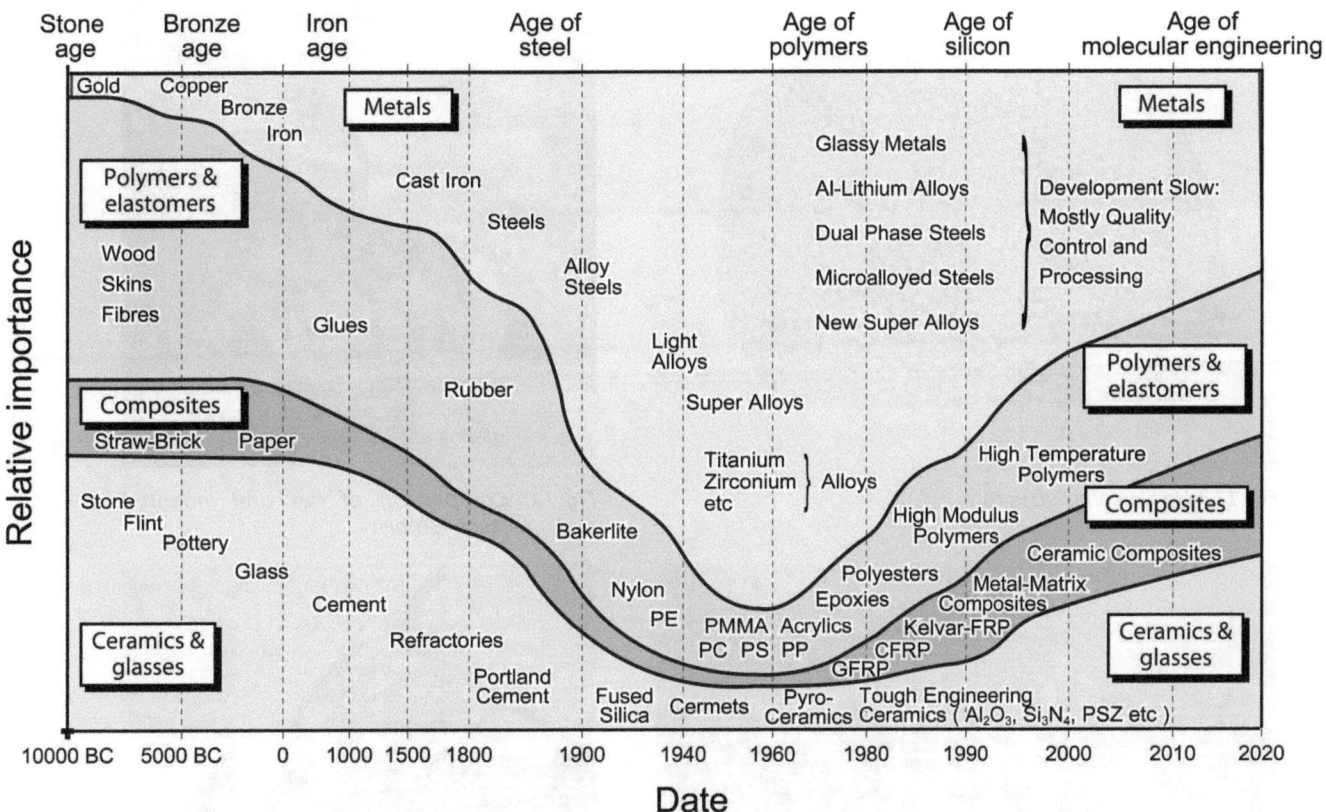

Fig.112 Illustration of the relative importance of materials at different times in the past, Mike Ashby, 1998.

Development

"Unprecedented in history is the development in new applications and the discovery of many new materials over the last 150 years: steel, aluminium, titanium and synthetic polymers, artificial ceramics, to name a few. Even for metals limits cannot be stretched endlessly... [Metals] reached their maximum contribution to constructions in around World War Two. From then on synthetic ceramics, polymers and composites began competing and will continue to do so for many years to come. The reason that application of metals is gradually decreasing is not that metal resources being exhausted, but that the most widely used ones, steel and aluminium, are no longer capable of meeting long term requirements of price and performance... change in the importance of weight. Now lightness, or performance per energy unit, is quickly gaining significance again, because... cheap energy is getting scarce."[174]

Structuring of material is a cheap way to make materials lighter. In the following paragraphs, different structuring methods and processes are described.

174 Ashby M.F. 1992 in Beukers, A. et al: Lightness, 1998, p.17

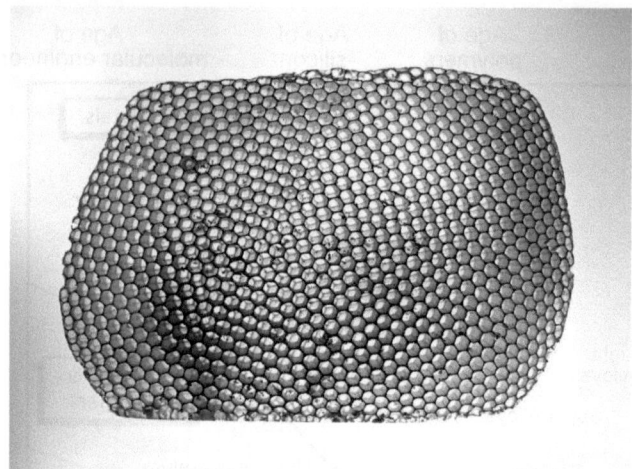

Fig.113 Honeycomb structure.

Fig.114 Paper honeycombs produced by glueing strips at intervals, then stretching the honeycomb to its final length.

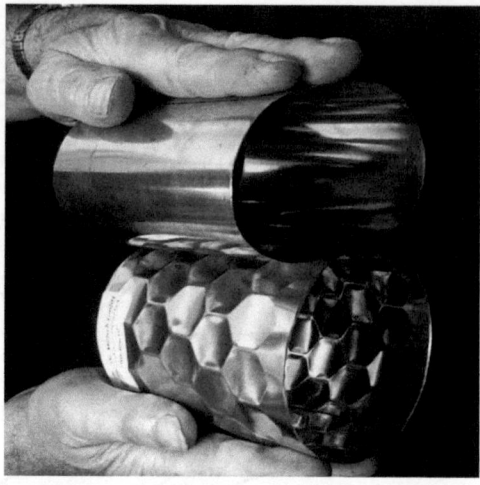

Fig.115 Comparison of flat and indented cans, Mirtsch GmbH.

Fig.116 Hexacan; hexagonal self-organised patterning of metal, Mirtsch GmbH.

Cellular structures

- **Honeycomb structure**

Honeycomb structures occur often in nature, as the hexagonal structure is the most densely packed structure in two-dimensional space. Technical honeycomb structures are made of plastic, ceramics, paper and metal. Processing includes: the cutting of hexagonal sheets and subsequent gluing, insertion of strips of glue between the sheets and subsequent stretching, or the use of moulds, especially when plastic is used. Honeycombs are used for the core of sandwich panels and composite designs. Due to their large surface area, they are suited for use in cooling machines and catalysers. They are also used as the surface layer of tyres and packaging.[175]

- **Honeycomb - vault structuring**

"Vault structuring uses the material's self-organization response to a very gentle and sustainable force to create a perfectly staggered three dimensional honeycomb structure.

It differs greatly from the traditional mechanical embossing and stamping technologies. The naturally occurring three-dimensional honeycomb structure shows typical properties of a self-organized process. The most important result of this natural process is an increased stiffness combined with a low weight. Vault structured materials are available in coils and sheets."[176]

The use of self-organised buckling to structure material enables industry to use thinner material, and thus eke out resources. Materials used include metal, paper and plastics. The product has spread into all kinds of applications: in architecture, roof and façade products of the Rheinzink company are structured in this way. Vault structured material is also used for ventilation technology, lamps, washing machines, in the car and packaging industries. Apart from the constructive aspect, the structure has the advantage of having its low weight uniformly distributed. [177]

---

175 Bürgin, T. et al.: HiTechNatur, 2000, pp.10

176 http://www.mirtschusa.com [08/2007]
177 http://www.mirtschusa.com [08/2007]

Foam structures

Nearly every material can be turned into a foam structure. The materials used for foaming are plastics, ceramics, glass and also metals. The size of the pores varies.

- **Plastic foams**

Plastics are usually produced from oil. They are categorised by their plastic forming properties in thermoplastics, thermoset plastics and elastomers. Only fusible thermoplasts can be recycled.

- **Polymers**

A polymer is a substance with a molecular structure consisting of a large number of similar units.

The foam structure is created by the mechanical addition of a gas (physical blowing agent) or with a chemical support substance (chemical blowing agent). Both methods can produce cell structures with open or closed pores.

- **PUR**

Polyurethane (PUR) is foamed in moulds, used as micro cellular structure and as soft-flexible foam. Polyurethane foam is applied in diverse products for architecture, furnishing and car manufacturing, as thermal and acoustic insulation and soft-flexible material.

- **PS**

Polystyrene (PS) is extruded and used in huge amounts for thermal insulation, also as packaging material and for disposable goods.[178]

- **Metal foams**

Aluminium is used for most metal foams. There are diverse processing methods: powder metallurgy, fuse-foam, and special casting processes...

The processing method determines the application of the material. Boards are used for absorbing high-energy impacts (bullet proof clothing and walls), reinforcing core for hollow sections (until now predominantly in the car industry), and for acoustic absorption. The problem of application of force to the foam structure can be solved by the creation of a compound effect with the solid metal. The strength of metal foams is higher than that of polymers. They are 100% recyclable, but the extraction from their ores consumes vast amounts of energy.

The boards can be produced at a maximum size of 60x60cm only. A wide range of companies produce metal foam for diverse products (e.g. Alulight of the Mepura company in Ranshofen).[179]

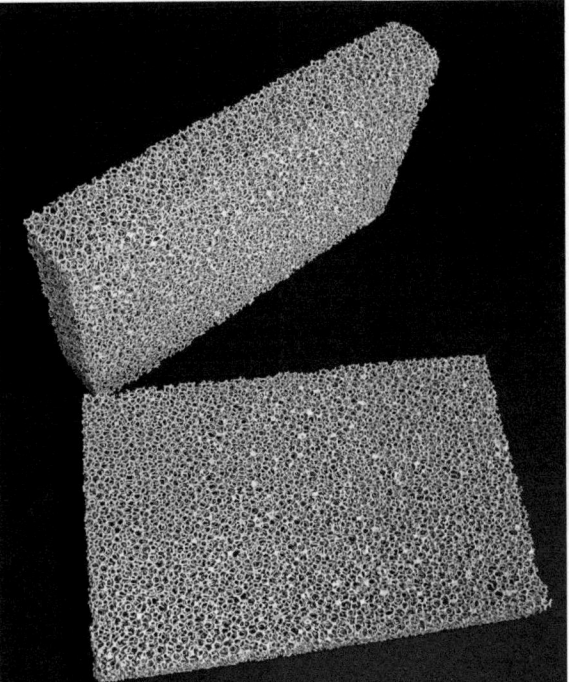

Fig.117 Metal foam plates, m-pore GmbH.

Fig.118 Foam structure.

- **Plaster foam**

In addition to being a structured product, plaster foam is also a recycled product. Plaster, collected during the cleaning of smoke gas of coal-fired power plants, can be processed efficiently into a foamed material, which is better than commonly used insulation materials with regard to ecological and health aspects. Chemically bound and therefore harmless isocyanates are used as propellant. The foaming process happens in normal environmental conditions, in contrast to the common foamed plaster construction materials, which are produced under pressure and at a temperature of 300°C to 500C°. The material is resistant to humidity and mechanically resilient.[180]

---

178 Stattmann, N.: Handbuch Material-Technologie, 2000, pp.72
179 Stattmann, N.: Handbuch Material-Technologie, 2000, p.44

180 Scientific American (Ed.): Spektrum der Wissenschaft Digest Nr.3 Moderne Werkstoffe, 1996, p.110

Fig.119 Honeycomb sandwich: decks made out of glass fibres with epoxy matrix, company Euro-Composites.

- **Foamed ceramics**

Foamed ceramics are open-pored structures made of ceramics (kaolin, clay, quartz and feldspar). A foamed structure of PUR is soaked in a ceramic suspension. During the burning process the ceramics are hardened and the PUR is removed, leaving behind the foam structure as hollow spaces. Different sizes of pores are possible. Foamed ceramics have the permeability of sponge and the strength of ceramics. The material is applied in filters.

Biomorphic ceramics use natural structures (wood, collagen) as the matrix for the ceramic structure. Chemical transformation processes make a ceramic material from the organic structure. This strategy provides the characteristics of both material and structure.[181]

Sandwich constructions

Sandwich constructions are laminated materials consisting of two covering layers separated by a light core. The distance between the outer layers defines the size of the geometrical moment of inertia. The principle is similar to I and H sections used in steel constructions, but the transmission of shear forces is low. Sandwich constructions are used in many areas of life: in the building industry (sandwich boards, gypsum plaster boards), the transport industry (ships, airplanes, trains, cars), packaging (e.g. corrugated card board), sports (snowboards, skis) etc. Covering materials are usually metals (aluminium), fibre reinforced plastics, wood and paper; the core consists of honeycomb cores, foams, light materials.

The core material absorbs the compression forces and transmits the bending force to the boards.

In natural sandwich constructions the core material is well adapted to the load.[182]

Biological sandwich constructions consist chiefly of fast growing plant material like straw, elephant grass, bamboo, jute and hemp. They are bound by a non-toxic inorganic substance which is flame resistant and water repellent. The orientation of the fibres provides additional strength against compression and bending forces.[183] Products currently on the market are Pacific board, Isoboard and Stramit, to name but a few.

---

181 Stattmann, N.: Handbuch Material-Technologie, 2000, p.58

182 Bürgin, T. et al.: HiTechNatur, 2000, pp.12

183 Stattmann, N.: Handbuch Material-Technologie, 2000, p.118

elastic properties of some fibres, natural composites and steel compared

material		density ($10^3 \cdot kg/m^3$)	Young's modulus ($10^9 \cdot N/m^2$)	Yield stress ($10^6 \cdot N/m^2$)	Yield strain (%)	elastic energy/weight (J/kg)
steel	0.2 carbon quenched	7.8	210	773	0.2	99
	piano wires/springs	7.8	210	3100	0.8	1590
animal	bovine bone	2.1	22.6	- 254	- 1.4	846
	ivory	1.9	17.5	217	1.2	685
	buffalo horn	1.3	2.65	- 124	- 3.2	1530
	sinew	1.3	1.24	103	4.1	1620
hardwood	birch	0.65	16.5	137	1.0	1050
	Wych elm	0.55	10.9	105	1.0	950
	ash	0.69	13.4	165	1.0	1196
	oak	0.96	13.0	97	1.0	703
	elm	0.46	7.0	68	1.0	740
softwood	scots pine	0.46	9.9	89	0.9	870
	taxus brevifolia	0.63	10.0	116	1.3	1100
natural organic polymer base fibres	jute	1.46	10-25	400-800	1-2	
	flax	1.54	40-85	800-2000	3-2.4	
	sisal	1.33	46	700	2-3	
	hemp	1.48	26-30	550-900	1-6	
	coir	1.25	6	221	15-40	
	cotton	1.51	1-12	400-900	3-10	
synthetic organic polymer base fibers	HM-Carbon (M40)	1.83	392	2740	0.7	
	HT-Carbon (T300)	1.76	230	3530	1.5	
	HM-Aramide	1.45	133	3500	2.7	
inorganic base fibres	S/R-Glass	2.48	88	4590	5.4	
	E-Glass	2.58	73	3450	4.8	

note 1: Northern hardwoods, sinew and horn are the basic structural materials for laminated composite bows and chariots from Mesopotamia and Egypt.
note 2: Taxus baccata is used for the medieval longbows.
note 3: Horn, a natural thermoplastic polymer was especially applied in the compression loaded areas.
note 4: Sinew, superior in tension, was applied mostly for strings and bow-reinforcement; more in general it was used as a shrinking (smart) robe to encapsulate and to connect different components.
note 5: Properties of natural materials are very variable, the figures shown are most averages in tension or compression (-) and collected from many different publications.
note 6: Yield stress is the lowest stress at which a material undergoes plastic deformation or failure.

Fig.120 Comparison of elastic properties of different fibres, natural composites and steel.

Fibres

Fibrous structures integrate fibres of any kind, of various lengths. Paper consists of layers of cellulose fibres and textiles have been woven of different materials since prehistoric times. The length of the fibres defines their usability, usually the longer the better. Nature offers numerous examples of fibres within complex structures. In former times the material for textiles consisted of short, naturally grown fibres, which had to be spun or braided (wool, hemp, cotton). Today, endless artificial plastic fibres are produced. Nylon, a bionic product, was manufactured by Dupont in the 1930s using natural silk as a role model.[184] Nylon can be extended by 20%, but has low stiffness.

Fibre reinforced composites

Nearly all load bearing materials and structures in nature are composites. Composite materials consist of at least two physically or chemically different phases with a separation interface (region where the matrix and the reinforcing material in the form of fibres touch).

The final material has characteristics that the components did not have. The matrix can consist of polymers, metals or ceramics. Other, metal, composites exist as well as the large group of fibre-reinforced composite, but these do not find applications in building.

Some characteristics are very important for construction with fibre-reinforced composites:
- **Anisotropy**

In contrast to the optimal fibre orientation in natural examples, which is adapted by growth to the load, technical composites show a very basic structure. The differences in performance of the layered material have to be taken into account:
- **Poor capacity to resist compressive forces**

Fibres can resist tensile forces very well, but bend and buckle under compression.
Several solutions have been found for this problem: prestressing of the fibres to avoid compression, introducing additional minerals (well bonded to the fibres), strong interconnections between fibres to provide stability against buckling and change of fibre orientation so that the compressive forces do not act parallel to them.[185]

184 http://heritage.dupont.com/touchpoints/tp_1935-2/depth.shtml [11/2007]

185 Jeronimidis, G.: Biodynamics, in: AD Architectural Design Vol 74 No 3, Emergence Morphogenetic Design Strategies, 2004, p.94

Fig.121 Hardhat constructed from eco-composite, Company Move Virgo, Cornwall 2007.

In natural composites the major proportion of fibres consists of natural polymers like cellulose, collagen, chitin and silk. In technical composites glass fibres, carbon and aramid fibres are embedded into polymeric matrices.

- **Glass fibres**

Glass fibres reinforce composite materials. They are used mainly with thermoplasts as glass reinforced plastics, GRP. Their production and processing is hazardous to health as they are extremely brittle. They are widely used, however, because they are comparably cheap.

- **Carbon fibres**

Carbon is similar to graphite. Carbon fibres are extremely good at resisting tensile stress (10x as good as steel) and very strong, but expensive and difficult to process.

- **Ceramic fibres**

Ceramic fibres are made of aluminium oxides or silica.

- **Metallic fibres and Bor fibres**

These fibres are not used in the building industry.

- **Organic polymer fibres**

Carbon- and Aramid fibres are organic polymer fibres. Product names include Kevlar and Nomex. They are very strong, their mechanical properties are almost constant between -200°C and +160°C, weigh little and are cheap. Photosensitivity is their big disadvantage as it reduces their good mechanical characteristics. In addition to all kinds of sports utilities and vehicles, textiles for the space industry and bullet-proof vests are made of these materials.[186]

Bio composites

"...Because of the inseparable bonds between fibres and matrix technical composite materials cannot be recycled."[187]

- **Plant fibres**

Plant fibres are an alternative to technical fibres, interesting from an ecological and an economic point of view. Plants are renewable resources; grass in particular produces a large quantity of biomass within a short time. Natural fibres are characterised by their tubular structure, and their blades have an enormous load capacity along their length. The fibres are bendable, have high tensile strength and easy to recycle. Plants like flax, straw, hemp, reed, banana fibres, pineapple fibres, coconut fibres, bamboo and china grass are all used in bio composites. The matrix consists of a biopolymer produced of sugar beet, potatoes, corn or cellulose made of waste paper. Cellulosedaicetate is commercially marketed under the name "Bioceta", is comparable in strength to epoxy resin, but softer and more plastic.

- **Bio composites**

Due to the fibres and matrix, bio composites attain performance characteristics that almost correspond to glass reinforced plastics, GRP's. Elements can be pressed, wound or extruded. A special feature is biodegradability: within 4 months microorganisms degrade 100% of the mass.[188]

3D fabrics

A 3D fabric is a semi-finished textile product, with a three-dimensional structure. 3D fabrics are soaked in polymer and then laid flat or bent in a mould. The form remains as it is after the polymer solidifies. The fabric's characteristics depend on the composition and the geometry; flexible as well as stable light materials can be produced. The material is used for the core of sandwich constructions.[189]

---

186 Stattmann, N.: Handbuch Material-Technologie, 2000, p.82

187 Stattmann, N.: Handbuch Material-Technologie, 2000, p.80
188 Stattmann, N.: Handbuch Material-Technologie, 2000, p.98
189 Stattmann, N.: Handbuch Material-Technologie, 2000, p.84

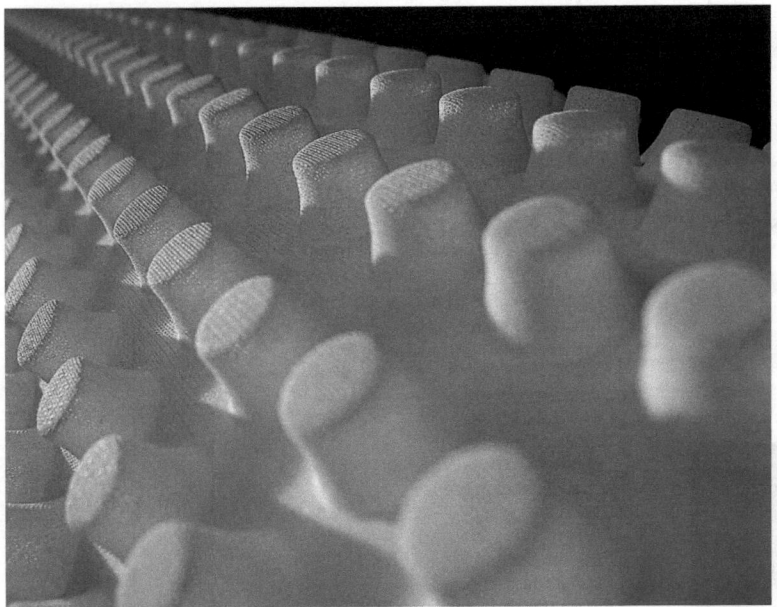

Fig.122 Fabric 3D-TEX, Height ca. 13mm, Mayser GmbH.

Fig.123 Elephant hawk moth.

Fig.124 Moth eye surface, the role model for anti-reflection nanostructures.

Fig.125 Moth eye, magnification 1:2000.

Nanomaterials

Nanotechnology is one of the key technologies of the 21st century. The term "Nano" embraces many different technologies, methods and materials. Nanocomposites consist of nanoscale particles, embedded in diverse matrices, e.g. in polymers, glass and ceramics. A functional layer can enclose these particles. With this technology it is possible to create a diversity of chemical, physical, electrical and optical properties. In architecture, functional surfaces are becoming more and more common. Examples of such applications are easy-to-clean (Lotuseffect) and photochromic coatings (darkens with light intensity).[190]

- **Anti-reflection surface**

Microscopically small gratings, narrower than the wavelength of light, cover the eye surfaces of moths. This structure influences the reflection and refraction of light. For the animal this delivers good night-vision because - with their eyes hardly reflecting anything - the available light and good camouflage are efficiently used. The artificial surface mimics the principle. It is produced by laser etching of plexiglass and foils, and is currently used as antireflection coating for optical glasses.

190 Stattmann, N.: Handbuch Material-Technologie, 2000, pp.106

Fig.126 Aerogal granulate, company Cabot.

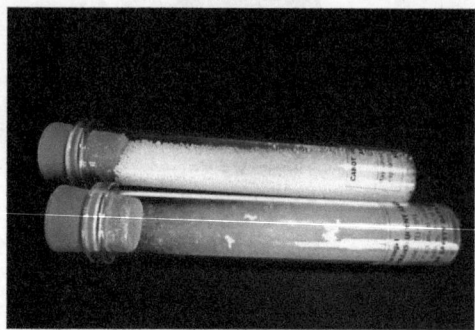

Fig.127 Aerogel in form of granulate.

- **Aerogel**

Aerogels were discovered in the 1960s. The basic materials are silicon-, aluminium-, iron-, titanium- and tantalum oxides. The most widely used is silicon oxide, chemically processed with carbon or formaldehyde. The gels are dried under high pressure and high temperature in a so-called supercritical state of the matter to avoid shrinking. The parts of the aerogel then create a three-dimensional network, similar to consolidated fog, which has particular properties:
- Extremely low density, high porosity and high strength
- Transparency (relative)
- Aerogels can be conductors of electricity, or insulators
- Superb insulation material, thermal conductivity being 0.012 W/m Kelvin (Styrofoam 0.024)
- Acoustic velocity is smaller than in air

The current hypothesis says that these effects result from a fractal structure of the material that is said to consist of 98% air. The state of matter is between fluid and gaseous.[191]

Because of the energy consuming production of the material is expensive. Therefore aerogel granulates are used. In architecture they are used for transparent insulation within double-glazing or boards of polycarbonate, and they are used in passive solar panels. Among others, the Swedish company Airglass and BASF produce aerogel.

Building inputs Material	Embodied energy MJ/kg
Kiln-dried sawn softwood	3.4
Kiln-dried sawn hardwood	2.0
Air-dried sawn hardwood	0.5
Hardboard	24.1
Particle board	8.0
Medium density fibreboard (MDF)	11.3
Plywood	10.4
Glued-laminated timber	11.0
Laminated veneer lumber	11.0
Plastics, general	90.0
PVC	80.0
Synthetic rubber	110.0
Acrylic paint	61.5
Stabilised earth	0.7
Imported dimension granite	13.9
Local dimension granite	5.9
Clay bricks	2.5
Cement	5.6
Gypsum plaster	2.9
Plasterboard	4.4
Fibre cement	7.6
In-situ concrete	1.7
Precast steam-cured concrete	2.0
Precast tilt-up concrete	1.9
Concrete blocks	1.4
Autoclaved aerated concrete (AAC)	3.6
Glass	12.7
Mild steel	34.0
Galvanised mild steel	38.0
Aluminium	170.0
Copper	100.0
Zinc	51.0

Fig.128 Process energy requirements (PER) for common building materials, Yeang 1999.

Ecological aspects

Ecological design means the creation of closed material cycles, and the minimisation of the total energy consumption. Ecological aspects have to be taken into account for the selection of materials in addition to constructive, aesthetic and economic aspects.
- Careful handling of resources, especially of non-renewable ones
- Ecological impact embodied in the material (as a consequence of production and transport)
- Potential reuse or recycling
- Toxicity (of the material or by-products)
- Energy balance, containing:
- Amount of chemically bound energy, which was needed for production
- Embodied energy for production and transport
- Energy for the operation and maintenance
- Energy to recycle the material[192]

The interrelation between material, construction and overall concept will be discussed later.

---

191 Scientific American (Ed.): Spektrum der Wissenschaft Digest Nr.3 Moderne Werkstoffe, 1996, pp.104

192 based on Yeang, K.: The Green Skykscraper, 1999, p.135

Fig.129 Hanging model of the Catherdral Sagrada Familia, Barcelona by Antoni Gaudi.

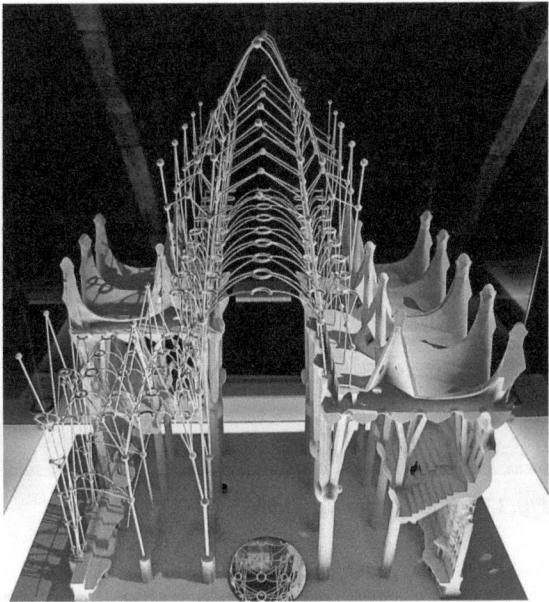

Fig.130 Exhibition model of the Sagrada Familia.

Use of natural processes - physical experimental formfinding

Much more surprising than the formal and functional transfer are the processes responsible for the formation of the phenomena that we are so fascinated by. The physical environment that technology shares with biology is responsible for many analogies occurring, even converging phenomena. Antoni Gaudi, Frei Otto and Heinz Isler all have found a simple way to make use of the phenomenon of catenaries. Under the continuous load of its own weight a chain connecting two points will follow a catenary curve - the minimum energy state of the system. Turning this upside down, an optimal curve for load transmission is found. This way, hanging models can be used to find organic forms in which only normal forces exist.

The form of the curve changes with the weight of the chain: in order to use this phenomenon for making analogue models one has to take into account all permanent loads. Gaudi and Otto were experts in using hanging models for formfinding. Nowadays computational methods replace the physical modelling.

- **Antoni Gaudi (1852-1926)**

Antoni Gaudi represents Art Nouveau in Spain, but developed a unique and independent technology and language of forms. He worked with hanging models to develop the constructions and forms of his projects. His largest and hitherto unfinished project is the Sagrada Familia cathedral in Barcelona (start of the building was 1883).

Fig.131 Main hall of the Sagrada Familia.

Fig.132 Shell hanging model, Frei Otto.

Fig.133 Soap film models, Frei Otto.

- **Heinz Isler**

From 1952 onwards Heinz Isler developed textile hanging models in parallel to the works of Frei Otto. These models were then soaked with a curing fluid (he also used simple water for ice models) to discover the optimal form for the flow of forces of shells. The self-generated shell geometry was then measured and transferred to plans manually. Isler's shells are made of reinforced concrete; the untensioned reinforcement ideally follows the main lines of principal stress. The elaborate boarding was used for more than one project. Isler developed and refined this technology for planning and building, although this has not yet been published in detail. The aesthetic quality of the shell, however, is lost with the insertion of façades.

- **Frei Otto**

With his Institut für Leichte Flächentragwerke Frei Otto investigated what he called synthetic analogy research. In experiments the physical laws responsible for specific phenomena in nature were modelled and explored.

Many "natural constructions" were investigated in this time, e.g. minimal surfaces, soap bubbles and foam, pneumatic and tensile structures.

Apart from his extensive publication work, which still inspires generations of engineers and architects, Frei Otto also managed to develop professional methods and applied them to big building tasks. For instance, the geometry of the Olympia park roof was derived from a physical model that delivered information on tensile forces and expected extensions.

- **Other phenomena**

Apart from gravity, other physical phenomena are used to define construction and to influence architectural parameters. Experiments with magnetism were carried out in urban planning, to visualise directions and design cities.

Some of the modelling strategies do not prove expedient for a particular task. Architects tend to use experimental results for formfinding, but also with phenomena not connected to the task as regards content, or whose appropriateness is the at least questionable. This method of using natural processes and phenomena to arrive at a formal pattern is very common, and is acceptable as another form of inspiration and creativity. The quality of the final project always depends on the adequacy of the application. But it is important to be conscious of that, and not mistake formal approaches with the scientific backing up of a design proposal.

With growing use of computational methods, natural processes can be digitally simulated and are available for the application to design tasks. The attempts to transfer physical laws or the dynamism of living processes onto built environment will be presented later in chapter 4.2 about the architectural interpretation of life criteria.

Ecological design and sustainability

The ecological and sustainable design movement began after the energy crisis in the 1970s. All kinds of efforts to reduce energy and resource input and processing in buildings have been experimented with on a 1:1 scale. As always, when one parameter is considered highly important in architecture, the other aspects necessarily suffer: the aesthetic side of ecological design has for a long time been neglected. The first low energy and passive houses did not become famous for beauty in public opinion nor did they appear in architecture magazines. Nowadays, basic rules of sustainable design are implemented into many national building regulations, and technology has delivered better and more aesthetic products (e.g. windows, façade panels...), but still the consideration of ecological issues is a challenge in design. Apart from basics in ecological and sustainable design, some iconic projects will be presented, which have managed to bridge the gap between aesthetic and other concepts on the one side and necessity of sustainability on the other.

Architecture and environment

The fundamental function of architecture is to provide shelter from changeable and often hostile natural environments. The built structure mediates between the environmental conditions outside and the interior climate, which is comforting for humans. Traditional architecture shows amazing adaptation to regional peculiarities of climate and resources.

In the 18th and 19th centuries a development of active systems started in Europe and the US to control environmental conditions within a building. This has lead buildings to be significant consumers of energy, and technology is a control medium. In the 1960s, the first critics of this development appeared. Victor Olgyay developed a model in which technology is merely given a role of fine-tuning at the end of the design process. Before that, the design followed the "bioclimatic" approach of a "design with climate". The energy crisis at the beginning of the 1970s made it clear that the critics have to assert themselves.

In general, we interpret the building as a system which interacts with its environment.

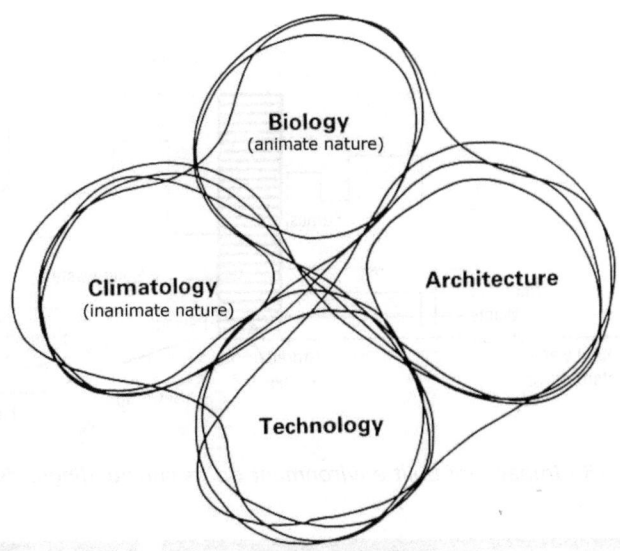

Fig.134 Model of environmental processes, Olgyay 1963, amended by the author.

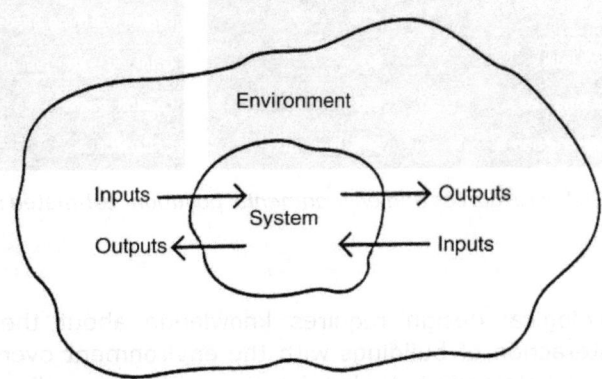

Fig.135 Model of a system and its relationships with the environment, Yeang, 1995.

Inputs and outputs

- **Interaction**

Different interactions (inputs and outputs) take place with the environment within the different building phases.
- Production: production and processing of materials, distribution, storage, transport
- Construction: construction, building site
- Operation: operation, maintenance, change, ecological measurements
- Demolition and recycling: demolition, recycling, reuse, restoration of the site

Designing and building should be interpreted as management of material and energy more than it is now.

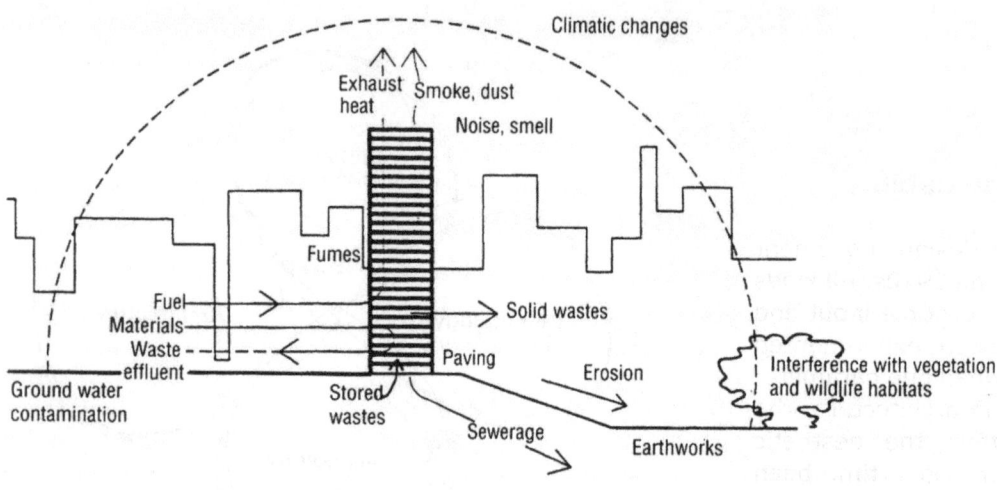

Fig.136 Impacts of built environment on its surroundings, Yeang 1995.

Global resources	
Resource	Building use
Energy	50%
Water	42%
Materials (by bulk)	50%
Agricultural land loss	48%
Coral reef destruction	50% (indirect)

Global Pollution	
Pollution	Building related
Air quality (cities)	24%
Global warming gases	50%
Drinking water pollution	40%
Landfill waste	20%
CFCs/HCFCs	50%

Fig.137 Global resources and environmental pollution, estimated share of the building industry AD 2001.

Ecological design requires knowledge about the interaction of buildings with the environment over the whole cycle of their existence, and the recycling of all inputs and outputs in other processes (closed loop cycles). Taking the interactions into account also means caring about changes of the environment by the mere existence of the new structure, and the consciousness of the dynamics of these interactions.

According to Ken Yeang, the ultimate measure for evaluation is the change in biodiversity (number of species and individuals, complexity of their interaction).[193]

- **Complexity**

When analysing interactions, one is confronted with a vast number of single variables. A holistic description is therefore impossible. What is crucial is the selection of the "right" variables. In his famous book "The Green Skyscraper" (1999), Ken Yeang states that what is crucial is the selection of the "right" variables and an open general system that functions as a grid for making diverse and complex interactions visible.

- **Building-related consumption of resources and pollution**

"Within our homes we create space for gadgets rather than for people... so whereas buildings consume half of all environmental resources, they either house the space where the other half is consumed or form the destination for the essential journeys required for human connection."[194]

The limitation of the human systems does not lie in the shortage of resources, but in the production of waste and environmental pollution. In densely populated urban areas (e.g. London, Hong Kong) the limits are visible. Apart from the fatal global impact of pollution, the survival of human civilisation itself is endangered by the poor health of populations. Architecture and the building industry produce a large amount of waste and pollution.

---

193 Yeang, K.: The Green Skykscraper, 1999, pp.99

194 Edwards, B.: Architectural Design Nr.71, Green Architecture, 2001, p.13

Principles	Measures		
	Buildings	Open spaces	Supply and disposal
Adaptation to natural and social location characteristics	• Integration into ecosystem depending on sun and wind • Zoning of ground plans • Minimum area consumption	• Minimal sealing • Few topographical changes • Maintain existing vegetation • Compact buildings	• Proximity to home, services and culture • Reduced personal traffic • Link to public transportation • Link to low-emission energy carriers
Energy saving	• Passive use of solar energy • Heat conservation • Heat recovery • Winter gardens + solar energy use	• Harnessing climate-regulating effects of vegetation and water surfaces	• Create closed cycles when possible • Waste-raw material • Rainwater • Grey- and cooling water • Waste heat-energy
Protection of resources and material	• Using environmentally friendly materials • Avoid toxicity • Low-energy production and processing	• Create green belt • Integrate parking into green area	• Substitute potable water • Avoid waste • Heat-power coupling • Minimise emissions
Creation of a high quality internal and external human environment	• Influence micro climate with building surface • Planted façades and roofs • Sun protection • Interior design • Ergonomic workplace design	• Enrich green area with plants and trees compatible with location • Create 'relaxation' areas • Stimulating environment	• Utilise surface water (rainwater) • Compost organic waste to improve soil

Fig.138 Principles and measures of ecological planning according to Klaus Daniels 2000.

Sustainability

"This word [Sustainability] describes economic cycles that do not harm the environment and do not consume more energy, raw materials etc. than are re-created in the natural cycle."[195]
In his book "Low-Tech Light-Tech High-Tech" Klaus Daniels defined sustainability and suggested basic principles for "contextual building design". To realise these principles, he suggests measures for building, free space supply and disposal.[196]

The list is a basic grid for sustainable design. Not all measures can be applied to all projects, but with these measures in mind, decisions can be taken more consciously. As Klaus Daniels differentiates, the approaches to sustainability can take different manifestations within the categories of high or low technology and high or low energy use.

195 Daniels, K.: Low-Tech Light-Tech High-Tech, 1998, p.18
196 Daniels, K.: Low-Tech Light-Tech High-Tech, 1998, pp.104

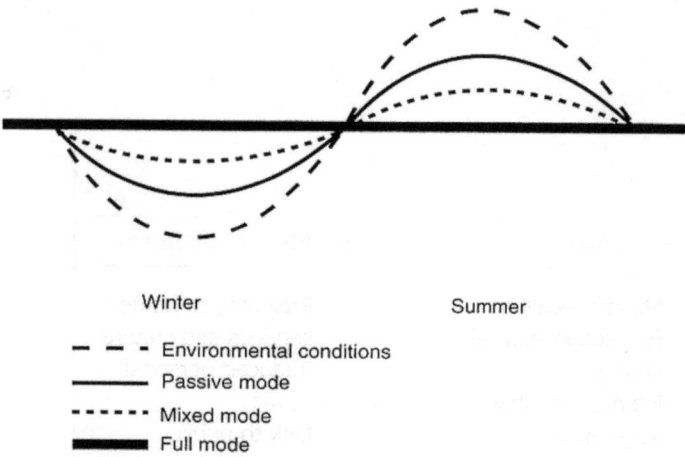

Winter　　　　　　　　Summer
– – – Environmental conditions
——— Passive mode
· · · · · Mixed mode
▬▬▬ Full mode

Fig.139 Comfort ranges of different modes, Yeang, adapted from Victor Olgyay's curves 1963.

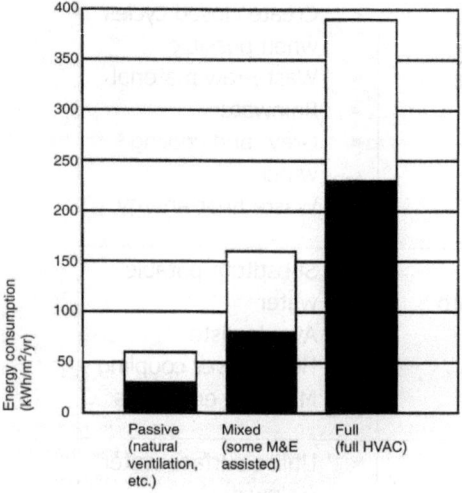

Fig.140 Energy consumption targets of various modes of operational systems, Yeang, 1999.

- **Ken Yeang, 1999**

Ken Yeang is a director in the big architecture company Hamzah&Yeang in Malaysia. He carries out research of the ecological design of intensive building systems, concentrating on large buildings in dense urban settings, preferably skyscrapers. He distinguishes between three groups of technical systems this kind of building is equipped with in its operating phase: passive-mode, mixed-mode, full-mode and productive-mode operational systems.[197]
A basic principle for ecological design is the preference of passive systems to that of mixed or full mode operational systems. In design, the levels have to be handled consecutively. Firstly the passive options have to be explored (e.g. orientation of the building), then the additional technical systems can be used (e.g. heating), and lastly the full technical solution applied (e.g. air conditioning). Production systems should be integrated whenever possible. The degree of connection between internal and external environment with respect to climate is a criterion for the decision, and the solutions depend on the climate zone of the site of the building.
In case of mixed mode and full mode, non-renewable energy sources (fossil fuels) should be excluded wherever possible. Some passive methods for intensive building types are:

- *"Built-form configuration and site layout planning*
- *Built-form orientation (of main façades and openings, etc)*
- *Façade design (including window size, location and details)*
- *Solar-control devices (e.g. shading for façades and windows)*
- *Passive daylight devices*
- *Vertical landscaping (i.e. use of plants in relation to the built-form)*
- *Winds and natural ventilation"* [198]

---

197 Yeang, K.: The Green Skykscraper, 1999, p.202
198 Yeang, K.: The Green Skykscraper, 1999, p.204

- **Spirituality as motivation for sustainability**

As humans depend on nature for survival, they have - everywhere in the world, developed in different cultural areas - different social and cultural constructions to ensure the conservation of the ecological balance. The term sustainability is established in the West, e.g. Europe and North America. The investigation of interactions is characterised by rational and scientific methods and by the idea of having or gaining "control" over nature. Ironically there is a negative correlation between ecological practice and theory. The societies most concerned with sustainability are at the same time those who act least ecologically (US, Germany, France). A major reason for the application of ecological principles is the expectation of higher economic profit.

Large areas of Africa and Asia have a low ecological impact per capita in contrast to the Western world. Sustainability in traditional societies is less measured but felt more: the traditional connection to nature is often of spiritual or religious character. The integration of natural processes is interpreted as a principle of attachment, from which an ethical system is derived. In built structures the attachment is expressed as a need for harmony with the cosmic order, or as harmony and balance between specific elements that the world consists of to maintain the flow of energy (e.g. Feng shui). The traditional building typologies are embedded into a natural cycle.

Apart from the connection of all things there is, in many rural societies, an unquestioned acceptance of natural rhythms. Changeability is a central issue in the Buddhist, Hindu and Taoist philosophies and is accepted also for built structure. The European principle of conservation and maintenance of historic monuments is, for example in Asia, considered rather exotic.

The third principle, which is lost with increasing individual isolation in Western societies, is social responsibility. What this all amounts to is that the access to sustainability in still traditional cultures is characterised more by a spiritual approach using a low tech direction, in contrast to the highly developed industrial societies' approach of low energy and high material.[199]

Fig.141 Fetish, village of the Dogon in Mali, West Africa, 2002.

199 Du Plessis, Crisna in: Edwards, B.: Architectural Design Nr.71, Green Architecture, 2001, pp.40

Fig.142 Model of "The Green Building", Future Systems 1990.

  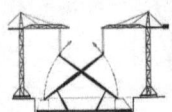 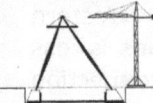 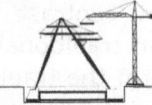 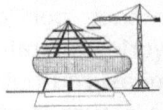

Fig.143 Process of building.

Green architecture

The term "green architecture" is established for projects implementing any natural aspect. Ken Yeang describes "green design" as including spatial approaches (a building occupies space and therefore modifies the environment) as well as the climatic approaches (buildings interact, "breathe" and thereby modify the environment) in a holistic point of view.[200]

In "green" high tech architecture, a high technology and low energy approach is taken. One of the iconic projects of ecological design was never realised, but symbolises a holistic approach to reach sustainability and architectural quality on many levels of the design. The study of the "Green Building" was one of the first taking into account different parameters and one of the first green concepts from high tech architecture for an urban setting.

- **Future Systems, Green Building, 1990**

"The Green Building is an experiment in reversing this tendency (complex control systems, high energy consumption): an attempt to attain 'green goals' without primitive architecture. The Green Building is an asymmetrical envelope supported by a tripod megastructure, which clears the site for use as public open space. Wind loads are resisted directly by the bending strength of the tripod legs, which are triangulated below ground level and anchored to large-diameter piles."[201]

Ove Arup & Partners calculated the construction and technical building systems.

200 Yeang, K.: The Green Skykscraper, 1999, p.54

201 http://www.future-systems.com/architecture/architecture_20.html [10/2007]

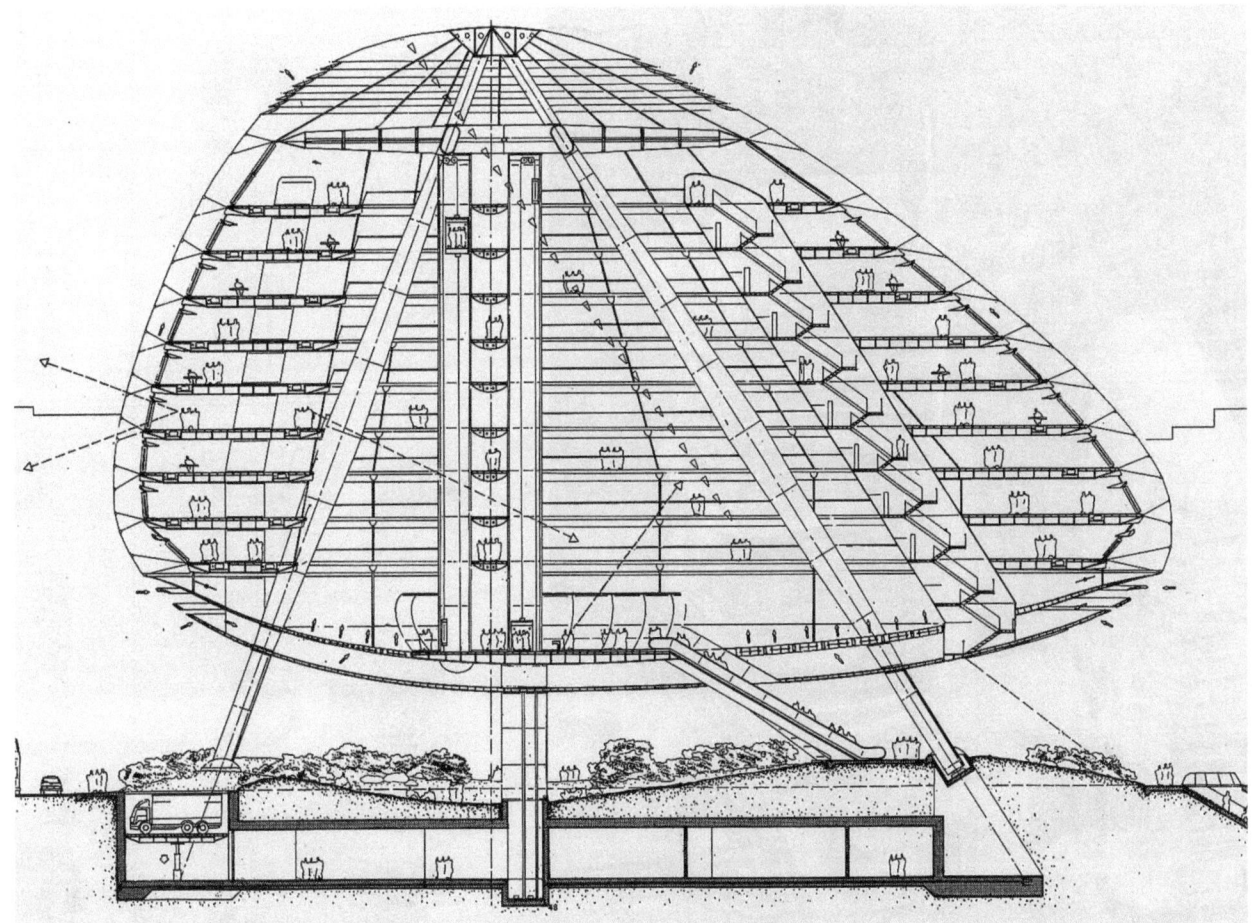

Fig.144 Section.

Future Systems, 1993, about the climate in the Green Building:
"Ambient air is drawn into the building at the base of the envelope which is raised 17m above ground level to reduce the ingestion of pollutants. As it warms, this air rises between the controllable inner skin of the building by stack effect, finally exhausting through louvers at its apex.
In addition to its ambient energy system, the fully glazed building makes maximum use of daylighting to reduce energy consumption. Solar glare and heat gain within the offices can be controlled by fabric blinds incorporated into the buildings inner skin. This skin also contains flexible plastic mirrors designed to deflect daylight into the innermost office spaces along a high-level horizontal path that is intercepted by ceiling mounted diffusers where necessary."[202]

Parameters considered in this project are land consumption, location factors, solar radiation, wind, energy consumption and passive energy systems, light, climate, ventilation, light construction, building technology and economy and quality of life. All of these factors should be considered in todays urban design.

Fig.145 First model, 1999.

202 http://www.future-systems.com/architecture/architecture_20.html [10/2007]

Classical approaches | "Natural construction"

Fig.146 Model of the EDITT Tower, Yeang 1998.

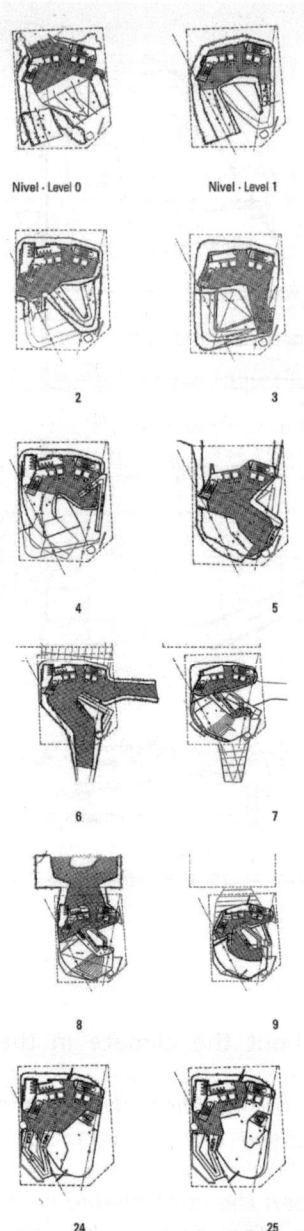

Fig.147 Floor plans of EDITT Tower, Ken Yeang 1998.

- **Ken Yeang, EDITT Tower Singapore, 1998**
"Can the skyscraper be 'green'? The proposition that the skyscraper and other high-density intensive building types can be designed to be ecologically responsible may be regarded with great suspicion... by virtue of their enormous size, skyscrapers consume huge amounts of energy and materials and make similarly extensive discharges into the natural environment, and are thus inherently ungreen."[203]

Ken Yeang is not just a theorist, but follows his own recommendations on the ecological design of environmentally responsive and sustainable buildings in his work. The EDITT Tower, Singapore, 1998, was designed as an exemplarily ecological building. The project was supported by the organisation Ecological Design in The Tropics (EDITT), the city's authorities and the National University of Singapore. The EDITT Tower represents an approach integrating the many parameters to reduce the negative environmental impact of skyscrapers:

- Response to the site's ecology
- Place making (vertical places, introduction and extension of street life)
- Views of the surrounding
- 'Loose-fit' (integration of flexibility to change usages)
- Vertical landscaping
- Water recycling and purification
- Sewage recycling
- Solar energy use
- Building materials recycling and reuse
- Natural ventilation and 'Mixed-mode' servicing
- Embodied energy and CO_2 [204]

203 Yeang, K.: The Green Skykscraper, 1999, p.18

204 http://www.trhamzahyeang.com/project/skyscrapers/edit-tower01.html [08/2007]

Fig.148 Transparent dome over Manhattan, RBF 1950.

Urban visions

The vision of an urban paradise arose again and again in history. In the 1960s, the belief in new technology as a saviour of humankind and solution to all problems, led to the design of new forms of urbanity; high dense structures which would nevertheless provide a good quality of life. The beginning of spaceflight triggered the idea of new settlements in space which were independent and capable of development.

- **Artificial paradise**

Another asymptote of lightweight design and environmental concern is the dome over midtown Manhattan, which Richard Buckminster Fuller visualised in 1950. The domes surface was calculated having just 1/85 of the total surface of the covered interior buildings, so was meant to reduce energy loss to the same ratio, and to pay off building costs within ten years. This was a brave statement, considering that until now, 50 years later, we have not managed to reach the size and transparency he suggested. Fuller also thought that his domes were going to be used for new settlements in the Arctic and Antarctica.[205]

"From the inside there will be uninterrupted contact with the exterior world. The sun and the moon will shine in the landscape, and the sky will be completely visible, but the unpleasant effects of climate, heat, dust, bugs, glare, etc. will be modulated by the skin to provide Garden of Eden interior."[206]

- **Mega structures**

A small group of architects designed urban utopias in the 1960s. In answer to an increasing lack of space due to increasing urban sprawl they designed tower cities, swimming metropolises and finally space cities: "Plug-In-City" Archigram, Peter Cook; funnel shaped cities of Walter Jonas; "City in the Air" of Arata Isozaki; "La ville spatiale" of Yona Friedman, "Arcologies" of Paolo Soleri to name a few. Even today, densification of urban settlements seems to remain the only sensible measure to meet the intensifying squeeze on land, the destruction of the remaining stable ecosystems and the shortage of resources.

"In nature, as an organism evolves it increases in complexity and it also becomes a more compact or miniaturized system. Similarly a city should function as a living system. Arcology, architecture and ecology as one integral process, is capable of demonstrating positive response to the many problems of urban civilization, population, pollution, energy and natural resource depletion, food scarcity and quality of life. Arcology recognizes the necessity of the radical reorganization of the sprawling urban landscape into dense, integrated, three-dimensional cities in order to support the complex activities that sustain human culture. The city is the necessary instrument for the evolution of humankind."[207]

---

205 R.B. Fuller in Krausse, J. et al. (Ed.): Your private sky, R.Buckminster Fuller, 1999, p.434, original in Fuller, R.B.: Utopia or Oblivion, 1969, p.353

206 R.B. Fuller in Krausse, J. et al. (Ed.): Your private sky, R. Buckminster Fuller, 1999, p.434, original in Allwood, J. 1977

207 http://www.arcosanti.org/theory/arcology/main.html [08/2007]

Fig.149 West Side New York City, Reiser and Umemoto, 2000.

- **Fluid - tectonics**

With increasing consciousness of environmental processes the interpretation of the city is changing. Not the static structures, but streams of activity and information now attract attention. The new picture of the city is that of an urban landscape consisting of fluid transformation and networks. In urban planning, time-related methods of design prevail. The method of "deep planning", suggested by Ben van Berkel and Caroline Bos, *"involves generating a situation-specific, dynamic, organisational structural plan with parameter-based techniques ... [incorporating] economics, infrastructure, program and construction in time."*[208] Jesse Reiser and Nakano Umemoto coin the term "infrastructuralism", and design a competition project for the New York West side. The conventional difference between infrastructure and architecture is dissolved; space is interpreted as "event space", which is subject to a fluctuating temporal use. In a differentiated spatial structure (in contrast to Fuller's domes) different scales and organisations are possible and thus enable the "local" within the "global". Many other examples for time-based urban design strategies could be mentioned.

208 AD Architectural Design Vol 70 No 3 , Contemporary Processes in Architecture, 2000, p.46

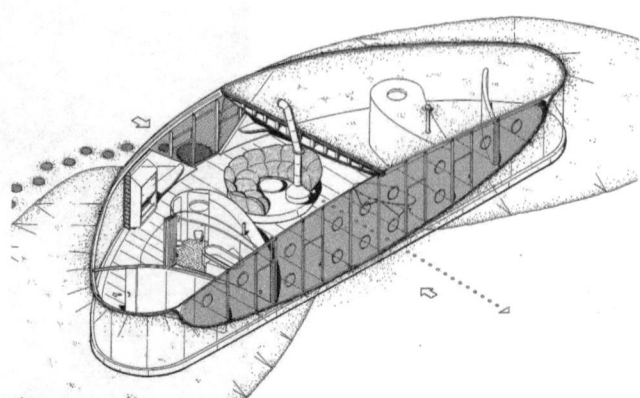

Fig.150 Axonometric view of project 222 by Future Systems, 1994.

Landscape

As mentioned in the background chapter earlier, architecture is considered a discipline dealing with the whole of built environment. Landscape scale is an issue, which is all too rarely considered in Western regional planning and politics. The destruction of (cultural) landscape by the growth of sites for industry and economy, as well as for housing, is an ongoing process. Populated areas expand into dangerous regions (unstable ground, unsuitable climate, danger of floods...), where safety can no longer be guaranteed; this fact being largely ignored and transformed into the business of non-sustainable technologies and insurance companies. The large-scale aesthetics of a landscape is taken for granted and not considered an object of design, "Gestaltung". So the interrelation between architectural planning and the landscape is usually not regarded an important issue, at least not in the direction of the landscape being influenced by the building. In contrast to built environment, e.g. cities where historical conservation forces rows of low-rise buildings to protect the cities' silhouette, the natural and cultural landscapes rarely enjoy such protection. The concept of how to deal with environmental landscape is tightly tied to the question of the interrelation of architecture and nature in general.

Several approaches are described in the following paragraphs.

Fig.151 House Kaufmann in Palm springs, Neutra 1946.

Invisibility

- **House in Wales, project 222, Future Systems 1994**

"Our objective has been to minimise the visual impact of the building and to site it in a way that makes the house appear a natural part of the landscape. The soft, organic form of the building is designed to melt into the rugged grass and gorse landscape, the roof and sides of the house being turfed with local vegetation. Views of the house are therefore only of grasses and the transparent glass walls outlined only by a slim stainless steel trim; an eye overlooking the sea. The surrounding landscape remains untouched with no visible boundary lines or designated garden area."[209]

Future Systems designed an invisible single-family house for a site at the coast in Wales. Apart from a main glass façade and the entrance, the house is subterranean - partly dug in, partly covered by earth and overgrowth. The furniture of the 150m² house was also designed by the architects. The sanitary system was delivered to the site as a prefabricated box, including all connections. Making architecture invisible is one of the solutions to avoid the visual destruction of the environment, certainly justifiable in an undisturbed coastal landscape, but in general this approach is questionable.

Connection of inner and outer space

- **Richard Neutra.**

Neutra stands as an example for many representatives of modern architecture who integrate landscape in their design.
From Richard Neutra's point of view the task of architects is to provide man with an appropriate environment. His "biological architecture" is based on sensual perception, and aims at harmonic cooperation of all the feelings triggered by it. The integration of landscape in his designs reminds one of the Japanese principle of the borrowed view (shakkei): architecture as a frame for the landscape and the landscape as a background for the staging. The warm climate makes the flow of space and the connection of elements much easier than in cooler climates, where insulation measures occupy more space.[210]

[209] http://www.future-systems.com/architecture/architecture_10.html [10/2007]

[210] Boeckl, M. (Ed.): Visionäre und Vertriebene, 1995, p.131

Fig.152 Spiral Jetty, Great Salt Lake, Utah, Robert Smithson 1970, © VBK, Wien 2010

Fig.153 The pavilion of the Netherlands, Expo 2000 Hannover, MVRDV.

Land art

The arts have always dealt with landscape, but the "land art" movement started in 1960s, using land space interactively for art projects. It never was a genuine "movement", but various artists who investigated space and landscape have been associated with it retrospectively, for example Michael Heizer, Walter de Maria, Robert Smithson, Richard Long, Dennis Oppenheim, Nancy Holt and others. Time is an important parameter for their projects. The sculptures of land art are thus in situ experiments and are subject to the laws of nature, slowly disappearing or degrading. What remains are documentation, remembering and influence. The artists work mostly with natural materials, and the importance of the specific site is characteristic.

- **Spiral Jetty, Great Salt Lake Utah, Robert Smithson, 1970.**

The "Spiral Jetty" of Robert Smithson is one of the most famous projects of land art. The spiral is 450m long, and about 4.5m wide, and was heaped up on the waterside of the Great Salt Lake in Utah in 1970. It then slowly sank below the water level, and later gaining renewed attention by re-emerging in 2004 - due to the long drought the water level of the lake had fallen to a record low.

Nature in architecture - integrated

Nature in architecture does not necessarily relate to natural construction, ecology, regionalism and event space, but can stand formally on its own. This is not about (artificial) reproduction of experiencing nature to fulfil the human desire for naturalness and originality, which is expressed also in life style, nutrition or medicine. The mapping or mimicking of nature, particularly in floral forms, has a long tradition, especially in all areas of interior design.

"... by choice, nature is robed, processed, delocalised, verticalised and its effects increased, multiplied, intensified and wrapped as a sandwich, dissected, enclosed and exhibited, artificially reproduced and infused with natural characteristics..."[211]

---

211 Krafft, S. in ARCH+ Nr.142, Architektur natürlich, 1998, p.22

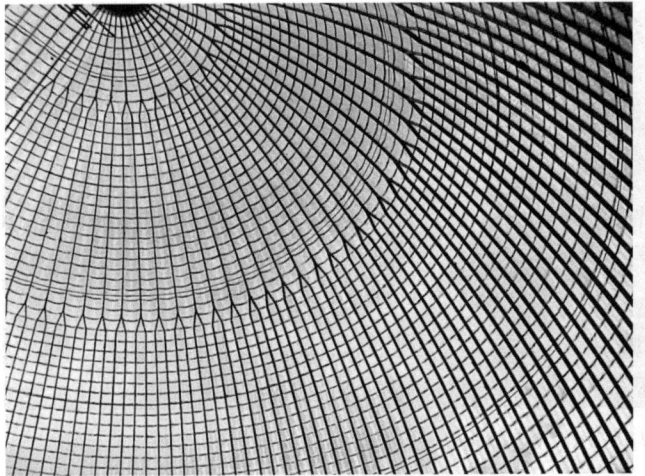

Fig.154 Palm house, Bicton Gardens, Bailey brothers, 1830, structure viewed from inside.

Fig.155 Palm house, Bicton Gardens, view from the garden.

Greenhouse renaissance

As at the beginning of the 20th century, when a series of botanical gardens, glasshouses and zoos were built in Europe to study and exhibit flora and fauna of foreign countries, we - at the start of the new millennium - find ourselves again in a founding era of projects intended to display the complexity of nature. One of the most innovative of old greenhouses is presented here, to compare and contrast with the very recent and highly innovative Eden Project.

- **Greenhouse of the Bailey brothers, Bicton Gardens, Devon, 1830**

This is the lightest design of a glasshouse imaginable in the 19th century, using steel and glass for construction materials. The small glass panes overlap scale-like and stiffen the steel construction, so the whole construction can be very delicate and transparent. Load bearing has to be done in a hybrid manner, and the differentiation and forking of the steel ribs appear to be obvious and natural. The overall form is not based on spheres, but is higher than it is wide, so that load transmission and the vault effect are improved. The massive back wall to the North provides thermal mass storage and is typical of the greenhouses of that era. The collection of thermal energy by means of building mass is an easy passive method of energy harvesting.

The function of greenhouses has changed since the 19th century. Then plants from foreign countries and climate zones were collected to be studied and classified, while today's collections and cultures can be established to provide against the rapidly increasing extermination of species.

By integration of phenomena from nature, sensory perception is intended to be enriched by the visual, acoustic, tactile and olfactory stimuli which have been lost in the urban environment.

This development has some side effects:
- Fusion of the natural and the artificial
- Decontextualisation
- Questioning of horizontality and landscape
- Materialisation of vegetation
- Intensification
- Link up with program (leisure industry)
- Loss of authenticity
- Hybridisation of technology with nature
- Inversion of the relationship between inside and outside
- Amorphousness
- Processes [Prozesshaftigkeit] in design methodology
- Loss of control or autonomy[212]

---

212 according to Oswalt, P. in ARCH+ Nr.142, Architektur natürlich, 1998, pp.74

Fig.156 Aerial view of the clay pit, St. Austell, Cornwall.

Fig.158 Interior space.

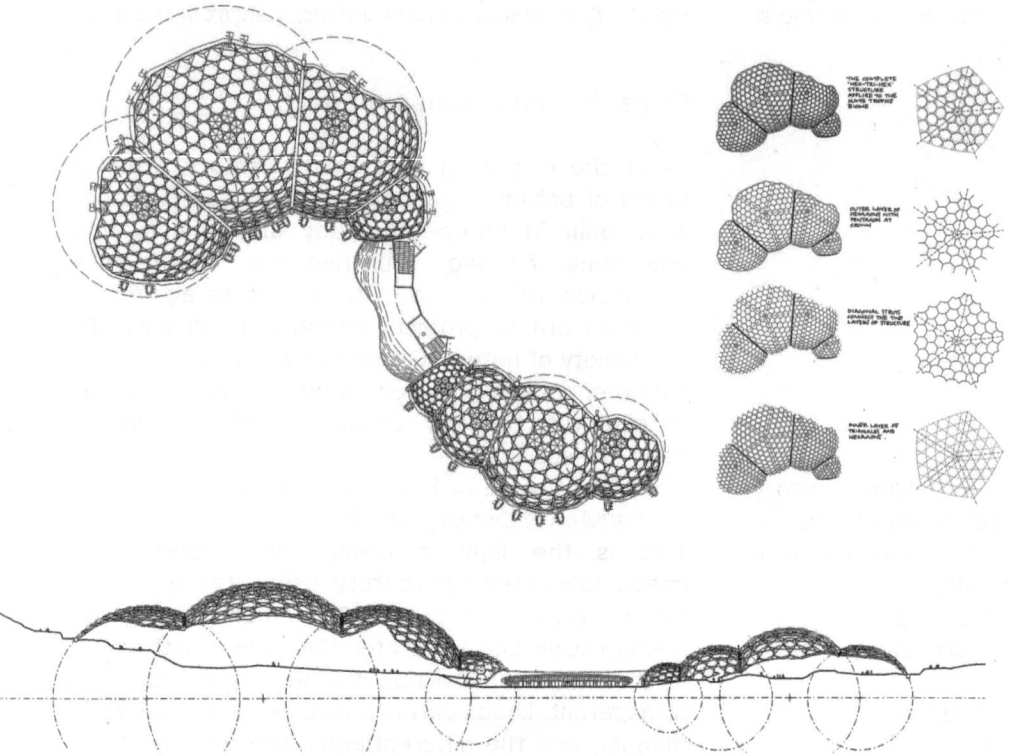
Fig.157 Floor plans, sections and construction layer of the Eden Project.

- **Nicholas Grimshaw & Partners, London.
Eden Project, 2000**

The Eden Project is the world's biggest greenhouse. Three different climate zones are cultivated: Tropical Wet, Mediterranean and Cornwall's own mild, temperate climate. The natural spaces of Rainforest, of California, South Africa and the Mediterranean are presented, to arouse the visitors' sensibility for ecological systems. The site for this mega project is a former clay pit. In spite of numerous adverse circumstances and with a tremendous commitment, Tim Smit, the client and founder of the Eden Project, managed to turn the vision of this project into reality. Planned as millennium project, it was opened one year late, in 2001.

The groundwork was gigantic, especially in contrast to the visible building, and was hampered and delayed by extreme rain and ingress of water in the first year. The whole site was drained, and water is permanently being pumped out (but reused for irrigation).

The construction spans across 124m at maximum, its form being based on intersecting geodesic domes. The light construction was a precondition to counteract the bad ground conditions. The structure is an interlocking system of steel pipes, with two layers: the outer layer based on a hexagonal structure; the internal layer consisting of triangles and hexagons. A trussed girder runs along the intersection lines of the domes.

Fig.159 Eden Project in building phase.

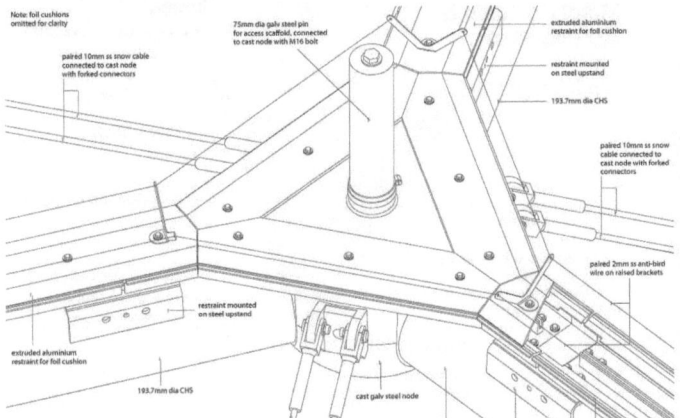
Fig.160 Node detail.

Fig.161 Close-up of the pneumatic ETFE cushions.

The cladding consists of light three-layered cushions of transparent polymer foil ETFE (Ethylene tetrafluoroethylene). The form of the geodesic domes allows the usage of sets of identical elements for cladding. The different sizes of the single domes are represented in the size of the construction elements: the diameter of the biggest cushion is 11m. The cushions along the intersections of the domes are customised. Construction and planning was done by Nicholas Grimshaw and Partners, London, together with Anthony Hunt Assoc. and Ove Arup and Partners.[213]

The extreme lightness of the construction is best visible in the interior space, and ventilation openings are situated on top of the domes.

In every respect, the Eden Project breaks all known scales. Diverse public and private organisations and banks did the financing. The hoped-for effect as a crowd puller has been achieved to an unexpected extent in spite of the remote situation in the South of Cornwall.

---

213 Edwards, B.: Architectural Design Nr.71, Green Architecture, 2001, pp.97

Fig.162 Water bear, © Dennis Kunkel Microscopy, Inc.

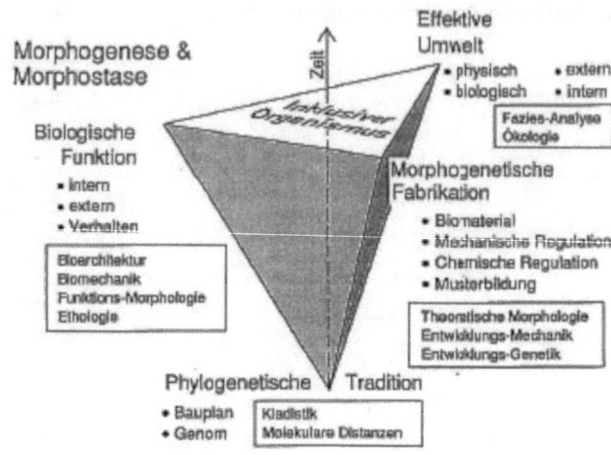

Fig.163 Morphodynamic tetrahedron, Kull 1996.

3.3 NATURE'S DESIGN PRINCIPLES

Looking at the "water bear" (eschiniscus blumi, phylum tardigrada), some of nature's design principles are obvious. The water bear is a tiny multi-cellular organism, which lives in moss in humid areas. The largest specimens reach a size of 1.2mm. Tardigrade means slow walking, bear-like in behaviour. If the environment becomes too dry, the organism can dehydrate, curl up and hibernate for up to 10 years.[214] This strategy ensures survival of extreme environmental conditions. The similarity of this small organism with the mammalian bear is striking - why does life show such similar appearance in such different dimensions? The question of the basic design principles of living organisms still keeps scientists busy, but some of the phenomena can be described. The principles found have been transferred into technological applications.

"In most cases many 'parameters' - simultaneously or successively - are responsible for the formation of an object in nature. Countless complex forms result from that, never static, but always subjected... to changes."[215]

Ulrich Kull et al. discussed the functional morphology of organisms and published the so-called "morphodynamic tetrahedron", which demonstrates influences of biological functions, effective environment and morphogenetic fabrication in the course of time, the phylogenetic tradition.[216]

214 Margulis, L. et al.: What Is Life?, 2000, p.169

215 Otto, F. et al.: Natürliche Konstruktionen, 1985, p.10
216 Kull, U. et al in Teichmann, K. et al.: Prozess und Form natürlicher Konstruktionen, 1996, p.34

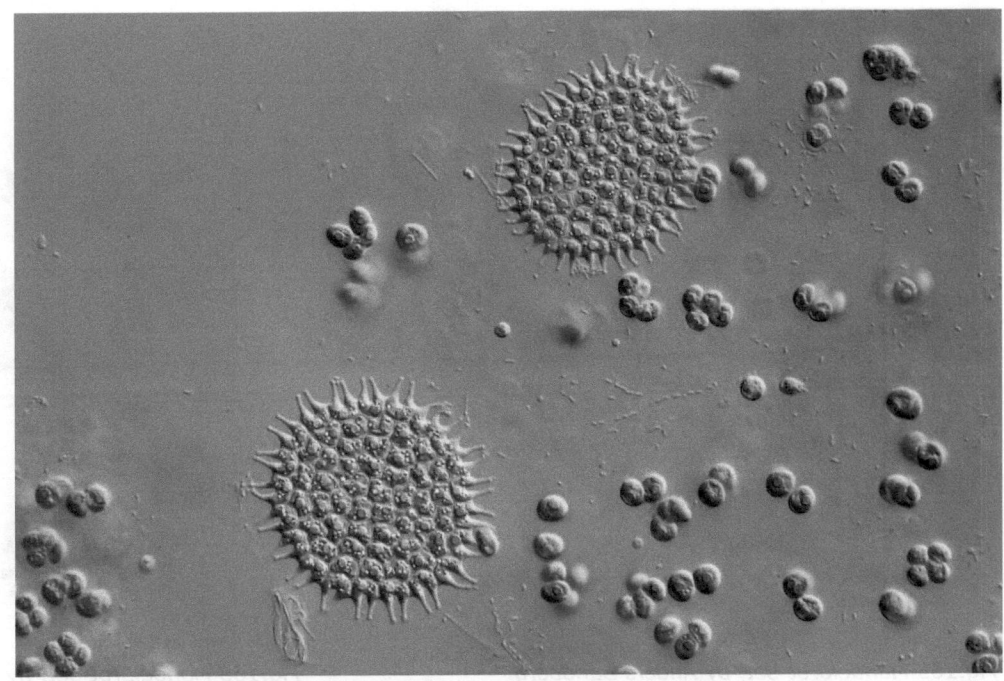

Fig.164 Green algae, Pediastrum, forming cogwheel-like colonies.

3.3.1 Physics

Life is subject to the same physical laws as inanimate nature. Erwin Schrödinger discussed quantum mechanics and the physical laws that life is based on in his famous book "What is life", first published in 1944. Lynn Margulies and Doris Sagan continued the discussion under the same title.

"Ilya Prigogine, a Belgian Nobel Laureate, helped pioneer the consideration of life within a larger class of 'dissipative structures'... a dissipative system maintains itself, and may even grow, by importing 'useful' forms of energy and exporting, or dissipating, less useful forms - notably, heat... This thermodynamic view of life actually goes back to Schrödinger, who also likened living beings to flames, 'streams of order' that maintain their forms."[217]

Schrödinger interprets life as entropy machines, islands of order within a universe of increasing entropy. The basic thermodynamic laws are responsible for some of the phenomena that make life so fascinating.

3.3.2 Energy efficiency

Organisms have to be efficient in terms of energy to survive. They do not waste energy and operate with high effectiveness. Werner Nachtigall states ten rules which are valid for the design of natural constructions:
*"All natural constructions and processes are optimised for energy use, and from this point of view all the following design principles of nature have to be conceived:
Integrated instead of additive construction
Optimisation of the whole instead of maximising single elements
Multifunctionality instead of monofunctionality
Fine adjustment with regard to the environment
Direct or indirect use of solar energy
Temporary limitation instead of useless durability
Complete recycling instead of waste accumulation
Integration instead of linearity
Development by trial and error"[218]*

The least possible effort is always taken in nature, to achieve a solution in a "practical approach". Organisms use what is available in their environment: availability of energy source, material, cooperation etc. Simple materials are used, and so information is needed for developing these to high performance structures.

---

217 Margulis, L. et al.: What Is Life?, 2000, p.16

218 Nachtigall, W.: Vorbild Natur, 1997, pp.21

Fig.165 Mandelbrot-set, by Anders Sandberg.

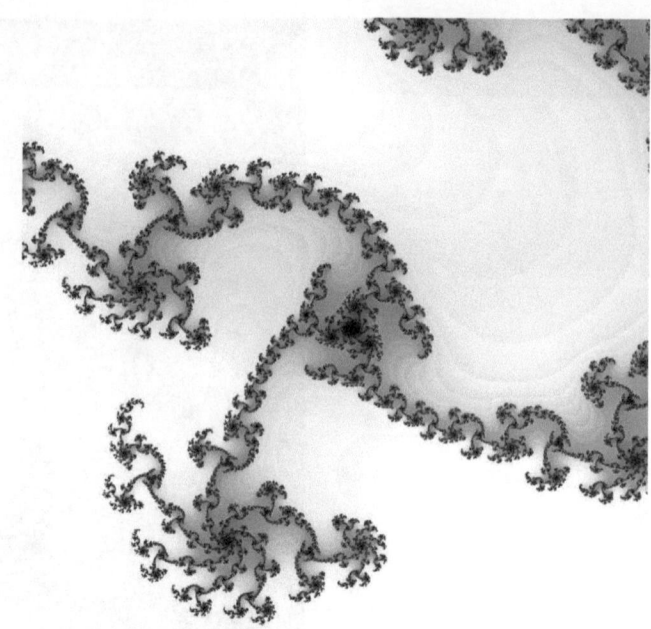
Fig.166 Mandelbrot-set, by Anders Sandberg.

3.3.3 Self-organisation and chaos theory

Self-organisation is a central concept of a systemic view of life. The term was coined in the 1950s. The theory of self-organisation is also called theory of non-linear (dynamic) systems (chaos theory). It is applicable to physical and chemical, biological, psychological and social systems.

"By 'self-organisation' we understand the ability of systems to develop and sustain their inherent order with no control from outside. The implicit ability of complex adaptive behaviour is a central characteristic of living systems. Furthermore, processes of self-organisation are also found in inanimate nature in fields of experience far from each other. Therefore, we can take the view of self-organisation as a concept bridging the gap between animate and inanimate phenomena in nature, presumably playing a key role for the understanding of life and consciousness."[219]

Preconditions for the emergence of order

"Order and chaos are two sides of our world which have a dynamic relationship. Chaos can change into order and order can transform into chaos. The destruction of order is the normal process, the creation of order happening spontaneously only under specific conditions."[220]

- **Nonlinearity and feedback**
The mathematical description of self-organisation includes nonlinear equations and behaviour of the systems is characterised by feedback loops.
- **Overcritical distance**
Self-organisation is possible at a specific critical distance of the system from the balanced state.
- **Openness**
Exchange of matter, energy and information with the environment.[221]

---

[219] Euler, M.: Selbstorganisation, Strukturbildung und Wahrnehmung in: Biologie in unserer Zeit, 30.Jahrg. 2000/Nr.1

[220] Ebeling, W. in Teichmann, K. et al.: Prozess und Form natürlicher Konstruktionen, 1996, p.24,25

[221] Ebeling, W. in Teichmann, K. et al.: Prozess und Form natürlicher Konstruktionen, 1996, p.24,25

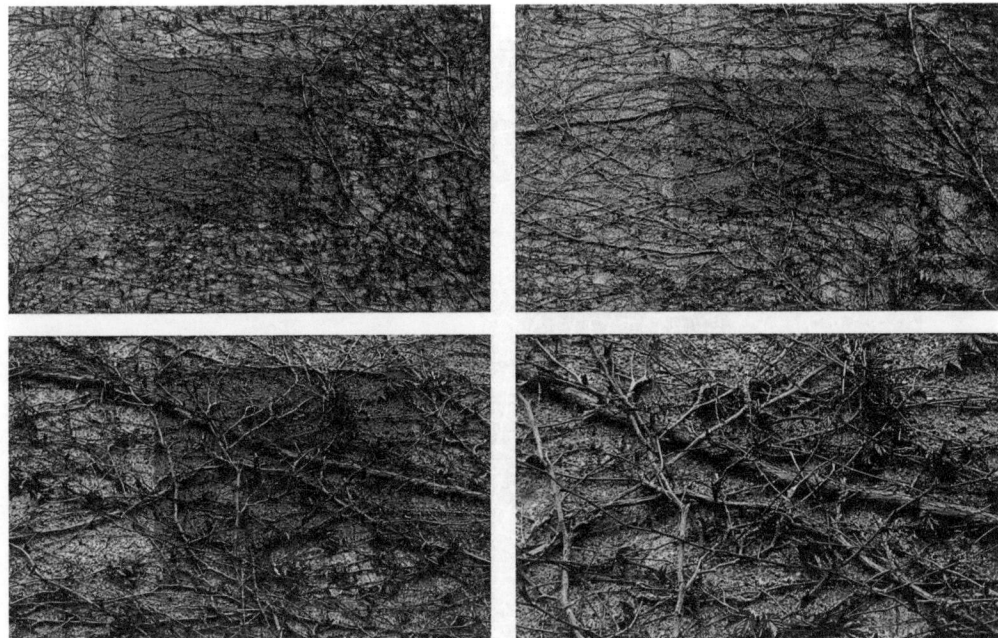

Fig.167 Wall with climbing vine at different magnifications; independence of the pattern from scale.

Characteristics of self organised systems

- **Entropy export**
- **Energy transformation**
- **Parameter or order**

Within the system, ordering parameters are responsible for the coordination of movements.

- **Stability and phase change**

Self-organised systems are stable for small disturbances. Overriding of critical parameters can lead to a collapse of the system. Phase changes or transitions are another form of instability; during transaction, strong fluctuations occur.

- **Breaking of symmetry**

Normally happens during formation of patterns.

- **Limited predictability**

Regular as well as irregular or chaotic structures can develop. Chaotic structures allow only limited statements about the behaviour of the system.

- **Historicity**

The specific developmental history of the system is unique.[222]

Conclusions

- The processes are not reversible
- Control is limited, but possible (experimental planning is possible)
- Erratic changes occur in chaotic systems
- The system is not arbitrary
- The system itself defines its borders
- Small changes can have big effects (the famous butterfly in Japan)

[222] Ebeling, W. in Teichmann, K. et al.: Prozess und Form natürlicher Konstruktionen, 1996, pp.24

Fractal geometry

Fractal geometry draws pictures of movements of dynamic systems in space. Fractals are geometric objects that repeat their structure in decreasing scales. Benoit Mandelbrot discovered them in the 1960s; the name derived from the Latin word fractus, broken.

Iterating specific equations in feedback loops generates fractal forms; the result of one step is the starting point of the next. Iterated fractal forms are currently used in all scientific areas to predict the behaviour of dynamic systems. Self-similarity is a characteristic of fractals, and for this reason fractal geometry is also called geometry between the dimensions.

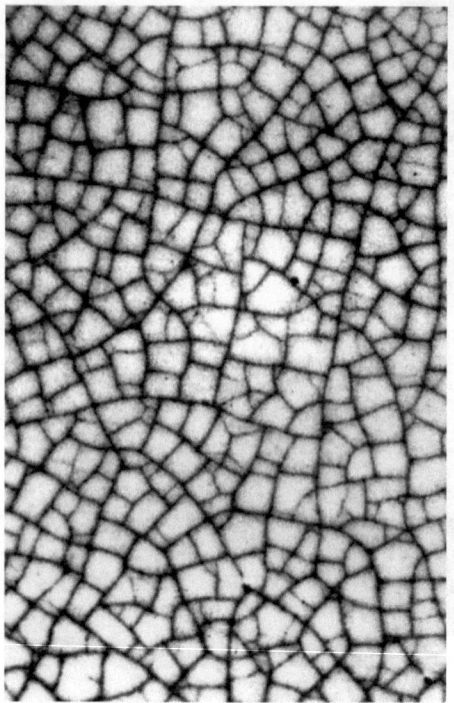
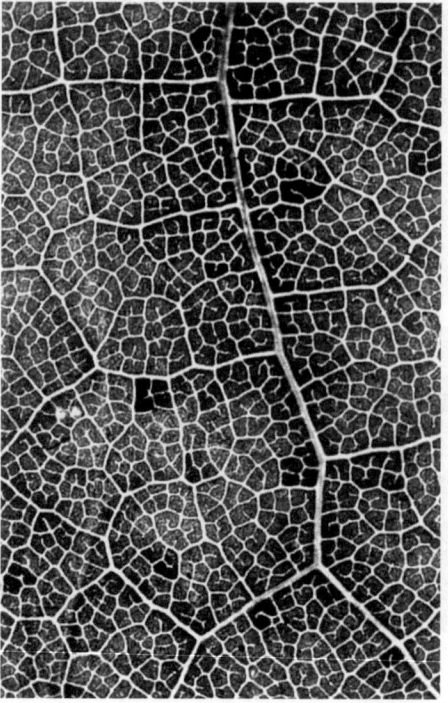

Fig.168 Cracks in porcelain.   *Fig.169 Maple leaf, part of vein system.*

3.3.4 Information

Information is an essential parameter in the design of organisms. During replication, organisms have to pass on information in the form of matter. This involves spending of material and energy for the production of plans that are as precise as possible. The control of information is crucial for the survival of organisms and species.

Information related restrictions in the design of living systems

The amount of information necessary for the construction of living systems is amazingly low compared to their complexity. According to Steven Vogel, a fertile egg for the construction of a human contains around 10^{10} bits of information. The scarcity of information is based on biological design.[223] Structures which are not genetically fixed are probably developed in processes of self-organisation, e.g. leaf veins or blood vessels in animals. Analogous fractal structures develop in unplanned urban settlements. Eda Schaur et al. investigated unplanned networks of paths.[224]

Strategies to minimise information

One of the strategies to minimise information is the use of structures that need little information and the multiplication of the available structural elements. The resulting geometries and patterns are referred to as "nature's numbers", providing an argument for the Creationist movement.

- **Use of fractal and simple structures**

In self-similar structures only a single rule is needed for constructing all subunits. A helix is also constructed from identical subunits, and this form is often used in nature (e.g. DNA, microtubules, microfilaments).[225]

- **Cellular structures**

Large organisms consist of cells. The rules for their assembly are simple, and the same rules can be used to assemble many types of cells into different tissues and organs.

- **Segmentation**

Segmentation is the use of identical subunits to form an organism, or part of an organism, when assembled. Segmentation is common with invertebrates.

- **Multiple use of information**

A single set of instructions can generate more than a single structure for the development of an individual organism (e.g. in humans the sizes of feet and hands are linked).[226]

---

223 Vogel, S.: On Cats' Paws and Catapults, 1998, p.25
224 Schaur, E. in Teichmann, K. et al.: Prozess und Form natürlicher Konstruktionen, 1996, pp.154

225 Vogel, S.: On Cats' Paws and Catapults, 1998, p.26
226 Vogel, S.: On cats' Paws and Catapults, 1998, p.27

- **Symmetry**

The initial meaning of the Greek word symmetry was harmony of proportions - a more general meaning than today's narrow interpretation. Many kinds of symmetries exist in nature. Radial symmetry (e.g. in flowers) and bilateral symmetry (e.g. in humans) extend the possibilities of reducing the amount of information required. Single mutations normally generate changes in both sides of the body, which proves that the information occurs only once. At the same time, the doubling up of important organs provides greater security.

- **Composition of small units**

Dense packing of small units is ubiquitous in nature in many scales, e.g. plant parenchyma cells, or seeds packed following geometric laws.

3.3.5 Minimisation and integration

Minimisation in nature is unrivalled by technological means: natural structures are "designed" down to the smallest molecular scales and beyond. Nanotechnology has just begun to mimic some of nature's smallest structures, but the tools available are still quite crude.

In nature's design, integration is the standard solution for any functionality needed by organisms. In the context of multifunctionality, elements are integrated into more than one system. Integrated functionality is possible at many scales. Sensing, for example, can be integrated into the material of the skin (sensing organs of insects are integrated into the hard shell of chitin) and sensing hairs on human skin are dispersed over the organism's surface. Other sensing devices have evolved to larger distinct and more complex organs, for instance eyes and ears. Dispersed or distinct sensing is probably determined by the relative importance of the signal. Sensing of temperature is crucial for every individual part of the body (to prevent freezing or sunburn), whereas seeing is important for the organism as a whole. But for other functions, e.g. locomotion, both dispersed and distinct elements are found in organisms.

Integration on the scale of a material is carried out by gradient design. There is no sharp difference of material as in technological designs, but gradual soft changes of material properties. Fibrous and composite structures are the main materials, which makes such subtle structuring possible. So-called gradient materials allow specific regional functionalities without an abrupt change of material. This is called nonhybrid design.

Fig.170 Fern.

3.3.6 Self-design, learning, smartness

Learning is one of the greatest achievements of the mind: the ability to deal with unknown situations and find new solutions is important for survival in a changing environment. Learning, on the other hand, also needs resources: a processing and storage unit, requiring a large amount of energy. Organisms usually avoid learning for as long as is possible: known processes can be dealt with by stored programs. The new is more strenuous to deal with, but is welcome if considered advantageous.

Talking about "smartness" instead of intelligence avoids the difficulties associated with its definition in technological contexts. Smartness implies the notion of intelligence on a small scale, reduced to single aspects or abilities of materials or elements. Organisms, or parts of organisms, can react to stimuli from their environment in many ways, e.g. by moving, communicating or growing.

Many smart structures in nature are passive systems, inherent in the structure of the material (e.g. seed pods opening in relation to humidity - seeds are released when the environmental conditions are good). These systems have evolved through phylogenetic development. The slow self-design of species in the process of evolution and natural selection is the most wonderful phenomenon, which has produced the diversity of the biosphere we can observe.

Fig.171 Translational, rotational, and bilateral symmetry in plants.

3.3.7 Nature's numbers

The position of mathematicians and physicists on symmetry in nature differs from that of architects and biologists. Ian Stewart stated:
" ...are the symmetries which we can observe in nature fragments of great universal symmetries... [?]"[227]

Symmetry

Symmetries and patterns in nature occur through the laws of physics and mathematics, and are deeply related to the formation processes in animate and inanimate nature.
Various kinds of symmetry can be distinguished:
- Reflection or mirror symmetry - simple mirroring
- Rotational symmetry - rotation at a specific angle
- Translational or glide symmetry - straight line movement
- Rotational and reflection symmetry - turning and mirroring
- Glide reflection - straight line movement and mirroring
- Ornamental symmetries - other combinations of basic symmetries, band and surface ornaments

Symmetry in inanimate nature, for instance, can be observed in crystal structures. Their exterior symmetry is based upon the internal symmetry of the regular alignment of atoms and molecules in a spatial grid, the crystal structure. Symmetry also appears in the laws of physics.
Translational grids exist in many forms: triclinic, monoclinic, rhombic, hexagonal... - this represents just a small selection of possible crystal structures. For humans, symmetry is responsible for the sensation of beauty. Symmetry in nature is not perfect; breaks of symmetry are ubiquitous and seem to be as important as symmetry itself. The mathematics of symmetry break explains pattern formation processes in animate and inanimate nature.

- **Flora**

Cone symmetry of trees (axis oriented according to gravity) and reflection symmetry of leaves are examples of symmetry in plants. In flowers, rotational symmetry is often combined with reflection, fivefold rotational symmetry being the most widespread variation.

227 Tarassow, L.V.: Symmetrie, Symmetrie!, 1982, p.109

- **Fauna**

Fivefold rotational symmetry is found in fauna as well, but only in invertebrate marine organisms. For most animals, a distinct front and rear is characteristic. The direction of locomotion is distinguishable. The plane of symmetry is once more determined by gravity. Bilateral (reflection) symmetry is characteristic for most animals. Symmetry is not restricted to matter; patterns of locomotion are also symmetrical.

Geometric orders - five platonic bodies

The investigation of the geometry of nature also leads to the five platonic bodies, regular polyhedra with faces of identical regular polygons that were described by Plato around 400BC and later by Euclid. Space allows the construction of only five regular platonic bodies:
- Tetrahedron
- Octahedron
- Icosahedron
- Hexahedron or cube
- Dodecahedron

Non-regular polyhedra are possible but also limited in number. Transformation changes and relationships between these bodies have been examined for centuries. Buckminster Fuller dealt extensively with the laws of formation of these forms and used the half-regular cuboctahedron, the "Dymaxion". In the so-called Jitterbug transformation he presented the platonic bodies as phases of a transformation process, a helical contraction.

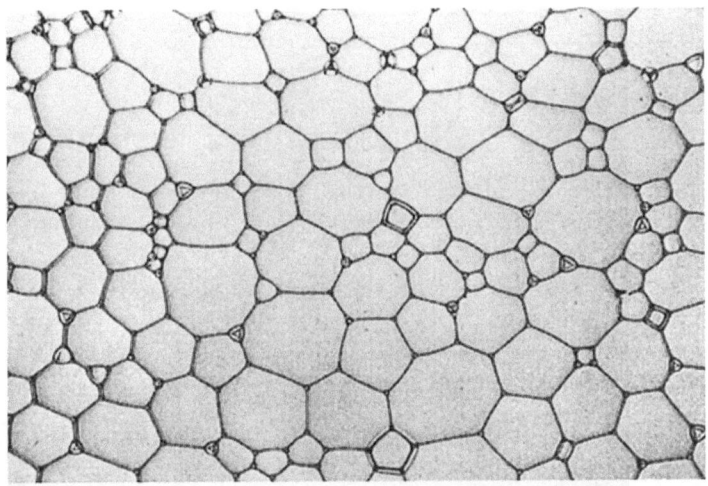

Fig.172 Floating foam, two dimensional packing.

Packing

Packing is an essential design criterion in nature. There are specific grids that frequently occur in space, for example:
- **Cubic face centred structure**

The hexagonal honeycomb pattern is the closest packing of spheres possible on a plane. Connecting the centres creates Y-shaped nodes with three connecting elements.

In three-dimensional space, the cubic face centred structure is the closest possible packing of spheres. This grid occurs in many natural elements, e.g. aluminium, gold, copper, nickel, platinum, silver and lead. The geometry of the closest packing of spheres in space results in the 120° and the 90° angle. Such packing is the variation with the lowest energy content.

- **Polyhedron foams**

If bubbles can move freely, they assemble in a self-organised way - as soon as they touch in close packing so-called polyhedron foams. Polyhedrons are created through packing from all sides. The boundary bodies are non-regular. Never more than three lamellas meet at an angle of 120° at an edge (Y-connection) and a maximum of four edges and six lamellas meet at a node, at an angle of 109°28'16". Otherwise the configuration is unstable and it rebuilds itself into a more stable form. These instable forms remain in materials that solidify rapidly. The fluid edges of the bubbles can be interpreted as a spatial minimal pathway network. Polyhedron foams are very stable and can transmit forces. The foams cannot be formed from identical polyhedrons: a non-regular polyhedron derived from the dodecahedron delivers an approximation to a polyhedron foam.

Classical approaches | Nature's design principles

Fig.173 Shell of a nautilus.

Fig.175 Rolled-up fern leaves.

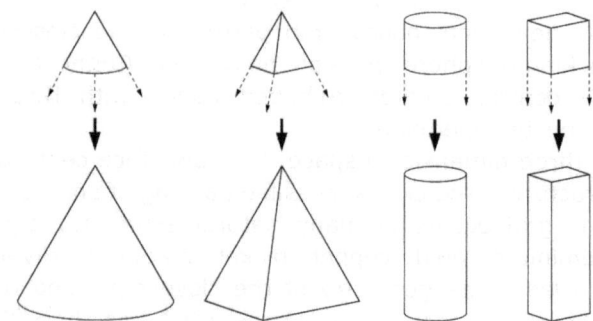

Fig.174 Growth and proportions.

Growth and proportion

Preservation of proportion during growth is a principle often found in animate nature. Growth happens during normal operation, which is why it is advantageous if the organism grows as continuously and constantly as possible. Maintenance of proportion is an advantage,[228] which could be the explanation for the diverse appearance of cone forms in nature. In plants, the tips of growing parts, the meristems, are almost always cone shaped. Juri Lebedew has investigated the principle of cones and stated that within flora the interrelation between two cones dominates - a lower one responsible for stability and an inverted upper cone responsible for growth and development.[229]

- **Spirals**

Scholars in the ancient world had already discussed the self-similar form of the spiral. In his famous book "On growth and form" (1917), D'Arcy Thompson states:

"In general terms, a Spiral is a curve which, starting from a point of origin, continually diminishes in curvature as it recedes from that point; or, in other words, whose radius of curvature continually increases."[230]

228 Vogel, S.: Cats' Paws and Catapults, 1998, pp.23
229 Lebedew, J.S.: Architektur und Bionik, 1983, p.46
230 Thompson, D.W.: On growth and form, 1992, p.748

Fig.176 Flower of marguerite.

The family of spirals is ubiquitous in animate as well as inanimate nature. Thompson stresses the difference to helical forms that have neither a point of origin nor change radius, but also appear frequently in nature. Lebedew commented on the stabilising effect of helical forms in contrast to cylindrical ones.[231] Architect Biruta Kresling investigated the geometry of the marine molluscan turbinate shells in order to derive a mechanical folding model.[232]

- **Kinds of spirals**

Most important in the large family of these curves are the spiral of Archimedes and the logarithmic spiral. The radius of the Archimedean spiral grows arithmetically, like a rolled up rope. The logarithmic spiral is also called the equiangular spiral, because the angle between the radius and the curve remains constant. This implies that the individual single parts of the curve are similar to each other, having the same proportion (the curve itself as well as the enclosed surface). Specific growth processes generate spirals in nature: the Nautilus, for instance, grows by adding material to its open rim. The size of the surface tension between water and the shell is determined by the angle of growth of the spiral, and this influences the growth process. Movement patterns also follow the equiangular spiral, e.g. the flight of insects using the angle between their flight direction and the direction to a source of light to navigate.[233]

- **Arrangement, growth and the Fibonacci sequence**

1 1 2 3 5 8 13 21 34 55 89 144 ...
$X_n = X_{n-1} + X_{n-2}$

The elements are assembled in spiral form in many flowers and infructescences. The resulting order and pattern is because of growth, and the growth of the single element is determined by internal and external conditions, e.g. nutrition. The sizes of elements differ, and relate to their respective position. The disc of the sunflower shows two overlapping families of spirals, one winding clockwise, and the other winding counter clockwise; some species have 34 spirals turning clockwise, and 55 spirals turning counter clockwise. These numbers appear as neighbours within the Fibonacci sequence. The number of petals is a Fibonacci number in most flowers.

The spiral also shows up in the growth of sprouts of many plants. The branching is determined at the tip of the shoot, in the meristem surrounding its tip, which grows slowly away from these so-called primordials. The spiral form is already recognisable in this phase. The obvious spirals are not necessarily most important to understand the process. The most important parameter is the angle that two consecutive primordials enclose if the tip is at their centre. This angle of divergence is almost exactly 137.5° in spiral growth. The reason for the evolution of this angle is that it enables the closest packed structure in space, as experiments with other angles have proved (multiples of 90° and 360° deliver the worst result).

---

231 Lebedew, J.S.: Architektur und Bionik, 1983, p.177
232 Kresling B.: The growing turbinate shell, 2004
233 Bürgin, T. et al.: HiTechNatur, 2000, p.112

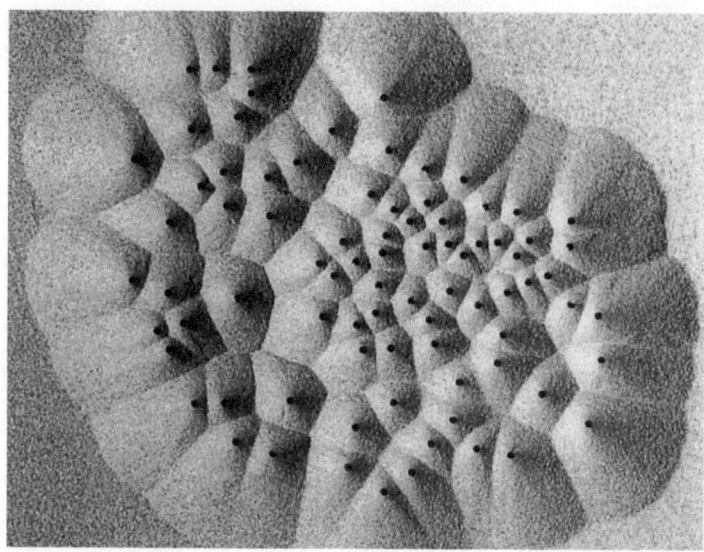
Fig.177 Sand model.

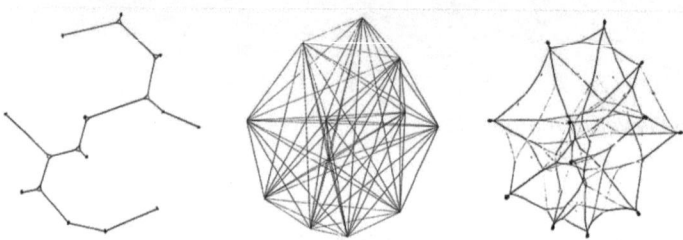
Fig.178 Pathway systems.

The branches are arranged in vertical rows of intersections, separated by free space. To fill the space efficiently, an angle of divergence is needed that is the most irrational multiple of 360°.[234]

- **Branches of trees**

Trees produce new branches throughout their entire life, so the total number of branches is always growing. In different experiments scientists tried to map the "real growth" in a model, or set of rules, and the Fibonacci sequence provides predictions of growth patterns.[235] So for the growth of trees it seems to be the principle of the most complete capture of space that is important (together with the principle of surface increase through structures limited in time - leaves - to maximise efficiency of photosynthesis, at least in moderate temperature zones). In real growth, these theoretical assumptions are influenced by a diversity of (local) environmental conditions, so that these rules are not discernible in full-grown trees. In other than temperate climate zones it may be important not to expose surface to the sun and for this reason other types of plants have developed, e.g. cacti.

- **Branching, supply and path networks**

The structure of networks, supply and path networks, in nature as well as in technology, is the subject of topology. In technology these networks have also evolved in a self-organised way, not in planned processes. For this reason, they can be investigated as a "natural" phenomenon. Frei Otto's group worked on this topic,[236] and Eda Schaur found out that the structure of unplanned settlements emerges through the interaction of two competing processes: the allocation of space and the ambition to have short connecting paths. This development is investigated by means of three different models or methods: minimal pathways - smallest total length; direct path systems - all points connected directly to all others; and the system of minimised detour, where both detours and total length are minimised. But none of the three models delivered structures similar to those actually existing. Foam and sand models finally delivered related structures. They mimic allocation of space, which seems to be more important for the generation of pathway networks in the investigated case rather than the pathway network itself.[237]

234 Stevens, P.S.: Patterns in Nature, 1988, pp.150
235 Stevens, P.S.: Patterns in Nature, 1988, pp.117
236 Teichmann, K. et al.: Prozess und Form natürlicher Konstruktionen, 1996, pp.150
237 Schaur, E. in Teichmann, K. et al.: Prozess und Form natürlicher Konstruktionen, 1996, pp.154

3.3.8 Structure and function

Structure and function are linked on all levels of biological organisation. One of the most amazing structures are birds, in particular their wings being constructed for optimal aerodynamic properties. Apart from the adapted form, the skeleton also has structural characteristics that contribute to the function of flight: the bones are strong but extremely light through a honeycomb-like structure. The muscles used for flight are controlled by neurons - cells with long extensions, which due to their specific construction, are suited to the transmission of signals. An example of functional anatomy on a sub cellular level is the cell organs responsible for respiration, the mitochondria, with their strongly folded and thus surface area increasing structure. The basic principle of connection between structure and function is valid for all "biomaterials".[238]

Fig.179 Seagull.

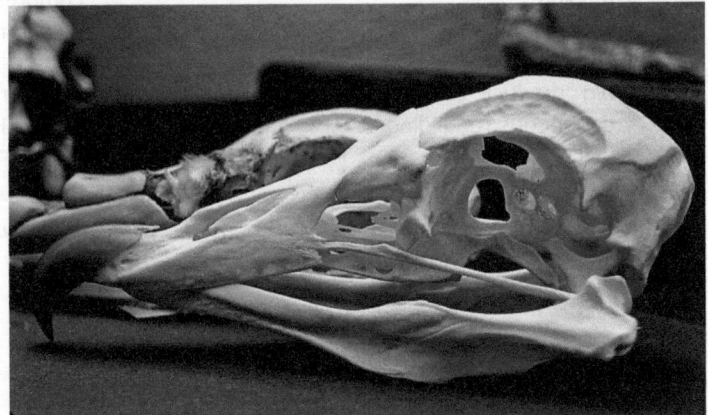

Fig.180 Skull of pelican (Pelecanus onocrotalus).

---

238 Campbell, N.A.: Biologie, 2000, p.9

natural construction	technical construction
round form	right angle
few parts, varied properties	many parts, homogenous
diffusion, surface tension, laminar flow	gravity, thermal conductivity, turbulent flow
strength	stiffness
toughness	brittleness
bending, twisting	sliding
flexible	streamlined
non metallic	metallic
tension	compression
-	wheel, rolling
submarines	surface boats
big "product" compared to factory	small products
continuous rebuilding	minimal maintenance
wet	dry

Fig.181 Comparison between natural and technical constructions, Vogel 1998.

3.4 PARALLELS, DIFFERENCES AND SYNERGIES BETWEEN DESIGN IN NATURE AND IN ARCHITECTURE

"There is a duality between engineering and nature, which is based on minimum use of energy. This is because animals and plants, in order to survive in competition with each other, have evolved ways of living and reproducing using the least amount of resource. This involves efficiency both in metabolism and optimal apportionment of energy between the various functions of life. A similar situation obtains with engineering, where cost is usually the most significant parameter. It seems likely, then, that ideas from nature, suitably interpreted and implemented, could improve the energy efficiency of our engineering at many levels. This transfer of technology, variously called bionics, biomimetics or biognosis, should not be seen so much as a panacea for engineering problems as a portfolio of paradigms."[239]

The quotation from Julian Vincent expresses the common base of design in nature and technology in terms of energy usage. Organisms have to be efficient in terms of energy to survive. They do not waste energy and operate with high effectiveness. Technology, and thus architecture, does not necessarily have to be energy efficient.

In general one could say that design in technology is based on the same laws of physics as animate nature, and that it is for this reason that similar problems, analogies, convergence and role models exist.

Julian Vincent and his group in Bath have compared problem solution in technology (engineering) and biology using principles of TRIZ as a base. They were able to prove that

"At size levels of up to 1m, where most technology is situated, the most important variable for the solution of a problem in is the manipulation of energy usage, closely followed by use of material... But in biology the most important variables for the solution of problems at these scales are information and space."[240]

To bring design in technology closer to that in nature, information has to be increased. This could be done, for instance, by differentiation and structuring of material and constructions.

The difference between design in nature and design in technology, and in architecture in particular, also lies in the difference of the design process as such. Steven Vogel compared grown and built structures and stated obvious differences, shown in the above table.[241]

---

239 Vincent, J.F.W. in: Beukers, A. et al: Lightness, 1998, p.44

240 Vincent, J.F.V. et al.: Biomimetics - its Practice and Theory, J.R.S., 2006, p.8

241 according to Vogel, S.: On cats' Paws and Catapults, 1998, pp.289

Design in nature is based on the powerful yet in some respects limited process of evolution and natural selection, which will be discussed in detail later on. According to evolution theorist Richard Dawkins, the process of evolution resembles a tinkerer rather than a real inventor, somebody who makes changes continuously, rather than somebody who creates something new.[242] Nevertheless, there have been a few and important "inventions", e.g. the development of species from water to land and flight. The number of fundamental plans in nature is quite limited, but the differentiation is breathtaking. Learning and intelligence is the only way for organisms to go beyond inherited capabilities.

Erwin Schrödinger compared natural evolution and technical making:

"It is illumination to compare this natural process with the making of an instrument by man... If we manufacture a delicate mechanism, we should in most cases spoil it if we were impatient and tried to use it again and again long before it is finished. Nature... proceeds differently. She cannot produce a new organism and its organs otherwise than whilst they are continually used, probed, examined with regard to their efficiency. But actually this parallel is wrong. The making of a single instrument by man corresponds to ontogenesis, that is, to the growing up of a single individual from the seed of maturity... The true parallel of the evolutionary development of organisms could be illustrated... by a historical exhibition of bicycles, showing how this machine gradually changed from year to year... Here, just as in the natural process, it is obviously essential that the machine in question should be continually used and thus improved; not literally improved by use, but by the experience gained and the alterations suggested."[243]

What Schrödinger did not discuss here was the primary invention of the bicycle. Design in technology is fundamentally different. Innovation in technology is not continuous and conservative, but revolutionary and really creative in a sense that phenomena emerge which do not have a predecessor in developmental history. Invention from scratch is possible.

Technological design exceeds the limitations of evolutionary design. On the other hand, evolutionary design incorporates more information, is refined to the maximum and conservative in a positive way. A synthesis of both processes should provide even more possibilities.

3.5 BIOMIMETICS IN CONSTRUCTION AND ARCHITECTURE

According to Juri Lebedew, the main components to be analysed by the field of "Architekturbionk" are:
"1. Functions with methods of analysis and analogy of the natural and built environment
2. Construction principles of organic nature
3. Development of form and harmony"[244]

Lebedew interprets the relationship between architecture, nature and humans in a triangular scheme as association. Architecture does not only protect humans, but also takes the role of an extended part of the human organism and interacts between humans and the environment.[245]

Lebedew favours the creative application of natural principles and discoveries in contrast to a mere copy of form.[246]

Biomimetics in architecture delivers identification of new and innovative fields together with a method of transferring ideas from nature's phenomena to architecture. As architectural projects are determined by so many parameters, the definition of "Biomimetic Architecture" as a new style or genre is not suitable. In addition to that, it is difficult to judge if an architectural project has biomimetic intentions. Often this is not obvious at all. Biomimetics can influence architecture in specific, but certainly not all respects. Therefore a methodical expression has to be preferred: biomimetics in architecture is a discipline to gain innovation in architecture by using natural role models, and the comparison between animate nature and built environment creates new insights.

---

242 R. Dawkins in Vogel, S.: On cats' Paws and Catapults, 1998, p.27, original in Dawkins, R.: The Blind Watchmaker, 1986
243 Schrödinger, E.: What Is Life?, 1967, p.114
244 Lebedew, J.S.: Architektur und Bionik, 1983, p.25
245 Lebedew, J.S.: Architektur und Bionik, 1983, pp.24
246 Lebedew, J.S.: Architektur und Bionik, 1983, p.225

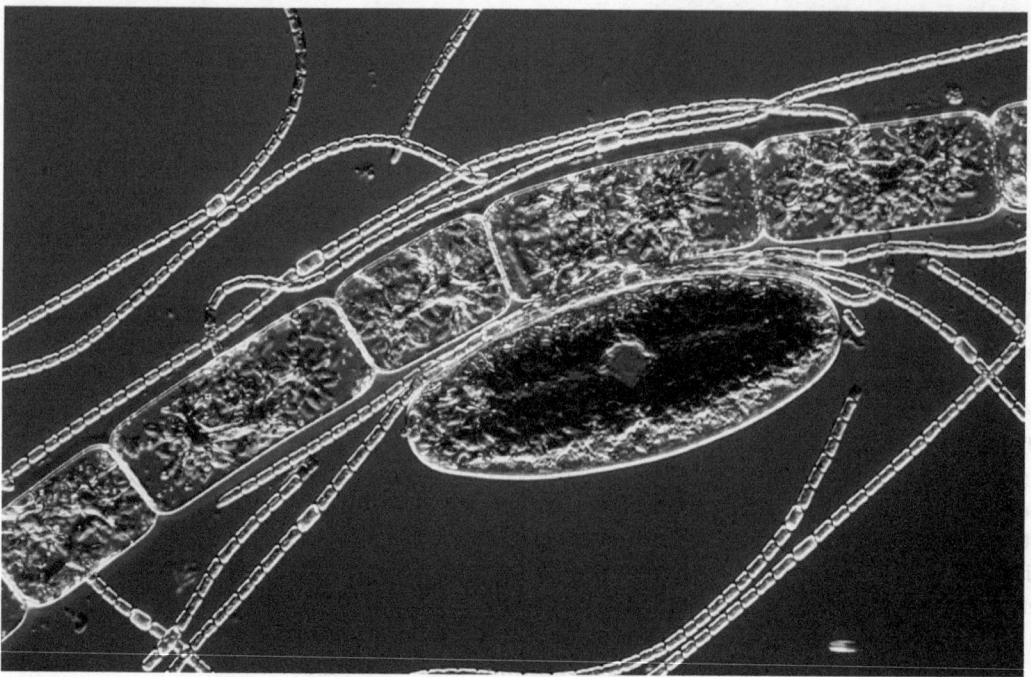
Fig.182 Cyanobacterium and green algae, © Dennis Kunkel Microscopy, Inc.

4 NEW APPROACHES AND APPLICATION OF BIOLOGY'S LIFE CRITERIA ON ARCHITECTURE

Considering the many movements in architecture dealing with the topic of nature, three common features can be observed:
- **All movements and approaches relate to life and living organisms**
- **Search for form is (still) the most important issue**
- **There is considerable confusion in terminology between the disciplines of architecture and the life sciences, because architects have taken over terminology from life sciences for their work (metabolism, evolution, genetic design...).**[247]

As a result, the discussion of life sciences terminology ought to be the next step to clarify the matter.

4.1 LIFE, BIOLOGY

The discussion of natural constructions requires a discussion of living systems and life itself.
What is life? There is no universal agreement on the definition of life.
Simple creatures like bacteria and algae are systems of enormous complexity, which offer many different fields of research. Even our best technical achievements are still far away from this complexity.
Especially in the 20th century, many scientists dealt with the question of when life begins, and tried to mimic spontaneous generation of life by experimenting with chemical models of the Earth's early environment, succeeding in the production of complex molecules.
"Despite much conjecture and intriguing research, it must be remembered that no life has yet been synthesized in the laboratory. The gap between chemical evolution and true cells remains unbridged."[248]

---

247 The same phenomenon happened the other way round in Information Technology, which has taken over the terminology of architecture.

248 Margulis, L. et al.: What Is Life?, 2000, p.75

Fig.183 Sand dunes in Egypt, showing characteristic ripples.

4.1.1 What is alive? Inanimate, animate and dead nature

The obvious importance of life in architecture leads to this old and famous question, which is surprisingly difficult to answer. Famous thinkers like Erwin Schrödinger[249], Humberto Maturana, Francisco Varela[250] and Lynn Margulis[251] have dealt with the question. During the last decades, different definitions have evolved in life sciences in this ongoing discussion.

To begin with, there is a set of classical criteria of life, all of which have to be shown by an organism in order to be called alive. The classical criteria are: order, propagation, growth and development, energy processing, reactions to the environment, homoeostasis - metabolism and evolutionary adaptation.[252] Order, as one of the most important features of life, involves: integrity, hierarchical levels, emergence and low entropy. Erwin Schrödinger defined living organisms as entropy machines.[253]

Self-organisation is characteristic for all processes in nature, and is important for the creation of living forms as well. All living systems are limited in space and time - there is as certain life span and a certain size which is favoured by organisms and these cannot easily be exceeded.[254]

The most recent definition of life comes from the discipline of Biosemantics; this proposes the semiotic character of life, characterised by communication and processing of signs on all levels of animate nature.[255]

Easier than the definition of life is the definition of not living and dead nature. Apart from animate nature, inanimate and dead nature is investigated in the field of biomimetics.

Inanimate nature

"Construction forms emerge out of themselves, according to their specific characteristics and in constant adaptation to external environmental conditions."[256]

There are typical patterns and emanations, effective forces depending on the scale of the system.

Inanimate nature is manifested in processes of self-organisation and tends to a state of minimal energy. Inanimate nature is only apparently permanent, but subject to a constant dynamic potential by permanent movement, synthesis and breakdown.[257]

---

249 Schrödinger, E.: What Is Life?, 1967, first published in 1944
250 Maturana, H.R. et al.: Autopoiesis and Cognition, 1991
251 Margulis, L. et al.: What Is Life?, 2000
252 Campbell, N.A.; Biologie, 2000, p.5
253 Schrödinger, E.: What Is Life?, 1967, p.71
254 Otto, F. et al.: Natürliche Konstruktionen, 1985, p.102

255 Hoffmeyer, J.: Biosemiotics: Towards a New Synthesis in Biology, European Journal for Semiotic Studies, Vol.9 No.2, 1997, pp.355
256 Thywissen, C. in Otto, F. et al.: Natürliche Konstruktionen, 1985, p.13
257 Thywissen, C. in Otto, F. et al.: Natürliche Konstruktionen, 1985, p.13

Otto Patzelt states that sand dunes emerge in a process that is characterised by positive and negative feedback effects. The wind throws up grains of sand - these grains throw up others - positive feedback effect - too much sand in the air slows the wind - negative feedback effect. The rhythm in time is expressed in a similar rhythmic spatial order of the sand-ripples.[258]

Architecture is part of inanimate nature, and thus subject to the respective processes:
- Self organisation: destruction- and deterioration processes, also diverse use of these processes in building, esp. in material processing
- Adaptation to external environmental conditions
- Basic patterns of manifestations
- Scale-dependency
- Tendency to energy minimisation: although validity for building is questionable
- Apparent permanence but permanent dynamics[259]

According to Lynn Margulis and the theories of Vladimir Ivanovich Vrnadsky, more and more of inanimate nature is integrated into processes of life. "The types as well as the mass of chemical elements in living bodies have increased through evolutionary time."[260]

Dead nature

Living nature produces individual forms. Dead nature evolves from it. Transformation processes reduce former living individual organisms back to a big uniform whole. Dead nature forms a considerable part of the whole of nature.

Dead nature is a considerable part of inanimate nature, and is most important for human economy and the building industry, e.g. oil, gas, coal, humus and lime.[261] Dead material is valuable because of its prestructured character. The natural transformation and recycling processes have not gone so far that the former living organisms are decomposed to the initial materials. So using dead nature in technology can be interpreted as step in an ongoing natural recycling process that human activity is part of.

Animate nature

"The shape alone is stable. The substance is a stream of energy going in at one end and out at the other. Life's purpose to maintain and perpetuate itself is understandable as a physico-chemical phenomenon studied by the science of thermodynamics. We are temporarily identifiable wiggles in a stream that enters us in the form of light, heat, air, water, milk... It goes out as gas and excrement - also as semen, babies, talk, politics, war, poetry and music."[262]

We can recognise life by its activities, but even in this age of scientific discovery the definition of life is still difficult. Different approaches to an explanation of the phenomenon of life exist today. The most general description is the following: living systems are open systems made of proteins and nucleic acids with the ability of self-synthesis of these substances.

4.1.2 The smallest unit of life: the cell

"The cell, the smallest autopoietic structure known today, is the minimum unit that is capable of incessant self-organizing metabolism."[263]

(Almost) all life is based on small basic modules, the cells. In contrast to technical constructions, laws of metabolism limit the size of cells. This is why big organisms consist of a diversity of single modules - cells. Frei Otto stated "Pneumatics is the building principle of life."[264]

From a construction point of view, cells are pneumatic structures, systems consisting of a soft fluid filling and a tensile and supple skin.[265]

Multicellular organisms are made up of a variety of these basic elements. Humans consist of around 10^{14} cells.[266] Cells specialise to build tissues and organs.

In the search for the beginning of life, many experiments have been made with so-called microspheres, an abiotic, soap bubble-like structure encasing a volume of air or fluid. The sphere is once more the form with lowest energy demand. Experiments with microspheres have shown the characteristics of cells.

---

258 Patzelt, O.: Wachsen und Bauen, 1974, pp.48
259 Otto, F. et al.: Natürliche Konstruktionen, 1985, pp.10
260 Margulis, L. et al.: What Is Life?, 2000, p.25
261 Otto, F. et al.: Natürliche Konstruktionen, 1985, p.18

262 Watts, A. in Margulis, L. et al.: What Is Life?, 2000, p.43
263 Margulis, L. et al.: What Is Life?, 2000, p.78
264 Nachtigall, W.: Das große Buch der Bionik, 2000, p.34
265 Otto, F. et al.: Natürliche Konstruktionen, 1985, p.84
266 Vogel, S.: Cats' Paws and Catapults, 1998, p.25

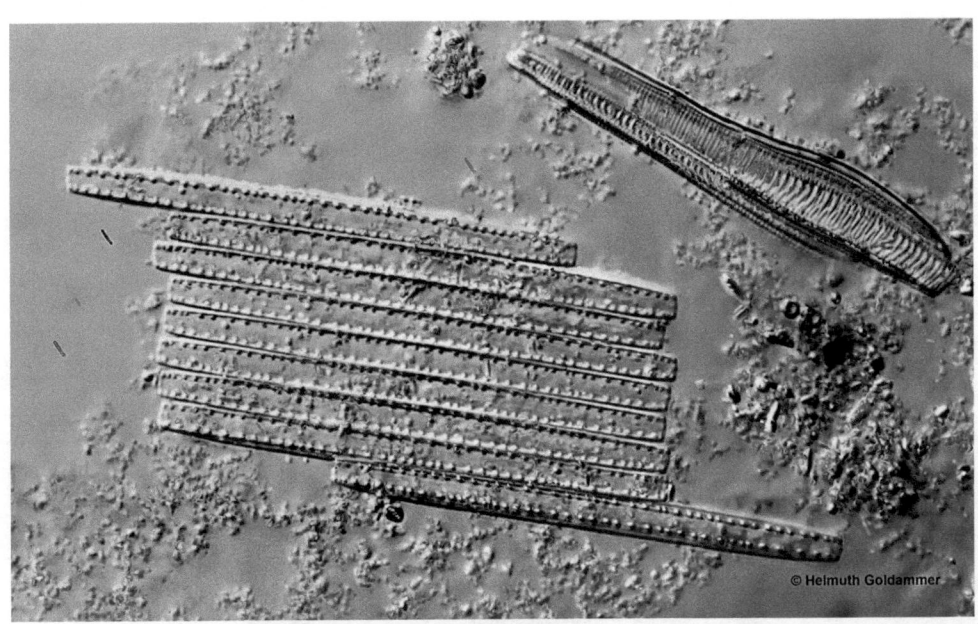

Fig.184 Bacillaria, alge forming motile colonies.

Internal pressure, for example, is large, and the smaller the microsphere the larger the pressure[267] (the maximum pressure that the skin may take per surface area always being the same). Cells have a semi-permeable cell membrane, or plasma membrane, separating the cell from its environment and fulfilling diverse other functions. The membranes consist of a lipid bilayer and integral membrane proteins. In fungi, bacteria and plants the cell wall is the outermost boundary.

Cells grow and divide, which requires low internal pressure. A water-filled pneumatic structure with a thin skin can divide easily internally, because there is hardly any strain within the water.

Cell division happens exclusively in a watery environment,[268] even if the organism itself, e.g. protected by skins, in eggs and abdominal cavities, generates this environment.

The organs in a cell are called organelles. There are two types of cells, prokaryotic and eukaryotic cells. Prokaryotic cells are usually protozoa, e.g. bacteria. Eukaryotic cells are usually assembled in multicellular organisms. A typical eukaryotic cell contains: lysosomes for storage, centrosome, microfilaments and microtubules for cytoskeleton and structure for cell division, mitochondria for respiration, flagellum for locomotion, nucleus containing DNA, ribosome for protein production, endoplasmatic reticulum "ER" for macromolecule production and transport, golgi-apparatus for macromolecule transport and modification, plasma membranes for separation from its environment and internal containers and vesicles for storage transport or digestion. The organelles float in cytosol, a semitransparent fluid in which other elements of cytoplasm are suspended as well as water, salts, organic molecules and enzymes.

The entire content of a cell is called cytoplasm. Plant cells contain additional elements for photosynthesis, the chloroplasts, a central vacuole for storage and a cell wall.

Lynn Margulis published the theory of endosymbiosis in 1970 in "The Origin of Eukaryotic Cells", explaining the existence of cell organelles, in particular the existence of chloroplasts and mitochondria, which have kept their own bilayer membrane and DNA - a result of the permanent symbiosis of bacteria - and to stress symbiosis as a driver for evolution.

There is enormous diversity of forms and specialisations, so that even the cellular nature may in some cases be no longer discernable. The combination and assembly of cells to tissues and organs requires adaptation of the typical characteristics in favour of another optimum on a higher hierarchical level - the characteristics of the tissue or organ being the determining factors for development.

267 Otto, F. et al.: Natürliche Konstruktionen, 1985, p.84
268 Otto, F. et al.: Natürliche Konstruktionen, 1985, p.90

Fig.185 Euglena, unicellular organism showing both plant and animal characteristics.

4.1.3 General characteristics of life

Different approaches to a definition of life have been taken over the past hundred years. The interpretation of life is influenced by the respective technological level of a certain time. Starting off with a mere listing of criteria we have now moved on towards a systemic view of this phenomenon. Nonetheless, important characteristics of life serve as criteria for a comparison with architecture and will be presented in the following.

Individual forms

"The necessity of thermodynamically isolating a subsystem is an irreducible condition of life"[269]
The partial separation of living entities from their environment by means of membranes is crucial for the processing of matter and energy.[270]
There is an obvious tendency to diversification and differentiation, so that life exists in an abundant diversity of individual forms. Five big kingdoms of organisms have been agreed upon: protoctista, bacteria, fungi (100,000-1.5 million species[271]), plants (nine phyla, e.g. flowering plants is estimated to have 250,000 species[272]) and animals (echinoderms, molluscs, craniates, coelenterates, etc. altogether an estimated 3-30 million species[273]).[274]

269 Morowitz, H. in Margulis, L. et al.: What Is Life?, 2000, p.85
270 Margulis, L. et al.: What Is Life?, 2000, p.85
271 Margulis, L. et al.: What Is Life?, 2000 p.175
272 Margulis, L. et al.: What Is Life?, 2000, p.195
273 Margulis, L. et al.: What Is Life?, 2000, p.151
274 Margulis, L. et al.: What Is Life?, 2000, p.119

Entropy

The first law of thermodynamics says that during any transformation the total energy of any system and its environment is constant: energy is neither lost nor gained. Energy - whether as light, movement, radiation, heat, radioactivity, chemical or other - is conserved. The second law says that in any moving or energy-using system entropy (degree of disorder and randomness) increases.[275]
Other forms of energy tend to convert to heat, and heat tends to disorganise matter. There is low entropy and a high degree of order in organisms. This is only achieved through permanent processing of solar and chemical energy.

"What an organism feeds upon is negative entropy. Or, to put it less paradoxically, the essential thing in metabolism is that the organism succeeds in freeing itself from all the entropy it cannot help producing while alive."[276]
Order is based on a hierarchy of structural levels, every level based on the one beneath it. Depending on the scale, different phenomena become important. Order is a fundamental characteristic of any architectural creation.
Organisms have found two different ways to get round the laws of thermodynamics: autotrophic organisms use solar energy to establish and maintain order; heterotrophic organisms use chemical energy by degrading high-molecular nutrition.

275 Margulis, L. et al.: What Is Life?, 2000, p.15
276 Schrödinger E.: What Is Life?, 1967, p.71

Living systems are open systems

Because metabolism requires exchange of matter and energy with the environment, living systems are open systems. Consisting of proteins and nucleic acids, they are able to synthesize these substances. Supply of energy and emission of positive entropy in the form of disoriented substance or thermal energy is necessary to keep entropy low.

Emergence

With a growing level of order, new characteristics evolve which did not exist on the level beneath. These characteristics are called emergent characteristics and result from the interaction between the components (synergism). Emergence occurs on all levels: with growing complexity, for instance, communication and sensing organs have to be used. A description of the levels is given below, with examples. Margulis states that even *"Mind may be the result of interacting cells."*[277]

Life is based on hierarchical levels of structure

Life is organised on hierarchical structural levels. Most processes in living systems take place on more than one level. The hierarchical structuring of matter is one of the most important characteristics of materials in biology.[278] Major levels are presented below, with characteristic examples of each level.
- Chemical level - atoms and molecules: cellulose fibres, chemical compounds
- Cellular organs and cell structures: chloroplasts for photosynthesis
- Cells: liverwort - cells walls and organelles are visible
- Organs and tissues: leaf structure, venation, supply system
- Multicellular organism with communication and sensing devices: moss animal
- Population, group of the same species within a limited area
- Community, ecosystem - complex interaction of organisms and inanimate environment: sweet water community

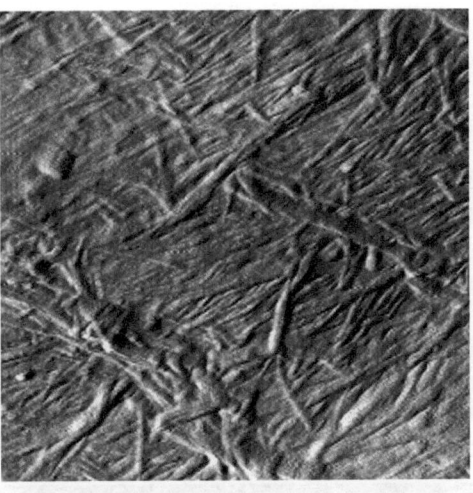

Fig.186 Cellulose fibres.

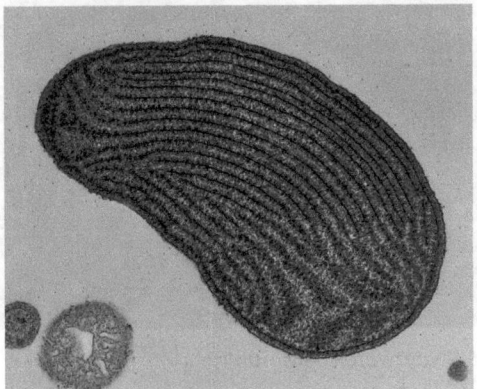

Fig.187 Chloroplast from red algae (Griffthsia spp.), © Dennis Kunkel Microscopy, Inc.

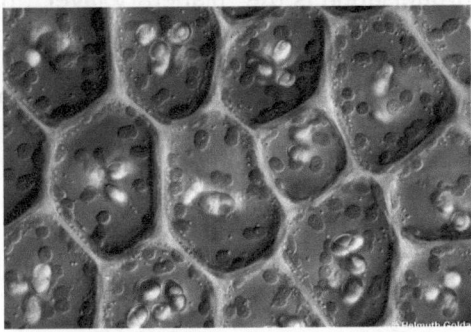

Fig.188 Liverwort.

Fig.189 Venation of Gingko leaf.

277 Margulis, L. et al.: What Is Life?, 2000, p.30
278 Campbell, N.A.; Biologie, 2000, pp.2

Fig.190 Fredericella, moss animal.

Fig.191 Sweetwater community.

According to Margulis, the highest level of organisation is the "planetary layer", the biosphere itself.[279]

"Life is not merely matter, but matter energized, matter organized, matter with a glorious and peculiar built-in history."[280]

Limitation

In inanimate nature, the size of constructions is limited through the size of elementary particles and the size of the universe.

In animate nature, structures from 0.01mm to 170m (giant sequoia trees) or several km² (fungi) exist. For most of the structures an existing size limit must not be exceeded, otherwise the whole system risks breakdown. The existence of most organisms has a time limitation. "Eternal" trees and long living animal colonies like coral reefs seem to be exceptions.[281] The giant sequoias on the West coast of North America are found to be more than 3,500 years old. For plants, strategies for not reaching a size limitation seem to be an effective way to achieve "eternal" life.

Bacteria, the very first organisms, do not die naturally. According to Lynn Margulis, ageing and programmed cell death evolved in relation to sexuality with protoctists.[282]

The relation between surface and volume limits the size of cells. The requirements of metabolism for supply are the limiting factor (0.1-10 micrometres)[283]. On a macroscopic scale the structural peculiarities of organisms are limiting their respective size. For example, the exoskeleton of insects limits their size under the respective atmospheric conditions.

The axiom of biogenesis (life is not created spontaneously, but only through propagation) does not agree with a definite "start" of life. As mentioned before, this paradox is still unsolved.

Scientists claim that life has started on earth 3.4 billion years ago, which is supported by the findings of stromatoliths - fossilised primitive microorganisms - in Western Australia.

---

279 Margulis, L. et al.: What Is Life?, 2000, p.14
280 Margulis, L. et al.: What Is Life?, 2000, p.43

281 Otto, F. et al.: Natürliche Konstruktionen, 1985, p.102
282 Margulis, L. et al.: What Is Life?, 2000, p.137
283 1 micrometre = 10^{-6} metre = 1 thousandth millimetre

Fig.192 Echinacea purpurea.

4.1.4 Classical criteria of life

The so-called criteria of life are defined as follows: order, propagation, growth and development, energy use, sensing and reacting, homeostasis and evolutionary development.[284]

These criteria substitute a definition. Living systems are supposed to display all of these criteria. This definition is incomplete, as it excludes for instance viruses that cannot reproduce independently but nonetheless cannot be called inanimate.

Order, or negative entropy

All characteristics of life develop out of the complex organisation of the organism itself. The existence of life depends on a specific level of complexity. The processes in living organisms take place in dynamic structures. Differentiation and change of structure and form are possible. All organisms exist in some kind of chemical order. Order often takes the form of ("natural") patterns.

As mentioned before, the order of living systems is not consistent with the second law of thermodynamics saying that with every transformation of matter and energy the universal entropy increases. Living systems use a trick to fulfil this physical condition: autotrophic organisms use sunlight as an energy source and create complex molecular material to store energy. Heterotrophic organisms internalise food and thus highly ordered material. This macromolecular material is reduced to lower levels delivering enough energy to survive. From this point of view, life is islands of order in an otherwise highly entropic universe.

Propagation

Organisms are capable of reproducing themselves through passing on genetic information. The axiom of biogenesis "omne vivum ex vivo" says that life can only arise from life. Cell division is the form of propagation in unicellular organisms to generate identical copies of individuals through so-called mitosis. Strategies of asexual propagation mostly in plants include budding and the generation of spores and seeds. Most animals, plants and some protists and fungi reproduce sexually. Offspring are produced by a male and a female cell containing complete sets of chromosomes. The strategy of sexual reproduction enables genetic recombination, leading to diversity in the genome, which is the basis for evolution.

Development and growth

Inherited programmes in the form of DNA together with RNA control growth and development processes and thus generate an organism typically representing a species. Development occurs in the timeframe of an individual (ontogenesis) and in the timeframe of a species (phylogenesis). The connection between ontogenesis and phylogenesis was discovered in 1866 by Ernst Haeckel in the so-called biogenetic law.

---

284 Campbell, N.A.; Biologie, 2000, p.5

Fig.193 Complex shape of pitcher plants.

Fig.194 Wave form of a scallop.

Growth

- **Growth by division**

Growth in nature relies on cell division and differentiation. Cells divide, assembling and building material for the living organism.

- **Growth and curvature**

A multicellular organism is not constructed by means of a plan defining positions of elements in space, but by means of growth rhythms and concepts, which are regulated chemically. Through growth the organic forms are generated in the way that addition and distribution of material imply. Where more material is generated, or growth is faster, lack of space leads to curvature and complex geometric forms. Depending on the speed of growth of border or central areas, saddle or bowl-like forms are created. Natural growth processes leave traces of growth on their products: to give an example, the growth phases of mussels are visible as ridges on their shells.

Concepts of growth

Only soft cells can grow, divide and reproduce. Any solidification hampers the process of growth, in particular if the organism cannot resolve the implemented material again (cellulose, chitin, silica acid, lignin). Hardening never affects the whole organism, and growth processes continue in soft parts. Organisms have developed different concepts to maintain stability and growth [285] which are conflicting parameters.

- **Growth on the edge**

Snails and mussels grow continuously until they die. Mineral substance is deposited on the margins of their shells.

- **Organisms with shells (exoskeleton)**

Growth is suspended after solidification, or it is necessary to complete or partly discard the shell or ecdysis. With crabs, spiders and insects, for instance, growth in stages is the solution. As they are particularly sensible and vulnerable in the phase without a hard shell, this is a dangerous period for the organism. Strategies of metamorphosis and pupa phases bridge the time of reconstruction.

- **Extrusion**

Hair, straws and tubes are extruded out of nozzles or grow similarly out of meristems. They can be compared to technical extruded sections.

- **Addition**

Solidified cells do not divide any longer, but do stay alive for a while. Soft cells in tips and margins, meristems, continue dividing and increase in size, then they also rigidify. This is how higher plants grow.

- **Budding**

Buds are pneumatic structures filled with fluid, which grow out of a solid, unchangeable organism. Elements grow within this protection, e.g. leaves and petals. When the elements are ready, the internal pressure breaks the bud and the completed elements unfold. Later, these elements or parts also rigidify. Folding of leaves and petals is created by growth - deployment through increase of pressure within the single cells.

---

285 Otto, F. et al.: Natürliche Konstruktionen, 1985, pp.90

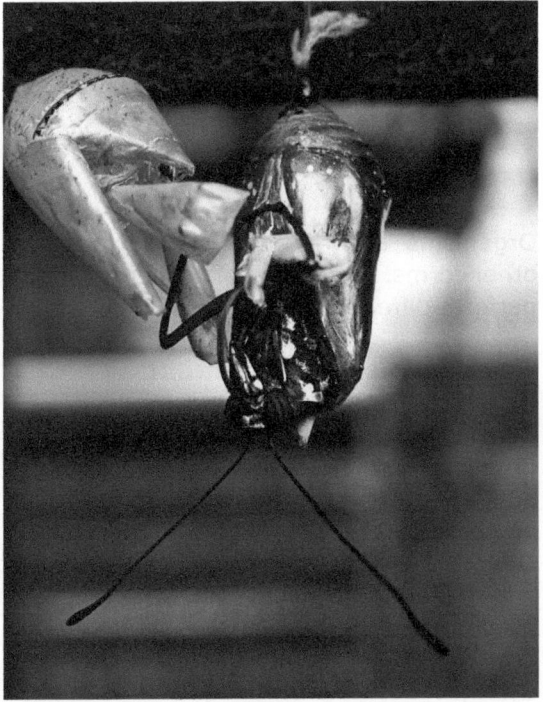

Fig.195 Eclosing butterfly.

Fig.196 Unfolding leaf.

- **Secondary growth in trees**

Trees grow on their exterior and solidify inside. The meristem layer is below the bark, between wood and bast. Growth in trees occurs by cell division in the meristem layer along radial directions and, as a result, circular sections are produced. The bark is torn in a pattern characteristic of the species.

- **Internal skeleton and growth of bone**

An internal skeleton reconciles growth and stability. Vertebrates have a skeleton system that grows and reforms continuously. This is a major reason for their success as medium-sized organisms. Bone substance is deposited between cells; a membrane encloses the bones. Bones grow inside a pneumatic structure which defines the form. Substance in bones is built up and broken down continuously, so that continued increase of size, reconstruction and diminution are always possible.

Fig.197 Different forms of bark.

Fig.198 Bee with pollen.

Fig.199 Mimosa leaf, closing when touched.

Fig.200 Ducks by a river, three females perfectly camouflaged.

Use of energy

Organisms absorb energy and transform it into other forms. They use solar energy or nutrients to perform different kinds of activities.

Reactions to environment

Sensing and reacting are vital for the survival of organisms and their species. All organisms have to adapt to their environment, therefore they have to be sensitive to external stimuli and process them.
"All living beings, not just animals but also plants and microorganisms, perceive."[286]

Homoeostasis, metabolism and autopoiesis

The internal environment of an organism is kept constant within certain limits by regulating mechanisms, in spite of variations in the environment. This regulation is called homoeostasis and is characterised by complex interwoven control cycles.
"Changing to stay the same is the essence of autopoiesis. It applies to the biosphere as well as the cell. Applied to species, it leads to evolution."[287]

Evolutionary adaptation

As organisms and their environment interact, life develops. As a consequence of evolution and natural selection organisms become more and more well-adapted to the environment, while at the same time shaping it.

---

286 Margulis, L. et al.: What Is Life?, 2000, p.27
287 Margulis, L. et al.: What Is Life?, 2000, p.31

4.1.5 Evolution theory

Evolution and natural selection are the most important concepts in nature.
The term "evolution" includes all the changes that life on earth has undergone since its very first beginnings to the diversity of today.[288]

Biological basis of evolution - the code

The biological basis of evolution is DNA, deoxyribonucleic acid. DNA is a long polymer whose structure resembles that of a ladder which has been bent to form a double helix. Sugar and phosphate molecules make up the backbones carrying the connecting four pairs of bases. The sequence of these bases, adenine, thymine, cytosine and guanine, forms the code. During the process of copying the connection between the pairs of bases is opened, and a particular segment is copied in the shape of a negative copy of RNA. Outside the cell nucleus, the information delivered by RNA is used for the production of proteins that in turn are responsible for certain processes and characteristics of the organism. Watson and Crick discovered the structure of DNA and in 1962 were awarded the Nobel prize for this achievement. The correlation between structure and function in biology is visible in DNA - the molecular structure defines the copy mechanism of genes.

The DNA of a human cell has around 3 billion pairs of bases. In humans, the whole of DNA is split into a diploid set of 46 chromosomes and about 20,000-25,000 genes.[289] A gene is a specific segment of DNA needed for the production of a RNA molecule. It is a part of the chromosome coding a distinct feature.

"The chromosome structures are at the same time instrumental in bringing about the development they foreshadow. They are law-code and executive power - or, to use another simile, they are architect's plan and builder's craft - in one."[290]

Charles Darwin, 1859, Great Britain

Published in 1859, Darwin's "On the origin of species", revolutionised the view of animate nature. Within a few years his theory was accepted by the scientific world, due to its logic and demonstrability.

Evolution theory: processes

Natural evolution is based on mutation, recombination and selection.

- **Mutation**

Mutations are small, non-directed changes in the genetic material, the coding DNA. (Genetic characteristics exist in genotype, but may not be expressed in phenotype - in the visible organism.)

- **Genetic Recombination**

Genetic recombination is the overall term for the production of offspring, who combine characteristics of both parents. Selection is the overall term for all external and internal criteria influencing the survival of an organism and its reproductive success.

Mutations deliver the necessary genetic variation, so that even in unforeseeable environmental conditions certain individuals can cope better than others with the altered situation. Recombination guarantees the passing-on of the genetic material.

- **Natural selection**

Natural selection is expressed in the reproductive success of different organisms, resulting from interaction with environment

- **Fitness**

If a specific feature is advantageous in a specific environment, organisms containing it have higher "fitness" than organisms without. This enhances their reproductive success and is thus likely to increase the frequency of the same feature within the population (population=group of individuals of the same species).

In this way, the existence of both the characteristic and the species is continued. The result of this process is visible in the striking adaptations of some organisms to their specific environmental conditions. Darwin described the adaptation of finches living on the Galápagos Islands, where geographic remoteness enhances the development of a unique flora and fauna, showing very specific adaptations.

---

288 Campbell, N. A.: Biologie, 2000, p.435
289 Nature 431: p.931-45
290 Schrödinger E.: What Is Life?, 1967, p.22

Evolution theory today

Genetics of populations

Within the past decades many discoveries and observations have increased our knowledge of the development of species, but the basic concept remains valid. Based on Darwin's theory, the principles of evolution are nowadays also applied to populations. The appearance and frequencies of features in populations are regarded in terms of mathematical models, taking into account gene drift, population densities, migration and isolation.

Limitations of natural evolution

- **Historical limitations**

"One of the most beautiful aspects of living things is that they bear within their very form the presence of the past... each body is the charitable gift of a biochemical museum, and each bacterial cell an unplanned time capsule."[291]

The environmental conditions triggering the development of specific characteristics of organisms may have changed or even disappeared by today. Therefore these characteristics are not important for survival any longer, but are still carried further as long as they do not become disadvantageous for the individual or the species. In this respect, evolution does not produce perfect organisms.

- **Adaptations are often compromises**

As organisms have to adapt to opposing or even contradicting parameters, most solutions are compromises. Every step of specialisation brings about the loss of other possibilities.

- **Non-adaptive steps of evolution**

The influence of chance in evolution is considerable. Catastrophes reducing a population to a small number of individuals can create an evolutionary bottleneck, which may initiate a decrease of fitness for the species.

Natural selection can only favour existing variations: out of all available variations, the ones best-adapted are selected, even though they may not be ideal.

Co-evolution and symbiosis

The basic concept of co-evolution is the mutual evolutionary influence of two species. Extended to the interconnectedness and mutual influence of all the decisions taken by organisms involving inanimate matter, the concept of co-evolution leads to a holistic concept of development of life and species. According to Lynn Margulis, symbiosis is an underestimated phenomenon, leading to new life forms through synergy of pre-existing forms.

"Great gaps in evolutions have been leaped by symbiotic incorporation of previously refined components..."[292]

Design of nature

Steven Vogel states: *"As a designer, then, nature is not only glacial in speed but lacking in versatility and erratic in performance."*[293] Schrödinger traced this back to a molecular level.

"The comparative conservatism which results from the high degree of permanence of the genes is essential... in order to ascertain whether the innovations improve or decrease the output, it is essential that they should be introduced one at a time, while all the other parts of the mechanisms are kept constant."[294]

As a result *"Fundamental innovation comes hard, and once achieved, it disseminates almost entirely within a lineage. To a remarkable extent the dazzling diversity in nature represents superficial features of systems of exceedingly conservative and stereotyped character."*[295]

To give an example: the diversity of appearance and mode of action of upper extremities of mammals is astonishing. From a construction point of view, the difference between these body parts is a gradual change of proportions, but the basic layout remains the same. This kind of similarity is called homologous.

---

291 Margulis, L. et al.: What Is Life?, 2000, p.81
292 Margulis, L. et al.: What Is Life?, 2000, p.9
293 Vogel, S.: On Cats' Paws and Catapults, 1998, p.31
294 Schrödinger E.: What Is Life?, 1967, p.22, pp.41
295 Vogel, S.: On Cats' Paws and Catapults, 1998, p.31

4.1.6 Ecosystems and cybernetics

Organisms are open systems permanently interacting with their environment. The science of the relations between organisms and their environment is called ecology.

- **Ecology**

(Greek oikos, house, and logos, science) Ecology encompasses abiotic factors (temperature, light, water, nutrients...) and biotic factors (organisms), which are linked to each other. On the level of ecosystem, the interaction between species and abiotic factors in a specific region is investigated. This interaction, and thus the system, is described in terms of cybernetics.

- **Cybernetics**

Cybernetics is the science of control and communication, as defined by Norbert Wiener in 1948.[296] By means of cybernetics, ecosystems can be described as interlinked control circuits that partly explain the behaviour of the biosphere. In contrast to many man-made systems (technological, economic, social forms of organisation), which (still) lack feedback and control mechanisms, ecosystems are genuinely cybernetic systems. They are self-preserving and fluctuate around specific set point values with small variation.[297]

4.1.7 Recent interpretations

As the current set of criteria is unsatisfactory for a definition of life, the search has continued until today. Important discoveries of the current time are projected onto the famous old question.

Entropy / energy

Thermodynamics is the science of energy transformation. According to Erwin Schrödinger, organisms represent entropy-machines. By degrading highly organised matter and energy they build islands of order and avoid the second law of thermodynamics, which requires entropy to increase continuously.

Chemoton theory

Tibor Ganti developed a theory of principles of life, which is based on chemical automaton.[298]
The importance of this model lies in the fact that it is the simplest existing model, which includes three essential processing units: metabolism, the membrane compartment and genetic information. The particularity of the model resides in the stoichiometrical coupling of the three autocatalytic subunits.[299]

Biosemiotics: signs and information

" ...when biologists and physicists talk about information, they talk about different kinds of things. While information as understood by physicists has no connection to values, relevance or purpose, biologists think about information in a much more everyday language sense, and in fact biological information always serves a purpose in the system, if nothing else it at least serves to promote survival. The point is that biological information is inseparable from its context, it has to be interpreted in order to work. "[300]

The new field of Biosemiotics investigates signs, communication and information management in living systems. It has been declared a paradigm for theoretical biology and is intended to connect life sciences with humanities.

"What we propose, then, is that the traditional paradigm of biology be substituted by a semiotic paradigm the core of which is that biological form is understood primarily as sign."[301]

---

296 Nachtigall, W.: Bionik, Grundlagen und Beispiele für Ingenieure und Naturwissenschaftler, 2002, p.428

297 Nachtigall, W.: Bionik, Grundlagen und Beispiele für Ingenieure und Naturwissenschaftler, 2002, p.428

298 Gantì, T.: The Principles of Life, 2003

299 Munteanu, A. et al.: Chaos in chemoton dynamics, 2007

300 Hoffmeyer, J.: Biosemiotics: Towards a New Synthesis in Biology, 1997

301 Hoffmeyer, J. et al in Kull, K.: Biosemiotics in the twentieth century: a view from biology, 1999

Fig.201 Bangkok, Thailand, March 2006

4.2 ARCHITECTURAL INTERPRETATION OF LIFE CRITERIA

The relationship between the fields of biology and architecture will be investigated by comparing them in the terminology of life sciences. Although the process taking place during this transformation of ideas is very interesting, it can only be covered briefly within the frame of this book. Different approaches are presented in the case studies section.

The comparison between organisms and buildings as architectonic units is obvious, but other dimensions will also be explored. As stated above, architecture is interpreted here as the built environment, which has always been influenced by nature and living systems. Urban design, building, processes and materials are all subjects of design and change that have to be treated individually because of their different scales.

The criteria of life are projected onto the field of architecture and illustrated by exemplary projects. Not all criteria of life can be found in all architectural categories and no quantification of life's criteria in architecture can be made.

Some of the criteria of life seem to be more important than others in analogies with architecture: order, growth, use of energy, sensing and reacting, metabolism and evolutionary development appear appropriate candidates for a transfer into architecture.

The translation diagram in figure 202 shows the outline of the approach.

The following criteria or signs of life are applied in architecture:

All organisms are **open** systems with the ability of self-synthesizing.

On a broad understanding, all life is the product of **self-organisation**. In architectural interpretation, self-organisation involves local independent decision making, planning and building without central control. In building, self-organisation processes are not yet used in a strategic way, but could provide more economic ways of building.

Information has to be processed to design and build. The flow of information is another interesting topic, as nature and technology use information in very different ways.

Limitation is thought of on all levels. Still, time-related processes are insufficiently considered.

Order - this is what architecture is about - exists on all levels, but the generation of order in architecture and biology differs greatly.

Propagation - we cannot yet build houses that propagate themselves[302]. There may be a propagation of ideas and building typology covered by the topic of evolution.

Growth is one of the most interesting aspects of life, but it is rarely applied in technological solutions.

Processing of energy is carried out at all levels of the built environment, and at all stages.

---

[302] There are already successful experiments in self replication, see http://reprap.org/ for the self replicating rapid prototyping machine of Adrian Boywer, Bath [12/2007]

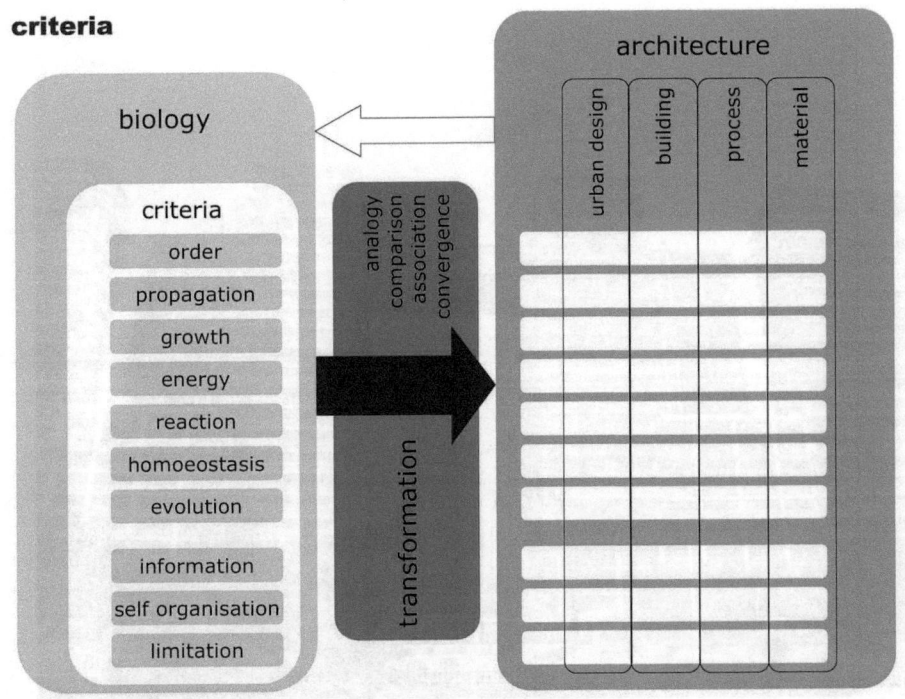

Fig.202 Translation diagram: biology and architecture, criteria and application fields, transformation and direction of information transfer.

Reaction to the environment is a rewarding topic in architecture - it opens up opportunities of what our future houses could deliver.
Homoeostasis and metabolism in architecture are becoming more and more important - internal environment is created and established.
Evolution - as in the notion of slow gradual development - is relevant in all categories, but the application of artificial evolution strategies is becoming increasingly available as a new method of design.

In the following, the selected criteria of life will be projected onto architecture. The meaning of the terms in a general and an architectural sense will be investigated at different scales; examples will be presented when available; and future projects suggested. A summary and look into the future is given in the last chapter: discussion and as yet unexplored fields.

4.2.1 Openness

"(openness of systems)... the degree to which a system operates with distinct boundaries across which exchange occurs capable of inducing change in the system while maintaining the boundaries themselves."[303]
As in living organisms, the conflicting basic requirements of enclosure and openness also exist in architecture, and both have to be met.

Openness in the sense of life sciences involves the exchange of resources, energy and information with the environment.
Openness as an architectural category also includes physical openness in terms of accessibility, visibility and permeability of any kind, but also metaphorical openness. In ecological design, the exchange of material and energy is given much significance.

Architectural "openness"

Architecture on all its levels can appear more or less open to human experience. This concerns issues of:
- The planning process, open to the input of future inhabitants or neighbours to encourage acceptance and identification
- Visible openness and transparency of urban structure and buildings
- Accessibility
- Orientation, e.g. "open floor plans" without dividing walls

Openness in terms of being in operation (e.g. commercial, industrial buildings) is signified by light.
Openness can be expressed by the appearance of the building as a whole being "open" or "closed", e.g. the Pavilion of Venezuela at EXPO 2000, which is an example of openness as well as growth of space. A more extended interpretation of openness also includes the matters of flexibility and change.

---

303 http://en.wiktionary.org/wiki/openness [02/2007]

Fig.203 Beijing, construction site for the Olympics, September 2006.

Openness of urban structures

The openness of urban structures is evident. Urban areas attract and absorb huge amounts of material and energy from their surroundings and produce amounts, just as huge, of waste and emissions, thereby influencing the environment. The working of large urban structures is a mystery on its own, which can only be mentioned within the frame of this project. Many of the supply chains are created in a self-organised and unplanned way (e.g. food for inhabitants), triggered by local needs. Planned supply normally includes water, electricity and other forms of energy, transported hundreds, or even thousands of kilometres. In some parts of the world this supply has not yet been established: in developing countries many methods of supply may be already available, but the disposal of waste may not - this is one of the most unsustainable phases of city development.

Infrastructure is needed to serve the purpose of exchange of energy, material, transport and information.

One of the most influential design criteria for cities is accessibility for people, who need space for locomotion and access. Just as streets are needed for individual transport, facilities are needed for public transport. Transport beyond a local scale occurs only if planned centrally.

Large building sites, e.g. for the Olympic games, involve the economy of an entire country and beyond. The recent building activities in China had global influences, affecting for instance the price and availability of steel in Europe.

The accessibility of urban areas depends on the "urban fabric", the relationship between built structures and space for traffic, the amount of supra-regional connections for individual transport as well as public transport and the structure of the networks. Accessibility is usually desirable, but some urban functions go along with restricted accessibility: intensive housing areas, pedestrian zoning, areas especially protected for reasons of conservation of cultural heritage or for security, to name but two examples.

Fig.204 Flower-shaped Venezuela pavilion, Expo 2000, Hannover, Fruto Vivas Architects and Buro Happold.

Openness of buildings

At a building level, architecture exchanges energy and material with the environment according to its life cycle. The accessibility of buildings is not unlimited: there is a distinct regime reflecting conditions of ownership and belonging. Accessibility (as well as possible emergency exits) in the form of circulation space is nowadays defined by building regulations, so that the actual limits are not reached. The closure of buildings and control of access are enabled through technological means. Openness of buildings differs according to the phase within the life cycle:

- **Construction phase**

During the construction phase, the exchange is obvious. The site is prepared; material is moved or removed. Accessibility is often restricted to working companies and people. Building material is transported and processed by using energy. Information is used to define the building. Expenditure is not measured in terms of energy but in terms of money.

A construction phase is always accompanied by disturbances in the neighbourhood. Emissions have mostly negative influences on the environment (sound, dust, traffic, waste, problems with land or water etc.).

- **"Life" of building or operational phase**

During the "lifetime" of a building, energy is necessary for its operation. With today's building technology (alternative systems: solar panels, photovoltaic, geothermal and wind energy systems) houses could be made independent of the large energy supply networks, and the so-called "Plus energy houses" can return a flow of energy. Water is supplied and disposed of constantly. Openness of materials to air and humidity is defined and worked with through characteristic values.

Constant air supply and ventilation is provided by passive means (permeable materials, passive ventilation systems), manual openings (windows, doors) and automated systems.

Electricity is provided for all sorts of amenities. Material may be necessary for maintenance. Apart from maintenance purposes, the exchange of material and energy while a house is being used is a fairly new achievement. Old traditional architectural typologies manage with no energy supply and thus without the amenities of our civilised life (which is one reason why they are endangered).

The most recent development is the exchange of information with the environment through sensory and control systems. A "smart" house has to have sufficient knowledge about its own current state and the ability to react to information coming from outside. Integrated systems manage and control information flow.

The impact a building has on its surrounding is varied. In terms of emission it is sewage, rainwater, gases, solid waste and heat. Ken Yeang and many other architects active in environmentally responsible design have dealt with the in- and outputs of buildings, already discussed in the chapter "Ecological Design and Sustainability."

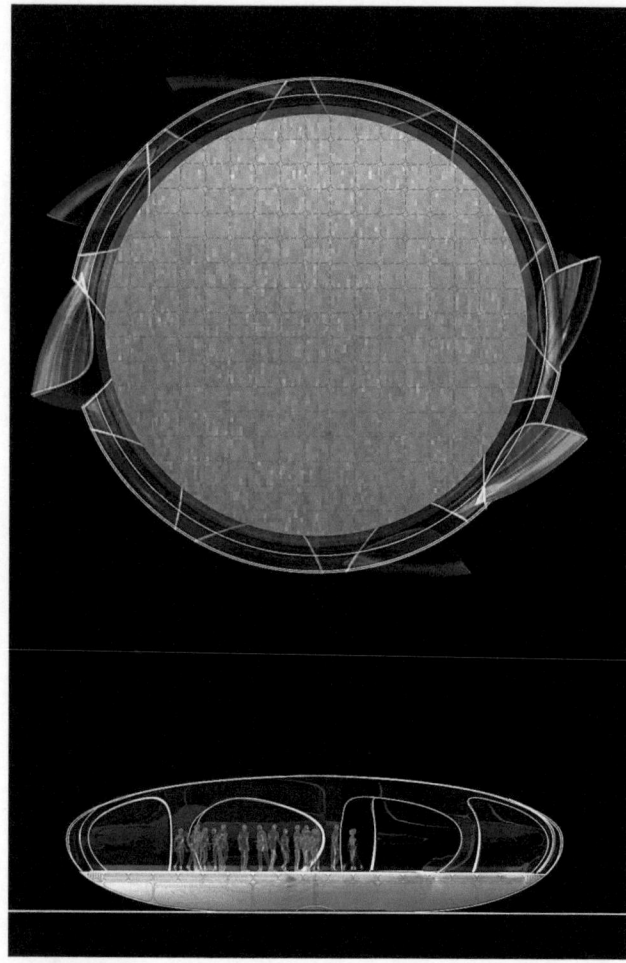

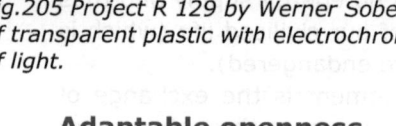

Fig.205 Project R 129 by Werner Sobek 2007, skin made of transparent plastic with electrochromic foil for control of light.

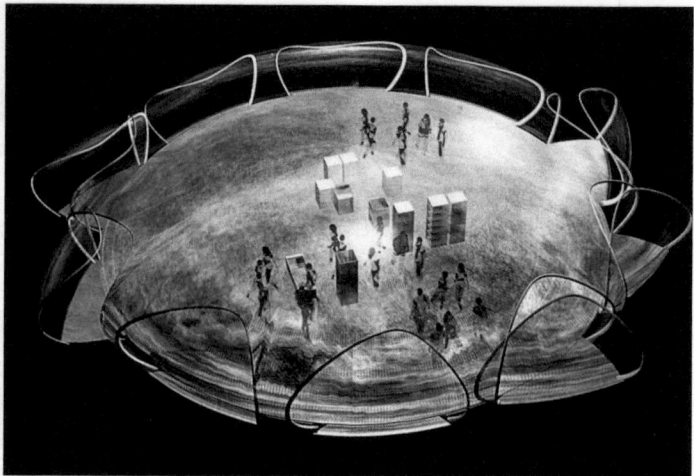

Fig.206 Visualisation of R 129.

Fig.207 Façade of the Institut du monde Arab, Paris, Nouvel 1987.

Adaptable openness

With nature as a role model, more subtle notions of enclosure and openness can be explored in architectural projects. Taking cell membranes as an example for openness, the adaptability and changeability of openness and closure is a striking feature. Openness in the notion of a semi permeable border and the adaptable permeability, for instance, of pores in skin, is attained in some fields by technological means. To give examples: temperature-controlled input of air into a ventilation system, or selective access control. Passive means would help to save energy and increase the reliability of the systems.

- **Light and vision**

Light and vision in architecture are usually closely connected. Few translucent materials, which allow light entry but block visual connection, are available on the market. New adaptable materials, e.g. electro-optical layers of liquid crystals or electrochromic materials in glass, separate the two parameters visual openness and openness for light.

- **Layers of closure**

Layering is one of the basic solutions for openness and enclosure: curtains, glass panes, louvres and blinds provide adaptable openness for windows. The principle of a diaphragm was used to achieve adaptable openness of the south façade in Jean Nouvel's famous Institut du Monde Arabe in Paris, (opened in 1987) reinterpreting the traditional moucharabiehs, gridlike window openings.

Ephemeral materials for enclosure include air and water curtains. These materials are used temporarily and require continuous input of energy. Air curtains are a standard in commercial buildings, to maintain inside temperatures in spite of permanent opening for access. Water curtains are a new development: an MIT project by the so-called "SENSEable city lab" was presented for EXPO 2008 in Zaragoza.[304] When forming the exterior walls, the water curtains can display messages and graphics.

---

304 http://senseable.mit.edu/ [10/2007]

4.2.2 Self-organisation

An introduction into the phenomenon self-organisation can be found in chapter 3.3 "Design principles in nature".
"Only in the cooperation of an enormously large number of atoms do statistical laws begin to operate and control the behaviour of these assemblées with an accuracy increasing as the number of atoms increases. It is in that way that the events acquire truly orderly features."[305]
"Self organization is defined as spontaneous formation, evolution and differentiation of complex order structures forming in non-linear dynamic systems by way of feedback mechanisms involving the elements of the systems, when these systems have passed a critical distance from the statical equilibrium as a result of the influx of unspecific energy, matter or information."[306]

In this way, order emerges from bottom up, in contrast to hierarchical systems. Local interactions of so-called "agents" (elements of lower complexity than the whole system) create global qualities.[307] The overall behaviour may be complex, but the local interactions follow simple rules. The system is maintained in spite of disturbances: regulation mechanisms initiate self-stabilisation, self-healing and self-regeneration etc. Efforts to predict the behaviour of systems were successful with the development of the so-called "multi-agent models", which simulate interactions between many and different components.
Self-organisation in the sense of thermodynamics occurs in architecture in the scale of material and structure. Self-organisation on a macro-scale is based on human activity: with no planning ahead or central planning, agents (humans or organisations) act according to their individual needs, thus creating an emerging, self-organised pattern.

Self organisation of urban structures

In unplanned or informal settlements, emerging settlement patterns result from self-organisation processes. Unplanned settlements are formed by the interaction of self-planning and self-organisation. Apart from selective environmental factors and topography, the structure of settlements is influenced by two phenomena: transportation and accessibility, as well as by allocation of surface area.[308] The similarity to fractal geometry is indicative of self-organisation processes.[309]
Within planned urban settlements, self-organisation is still possible wherever space is not centrally controlled, and some freedom remains where self-organisation can come to the fore - unregulated space is locally organised to meet individual demands (connecting paths in parks, appropriation of space...).

Integration of self-organisation in the design process

Self-organisation processes are integrated in software for form generation, for urban design as well as for buildings. Self-organised design strategies exist, for instance, in evolutionary design approaches, which will be discussed later.
On the building level, the participation of users or occupants in planning processes could deliver an element of self-organisation in the creative process. The future users' wishes are integrated into the planning process (whether processed manually or by computational methods) and generate the composition of the final project by defining floor layout, façade patterns etc. Eda Schaur suggests new methods of urban design and settlement development that connect conscious planning with processes of self-organisation in sub areas.[310]

---

305 Schrödinger E.: What Is Life?, 1967, p.10
306 Schweitzer F. et al.: Communication and Self-organisation in Complex Systems, p.4, original in: SFB 230 (1994): Evolution of Natural Structures, Proceedings of the 3rd International Symposium (Mitteilungen des SFB 230, Heft 9), http://intern.sg.ethz.ch/fschweitzer/until2005/papers.html [11/2007]
307 Schweitzer F. et al.: Communication and Self-Organization in Complex Systems: A Basic Approach, in: Fischer M. M.; Fröhlich J. (Eds.): Knowledge, Complexity and Innovation Systems, 2001, pp. 275-296, http://intern.sg.ethz.ch/fschweitzer/until2005/papers.html [11/2007]

308 Schaur E. in Teichmann, K.; Wilke, J. (Eds.): Prozess und Form "Natürlicher Konstruktionen" 1996, Zur Phänomenologie und Evolution ungeplanter Siedlungen, p.154
309 Mast, H. und Schaur E. et al. in Teichmann, K.; Wilke, J. (Eds.): Prozess und Form "Natürlicher Konstruktionen" 1996, Struktur und Konstruktion von Siedlungen und Städten, pp.150
310 Schaur E. in Teichmann, K.; Wilke, J. (Eds.): Prozess und Form "Natürlicher Konstruktionen" 1996, Zur Phänomenologie und Evolution ungeplanter Siedlungen, p.156

Fig.208 "Twisted" building part of the BMW museum in Munich by Coop Himmelb(l)au, 2007, © Hélène Binet.

Physical self-organisation of material and structure

Experimental formfinding analogies work with self-organisation based on some kind of physical information. Models that visualise directions and centres in urban design with iron filings and magnets make use of magnetism, Gaudi's hanging models use gravity. Frei Otto's soap film models use surface tension and the principle of minimum energy surfaces. All these examples integrate self-organisation into a design or development method. The industrial production of building materials can integrate physical self-organisation processes. The hexagonally structured sheet metal of Frank Mirtsch's company is an excellent example: the rigidity of semi-finished material can be considerably improved by hexagonal structuring, created by a self-organized vaulting process.[311] This patented production method works on the basic principle that thin sheets fail under pressure. No moulds are required for the generation of the hexagonal pattern, since the pattern is the minimum energy formation, which the material forms "automatically" under certain circumstances. The patterning of material is a vast application field for self-organised processing. For example, change of length or other physical properties during a drying process is exploited for the surface of ceramics - the emergence of crack patterns is controlled only by controlling the physical conditions under which the phenomenon takes place. These methods and strategies are often found in traditional craftsmanship and are especially popular in Eastern arts and crafts.

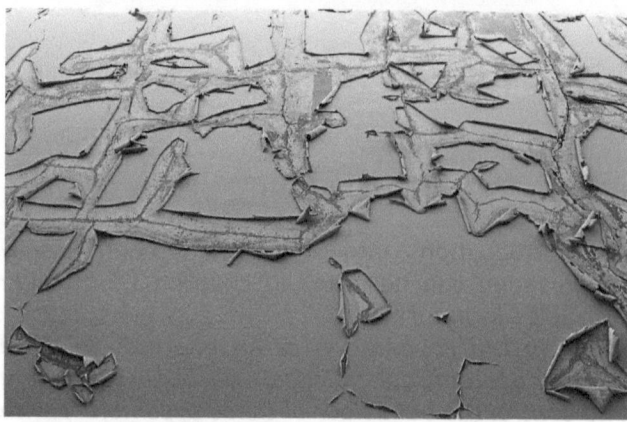

Fig.209 Self-organised pattern - cracks in a colour layer.

Fig.210 Foam structure of the watercube stadium in Beijing, PTW architects, Syndey.

Fig.211 Watercube stadium, building phase, September 2006.

The use of self-organisation processes is ubiquitous in modern art, which appreciates the interrelationships between planned creation and self-organisation (Jackson Pollock, Hermann Nitsch).

311   http://www.woelbstruktur.de/Hauptframe/woelbframe.htm, [03/2007]

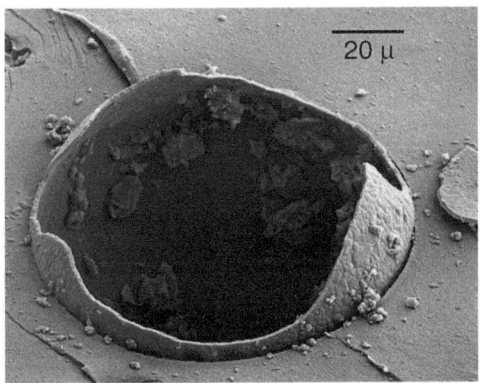

Fig.212 ESEM image showing ruptured microcapsules.

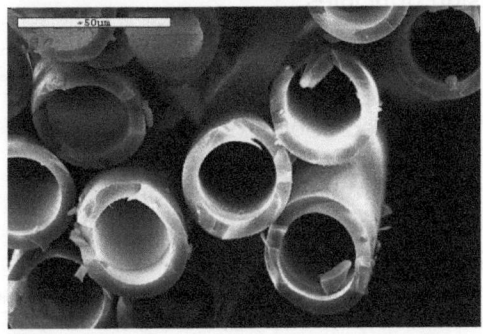

Fig.213 Hollow glass fibres, which contain polymer to fill micro cracks within the structure.

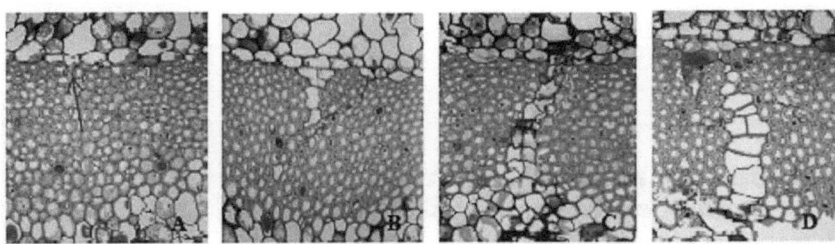

Fig.214 Different phases of fissure repair in the twining vine Aristolochia macrophylla.

Apparent self-organisation

Apparent self-organisation exists when "patterns of nature" are used with presumably self-organised creation. Apparent self-organisation is common on the material level (e.g. the rusty appearance of Corten-steel which was made popular by Richard Serra's sculptures), but also for architectural constructions. Self-organised forms are used in deconstructivist architecture. The deconstruction process creates mashed forms, heaps of material or crumpled-up shapes which seem to be randomly organised - in reality the technical effort for realising these effects is considerable.

Another quite formal approach is the scaling-up of self-organised natural structures to architecture projects, e.g. the use of crystal shapes as models for skyscrapers. PTW architects took the structure of foam as a role model for the design of the Olympic swim-stadium in Beijing. The geometric structure follows a mathematical model of 3D-foam, the edges being transformed into a large-scale steel construction. The basic idea of the foam was adapted to the needs of the architectural structure and function.

Self-healing

Self-healing is one of the most valuable characteristics of living organisms: the ability of self-healing allows economical use of material. In technology, the repair of broken elements is a standard procedure, and the failure of elements is taken into account as well. In contrast to nature, detection and repair of broken elements does not usually happen in a self-organised way. Self-organised healing and repair of elements is especially interesting in cases where local failure would lead to a total system breakdown, as in airplanes, space technology, or in pneumatic structures, which rely on air pressure to maintain structural integrity.

Self-organised crack polymerisation is experimented with for use in space applications. Hollow glass fibres filled with polymer matrix are integrated into composite materials. The breakage of the fibres, and the following release of the sealing matrix allow the break to be found and repaired all at the same time.[312] Similar systems were investigated for concrete, using polymer filled microspheres.

Thomas Speck has investigated the self-healing characteristics of climbing plants, and developed a self-healing membrane material for use in pneumatic structures. The inside of the pressurised membranes is covered with a self-expanding foam layer. In case of a fissure, the foam extends into the space and seals the opening.[313]

---

[312] Trask, R. et al.: Enabling Self-Healing Capabilities – A Small Step to Bio-Mimetic Materials, ESA 2005, pp.6

[313] Speck, T. et al: Self-repairing Membranes for Pneumatic Structures, 2006, pp.117

Self-organisation versus planning - conflicting approaches?

If self-organisation processes are used as a design tool, they separate the planner from the design, as any tool would do. This separation results in a certain loss of control for architects. It is not the design itself that is controllable, merely the boundary conditions, which serve the self-organisation process. This may lead to a certain loss of responsibility for the designing architect, but when used appropriately it can also bring about improvement.

Referring to the hierarchical levels that exist in nature as well as in design, the conscious choice of level is important for successfully implementing self-organisation in architectural design. Some levels of design cannot be influenced locally; on the other hand, some local decisions are difficult to overrule by a central plan. The introduction of self-organisation processes makes sense in connection with a clear definition of sphere of influence.

On an organisational level, self-organisation structures can achieve higher levels of performance than traditionally managed organisations.[314] This advantage could also be exploited in architecture. In European countries, building regulations tend to influence the fine details of buildings, which is annoying for planners as well as builders, clients and the controllers themselves. On the other hand, to give one example, growth of commercial buildings on former farmland in Austria seems to be out of control. Building without unnecessary regulations, but within a stronger framework of regional and urban planning, would enhance the quality of the built environment, preserving spaces which serve other functions rather than merely the bare necessities of industry and the economy.

Self-organisation systems are failure proof as there is no active control system which could possibly fail. The use of such systems in building operation context, e.g. in building automation, is an interesting field.

All in all, the differentiation between planning and local self-organised decision-making depending on information content of a local agent would reduce the effort required to generate high quality built environment.

4.2.3 Information processing

Information is *"Facts provided or learned about something or someone"* or *"What is conveyed or represented by a particular arrangement or sequence of things"*[315]

Information processing, and the resulting strategies in biology are described earlier in chapter 3.3 "Nature's design principles".
Information processing is also crucial for all phases and sizes of built environment. In planning, information has to be available as the basis, and information is provided about the planned project. For building, information has to be passed on to all relevant participants: administration, builders, clients, consultants and so on. After the completion of a building, information still is important for the operational phase. Facility management relies on representation models of the built environment to control supplies, comfort and security. Architecture itself informs the observer, on one hand representing something within the environment and on the other hand representing the environment itself.

Beyond the individual building, information is available in the global "knowledge base" of existing projects and techniques as well as planning and building technologies. Innovation aims at the increase of possibilities and thus knowledge-based information for further use.

Just like the genotype in form of DNA in biology, information consisting of plans for buildings is useless if not translated into reality. The information about how to translate drawings into reality relies on the ability of humans to think abstractly, and requires a sound knowledge base.

---

314 Bennet, J: Organisational Strategy in Clements-Croome, D.: Intelligent Buildings, 2004, p.237

315 New Oxford American Dictionary, program version 1.0.1, 2005

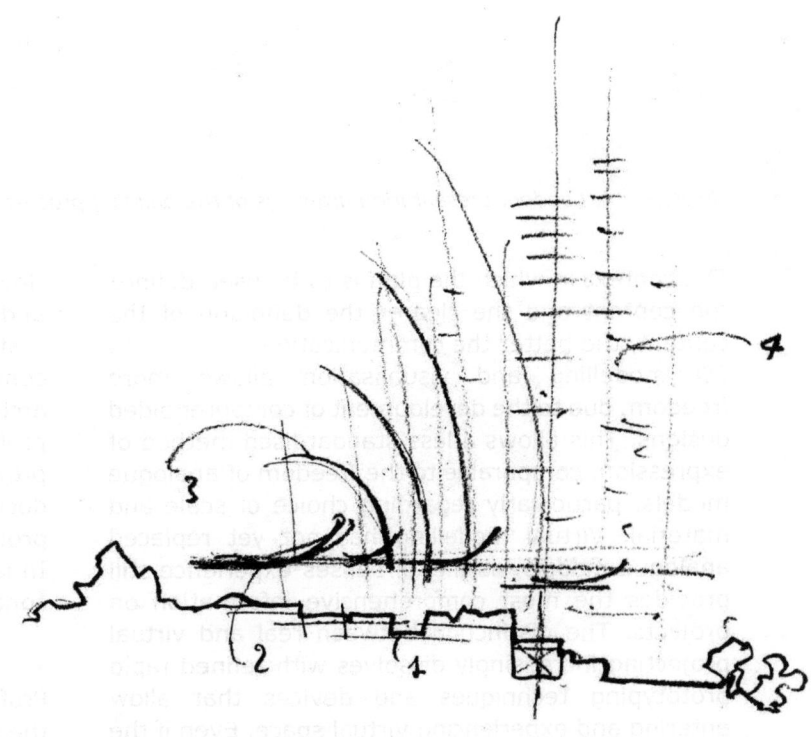
Fig.215 Sketch of Renzo Piano for the Centre Kanak, New Caledonia.

Information transfer in architectural projects

- **Collecting information about site**

Site plans are usually available for planning purposes. They inform about topography, perhaps ground condition, installation, vegetation, existing natural structures, existing built structures, property lines, boundaries and building regulations.

The "character" and atmosphere of a site are intangible and difficult to grasp - usually architects collect what they feel to be important. No "character" maps are available, there is no defined standard and too many parameters influence what we feel to be "characteristic". So architects rely on their own experience and work with visual tools like photographs, sketches, drawings, aerial views and film. Measured environmental parameters, physical information on sound, light, temperature, wind speed, earthquake forces and so on are provided by different organisations.

Apart from a regulatory expression of building density and function, standardised information about local conditions in terms of qualities, needs, future projects, wishes and visions is not available. In Western countries, studies on specific topics in specific areas have been published especially for urban regions, but there is no general system, no integrating surface to gather, process and facilitate access to this information for planners, let alone the public.

Information about a site in a national or even global frame is usually not accessible. Information about the biodiversity of a site is not available, either because surveys don't exist, or because the separation of disciplines is still too rigid.

The environmental impact of a project above a certain size has to be studied, at least in Western countries. But the assessment of such environmental impact is avoided whenever possible, as it costs time and money and might uncover environmental problems and prevent or hamper the whole project.

- **Architectural language**

Communication of information in architecture is generally restricted to visual information transmitted as plans depicting abstraction of built structure in scaled drawings. Vertical and horizontal sections are standardised means (apart from language and measures the standards differ only slightly internationally), and 3D visualisation is used for further understanding. This simple kind of Cartesian order can deliver very complex buildings. Real analogue models are still standard, but can nowadays be produced automatically using digital data and rapid prototyping techniques.

The language of architectural drawings can vary considerably between artistic expression (e.g. at the competition stage) and technical drawing, according to the scale of the project, the stage of the design and use of plans.

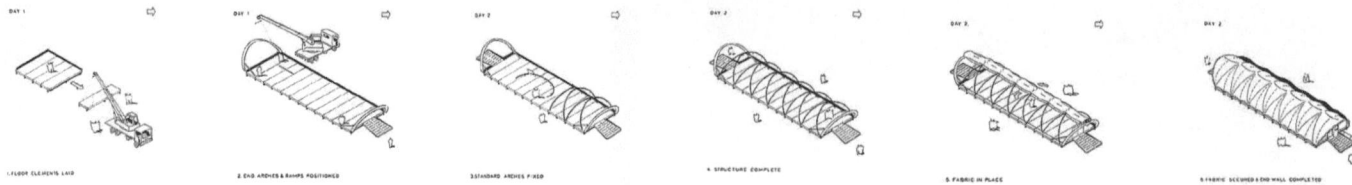

Fig.216 MOMI-tent, London, presentation drawings of the building process, Future Systems.

The context in which the plan is to be used defines the content and the clearer the definition of the content, the better the communication.

3D modelling and visualisation allows more freedom, due to the development of computer aided designs. This allows a less standardised method of expression, comparable to the freedom of analogue models, particularly regarding choice of scale and material. Virtual modelling has not yet replaced analogue models, as the all-senses-experience still provides the most comprehensive information on projects. The distinction between real and virtual projecting increasingly dissolves with refined rapid prototyping techniques and devices that allow entering and experiencing virtual space. Even if the reproduction of an envisioned reality is possible, it is abstraction that provides the key to any expressive representation, as the specific spatial qualities have to be stressed.

- **Information flow during project planning and building**

We have now reached the level where digital information is passed on to all collaborating partners during the entire planning and building process. Digital data are made available on websites, access can easily be controlled and distribution monitored. Further development includes automation techniques in the building industry, where digital data directly inform element production.

Usually a final stage of element assembly is depicted in an architect's plans. Process information is seldom communicated (but always has to be taken into account). Usually this information is implied or passed on orally. New computational tools, e.g. parametric modelling, allow architects to leave the space of exact metric description and move towards a domain of rule based and relational definition, which comes closer to information processing in nature.

Architect's drawings are used as a base for all other information that has to be passed on during the building process. The various consultants, e.g. construction engineers, building physicists, HVAC experts (heating, ventilation, air conditioning), and others draw their own plans based on those provided by the architect. The execution plans of building companies, which they usually draw themselves, are focussed on their respective needs. The planning architect and the client control the information, advancing step-by-step.

However, linear approach cannot be maintained under time pressure. When the schedule demands that the information be spread simultaneously, control and integration has to be carried out by the architect. Integrating computing tools for different professions working on a common architectural project simultaneously are becoming common. The documentation of the whole planning and building process is an important part of the architect's work. In large projects, specialised companies for project control carry out these tasks.

- **Friendly communication**

Professional communication is very important for the successful implementation of a project, including the orientation of all partners towards a common objective upon which all of them must agree. Otherwise the planning and building process could turn into a continuous fight and could possibly end up in chaos. Formal rules have still to be followed and provisions made for the inevitable mistakes. Once established, successful communication between partners usually leads to follow-up projects (client-architect, architect-consultant, architect-company cooperation...), but this is not regarded to be of direct inherent value (within a bidding process for instance).

- **Administration of architecture**

Facility management is a business that helps to control the operational phase of buildings. FM was introduced with the increased sizes of buildings, and can be interpreted as technological progression from the classic caretaker. The administration of space as a resource, supply, the functioning of technical systems and security issues are all included in FM. Modern computational systems are able to integrate large databases; full management of all building operations is, however, applied only to high density building types.

Maintenance can include services from cleaning to renovation. The supply of the building with resources and energy is controlled, as is the internal environment, temperature, light and air humidity. Information about the position of moveable elements of the building itself; elevators, doors, windows, etc., is also represented in a digital model as needing control. Access to buildings is restricted to authorised persons, by card access systems or biometric identification.

Information about user activity is gathered by automatic surveillance, analysed by humans. Software recognising human activity is being developed and will soon be refined enough for automatic surveillance.

When information is used to control a building largely without human interference we speak of "intelligent building systems". The notion of intelligence will be discussed later.

- **Backflow of information**

The "working" of a project is defined by economic success and general acceptance - from users, companies and the public. Evaluation of projects is not standardised in architecture, and depends on the culture of the architecture firm, so there is no standard feedback of information and validation from a built project to the planning architect. The success of a project cannot itself be measured, but it is still defined by economic success, sales, renting, and its overall performance and acceptance. The imitation of architectural innovation is not necessarily a sign of success, but it shows some kind of public interest. The reluctance of architecture as a discipline to evaluate its results may originate in the complexity of the task and in the proximity of the profession to the arts, which also evade general evaluation.

- **Processing problems**

The apparent infinite accuracy of digital data is tempting, but this accuracy is not needed in planning and not executable in building. At the same time, a freedom of "fuzziness" is lost, which was, for instance, given by a very soft pencil within the design process while with digital data processing, measurements tend to become fixed too early within the design process.

Different software languages still pose problems. Transfer of data is still prone to losses, and with growing complexity and specification of the programmes compatibility decreases.

Contract and coverage matters are quite influential, and hamper innovation; overregulated and inflexible systems make the implementation of new ideas difficult. The documentation and monitoring of processes (e.g. discussions, advance of works) are still done in a traditional way and take much time and energy (Protocols to control agreements, for example, could be automated by using video recording and speech recognition).

- **Representation**

Architecture also represents information about the environment. Built structures have inherent information, which they express by their appearance. They are responsible for the "atmosphere" of a space. Landmarks usually signify a specific place, that can be at the centre or entrance of something. Identification with landmark buildings is extensive in some cases, when buildings represent cities (Eiffel Tower) or even continents (Sydney Opera House). Religious buildings carry information about the population's relation to god or their view of the world. All buildings reveal something about the owner, through a number of aspects - style, form, structure, size, material etc. In the extreme, architectural projects represent their owner(s) (and may thus enhance corporate identity).

- **Information by architecture**

Architecture expresses information about space: it points out entrance, closure, openness, orientation, passage, it can show the way and frames space.

Information systems are also integrated into architecture. It is not only the space that informs, but signs and orientation systems as well. Nowadays technological equipment, screens and loudspeakers are not only added retrospectively, but are already integrated into new buildings to provide information.

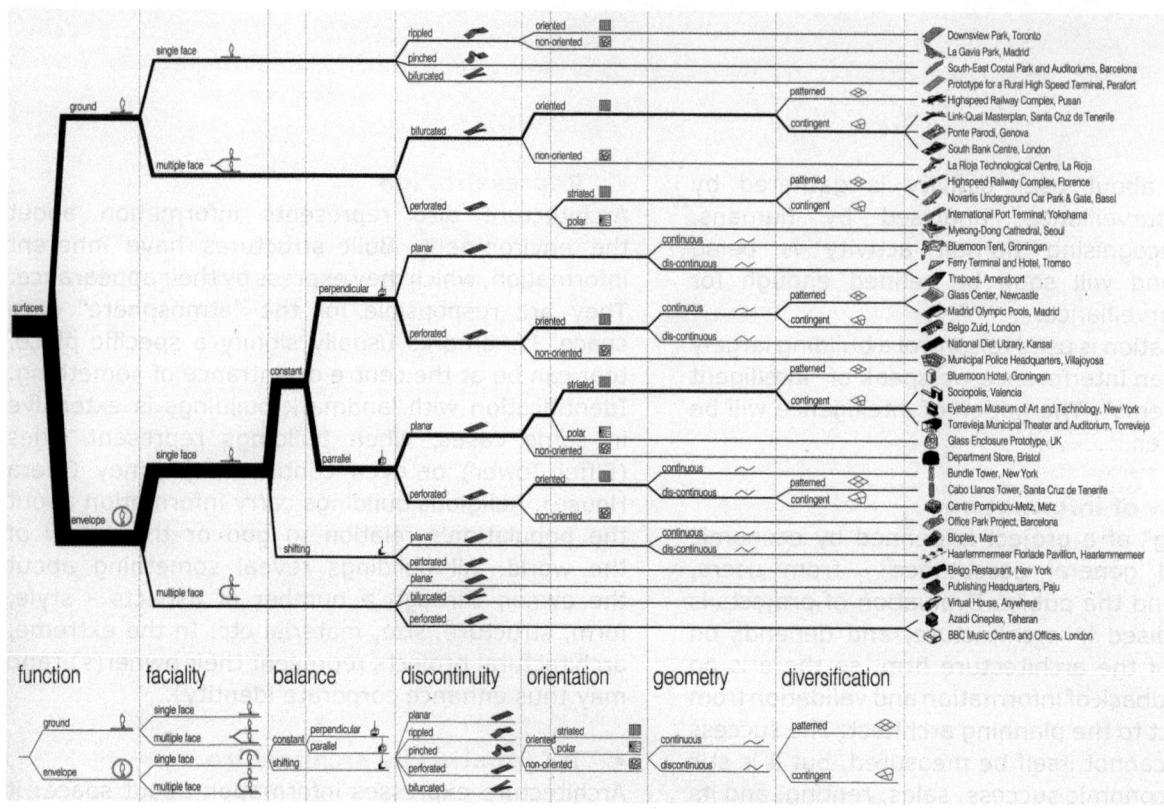

Fig.217 The phylogenetic tree interpreting the projects of Foreign Office Architects as evolutionary development.

Information transfer beyond the project

Hillier defines "architectural competence" as the systematic intent of intellectual choice and decision making aiming at innovation rather than cultural reduplication, in the sense of Margaret Mead's *"transmission of culture by artefacts"*.[316] As was stated earlier, and will be further discussed in the case study about Nias traditional architecture, the difference between "proper" architecture and traditional building development cannot be defined as an approach of different quality. "Western" architecture works with a wide range of possibilities and solutions, of which architects may not be fully aware. The available options (structures, materials, technologies, energy) allow a choice of structures and materials wider than ever before. In traditional building, the resources were limited, and innovation occurred for pragmatic reasons, within the boundaries of the available options and in an empirical way.

In this understanding, innovation increases the available knowledge and information pool, which becomes the base for further design.

---

316 Hillier, B.: Specifically Architectural Theory, 1993, p.10

- **The transfer of information from one project to another**

In traditional building, information is stored and passed on through learning. Craftsmen are living conservationists of building culture. Traditional architecture relies on oral and empirical tradition, and usually has no written documentation. Apprentices are introduced to the "art" of building. Measuring systems are often taken from the human body. Additional variation in design is possible through the combination and multiplication of certain features. Innovation occurs because of new challenges, unforeseen influences. Technological enhancement, better tools and materials bring new opportunities. Changes in society, culture or in the natural environment, even catastrophes, trigger changes in building tradition, made possible through the inventiveness of the builders in a trial and error process. The strong need of humans to present their individuality is another driver for differentiation in design.

In architecture as practised today, information exchange between architectural projects is dynamic and occurs in various ways. The basic information, needed to design a building is provided by the education of architects and builders at universities and academies. Information about buildings is published in magazines, books and the Internet.

Companies and architecture firms have their own building cultures, collections of solutions, products and experience, which they use as a knowledge base for their work. Information collected within organisations forms a "company heritage", an important asset which is handled with respective care. The further development of projects in an office comes close to phylogenesis.

Certain building functions requiring high security standards have special reasons not to give away information about the building: particular public buildings, banks etc.

There is usually no direct flow of information in the sense of copying plans. Architects who intend to design specific and high quality solutions do not usually appreciate the transfer of their plans onto another project. Nonetheless projects have been copied, especially when they share the same function and are built by the same organisation. In housing, the use of very similar or even identical floor plans and layouts is not uncommon, and makes sense if they are good (which unfortunately is only apparent much later, when adaptability to a changing environment is required). Copyright issues of architects' plans together with the uniqueness of the projects create an atmosphere of confidentiality.

It is difficult to access plans of a project if one does not have a personal connection with one of the companies involved. Considering that architects are paid for ideas, and that these plans represent an asset, such secretiveness is understandable to a certain degree.

- **Information flow between architectural categories**

Flow of information is shown in the rough categorisation of architecture described in chapter 2.1. Traditional architecture contains a lot of useful information about local conditions and can therefore serve as a pool of knowledge well adapted to an individual site. It is for this reason that many famous modern architects have investigated local building traditions, and have taken local conditions into account. The implementation of new ideas usually happens in one-off and exemplary design projects. As said before, information - the particular technology or product - then slowly spreads down to the mass market. Innovation coming from a mass market usually reduces quality, finding new and cheaper solutions (mostly materials) for the same tasks. Nonetheless standards (e.g. price per m^2) are defined by quantity, and cheaper ways of building may also be regarded as more efficient and a step forward in development.

- **Integration of information**

"Already, buildings are under pressure to provide the sophistication we have become accustomed to in the design of vehicles and media. The potency of cinema as a worldwide cultural force, even, or perhaps especially, when it is based on the suspension on reality by special effects, challenges architecture to find a voice that is as powerful and eloquent."[317]

Input of other disciplines and new technologies, other sources than the development in architecture itself, has always influenced architecture. New products are integrated into built environment, and materials and structures initially developed for outer space, aviation or the car industry etc., influence architectural design. The interdisciplinary exchange of information depends on the commitment of individuals or companies interested in innovation and able to translate information and adapt results coming from other disciplines. This process of evolution in architecture will be discussed later.

317 Novak, M.: Transarchitecture, http://www.heise.de/tp/r4/artikel/6/6069/2.html [12/2007]

Fig.218 The front representation space of the King's house in Onohondroe, Nias, Indonesia, in August 2005. The decline is clearly visible; valuable parts and carvings have already been sold. Meanwhile, the back part of the house has collapsed.

4.2.4 Limitation - space and time, death as renewal

As in living organisms (described in chapter 4.1 on the general characteristics of life), limitations in space and time define the existence of architecture. The spatial extension of built environment refers to a particular scale. Time is the fourth dimension of the built environment. Although considerations about urban development and life cycle are not uncommon, aspects of time have not yet influenced architecture in the way they could. Time related visions, and thus processes in the course of the generation and existence of architecture, provide essential information on the future of cities and buildings. When they are taken as a planning base they can enhance architectural quality (management of large building sites - how to preserve quality of life for the environment, taking into account a changing environment when planning, technological development for future projects, etc.)

Phases of a building's existence, or "life cycle", are: design and planning, production of materials, distribution and transport, building and assembly, operation, change and reorganisation, death and recycling.

Rituals like laying the foundation stone or a celebration when the primary structure is finished emphasize the start and progress of building and also have a social function. Opening ceremonies are also common. In contrast to this, the end of a building is not regarded as significant.

The introduction of time into design processes connects architecture to evolution, and implies the interpretation of the existing building as a stage in the history of a possible development. Of course, all suppositions about the future are based on knowledge of the present - therefore it is impossible to make predictions.

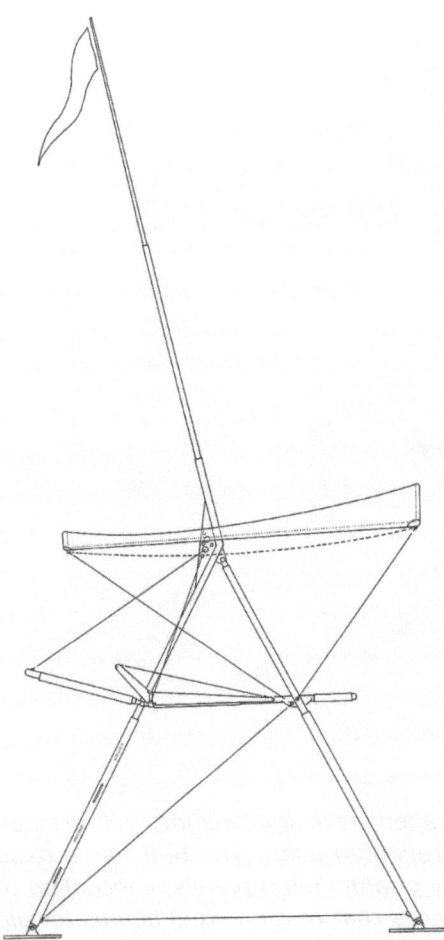

Fig.219 Elevation.

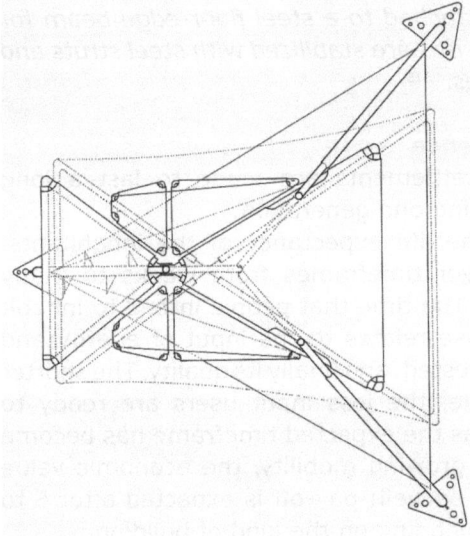

Fig.220 Floor plan.

Fig.221 Point Lookout on Frazer Island, Australia, 1993.

Timeframe of existence

Apart from other contextual factors the life of buildings is limited by the durability of component elements, which again may have different life spans. In the selection of building materials, durability is just one parameter besides availability and ease of processing. Solid buildings, consisting of massive material, are naturally most durable (take the temple remains on Malta, which are estimated to be 5000 years old, or recent examples like the concrete bunkers of the "Atlantic wall"). Brick buildings can last centuries. Solid buildings of adobe - dried mud, mostly shaped as bricks - commonly used in hot and dry climates, have a lifespan of 30 and in some cases of more than 100 years, depending on the composition and the climate. Lighter frame structures are generally less durable, not because of the material of the load bearing parts but the poorer durability of covering materials. Wood can survive centuries if maintained properly. Natural covering materials like grass, palm leaves and straw have to be replaced regularly. Metal structures containing iron are very durable in theory, but are susceptible to rusting. In general, humidity and infestation by vermin are the main destructive factors for frame structures.

Depending on the rhythm of life, built structures have different degrees of permanence.

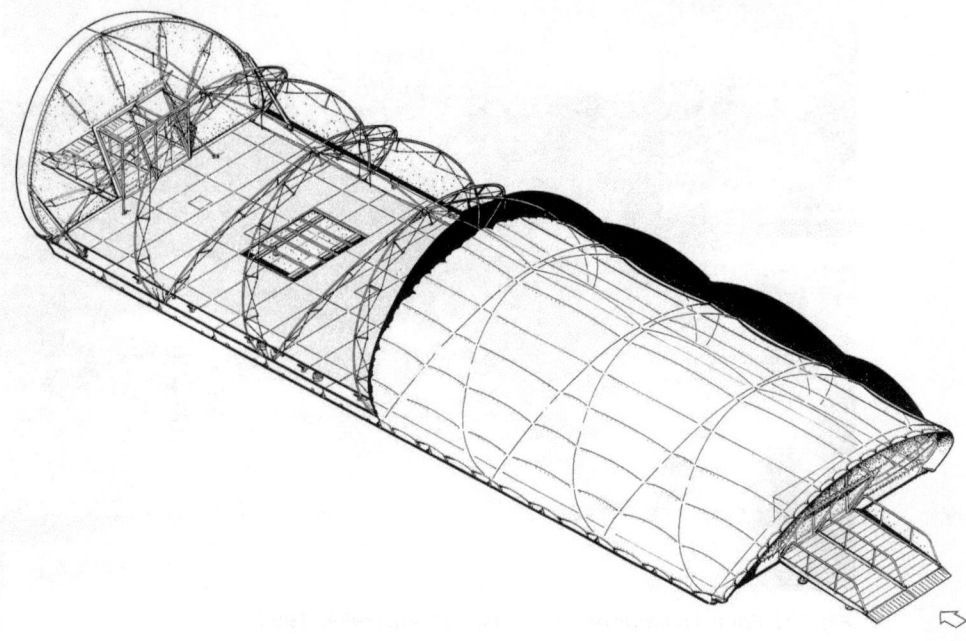

Fig.222 MOMI tent, Future Systems 1991, isometric view, assembly of the construction.

- **Temporary architecture**

A nomadic way of life requires shelters that can be easily transported, so tent-like structures prevail. All sorts of temporary structures are also found in modern architecture, especially for sports and recreation. The projects of Richard Horden and Future System illustrate the basic principles which are important for temporary structures: material, lightness of all components, easy and fast assembly and disassembly. Horden states:

"We interpret architecture through the phrase 'touch the earth lightly'. This has many meanings, from physical lightness to the effect of light on form, aesthetics of light and colour, lightness of touch, lightness in relation to energy and the environment and most of all 'social lightness'."[318]

Jan Kaplicky and Amanda Levete designed a mobile tent for the British Film Institute, which opened in 1991. It was located in front of the National Film Theatre with a view of the Thames. Six workers can put the MOMI tent up or take it down within two days. Ove Arup & Partners did the structural engineering. It is typical of Future Systems projects that the process of building is an important part of the design.

"It was envisaged as a lightweight, yet dramatic structure for repeated use... The tent has a raised floor assembly of aluminium panels, concealing the electrical services that rest on steel beams levelled by jacks. Its white Tenera fabric membrane is stretched between pairs of inclined arches formed with 32mm GRP rods, with a braced inclined arch at either end attached to a steel floor-edge beam for stability. The ribs are stabilized with steel struts and tension cables."[319]

- **Permanence**

Permanent settlements are made to last a long time, exceeding one generation.

Related to the life expectancy of the inhabitants, buildings cover timeframes from weeks to many generations. The time that people intend to inhabit or use a house relates to the input of energy and resources invested, and finally its quality. The shorter the timeframe, the less input users are ready to supply. Just as the expected timeframe has become shorter with growing mobility, the economic value has changed as well: pay-off is expected after 5 to 15 years, depending on the kind of building.

Change and refurbishment is a normal procedure in European architecture, as the large part of our built environment consists of massive buildings. The integration of change procedures into the "normal" life of buildings and inhabitants depends on the scale of measures and the rhythm of renewal.

---

318 Horden, R.: Light Tech, towards a Light Architecture, 1995, preface

319 http://www.future-systems.com/design/design_13.html# [10/2007]

Fig.223 Steel reinforcement prepared for recycling, Nias island 2005.

Fig.224 Space for One Summer, student's project in Rotis, Germany, summer workshop TU Berlin, Prof. Fink 2000.

Normally the usability of a building is restricted during refurbishment phases, but regular and short-timed maintenance measures do not affect use. Keeping a building in good repair implies resource and energy input by regular maintenance measures. There is a rhythm of renewal in different building parts and materials.

- **"Death" of buildings**

The "death" of a building arrives when it is not used or has become unusable. The reasons may vary from structural failure to lost suitability of any kind: the purpose of the building may be lost, the architecture may no longer be appropriate to the needs of the users, migration or other internal or external issues can lead to the abandonment of a building. This is a process which may take a long time, but may sometimes be stopped by complete refurbishment. Death of buildings is not necessarily a sign of decline, as in many cases renewal requires the removal of old structures.

- **Recycling**

The significance of any object is contextual. Dismantling a building deprives its components of their significance, even if they would still be usable. Huge amounts of building components are wasted all the time, because recycling does not (yet) pay for building companies. In poorer regions recycling is a matter of necessity (e.g. recycling of steel reinforcements on islands hit by Tsunami, or of the copper from houses robbed of copper wiring in devastated living areas).
The reuse of hewn stones for new buildings has often occurred in history.

Fig.225 Assembly principle of recycled PET bottles.

Architects are experimenting with the reuse of material, especially plastic. The design project "Space for One Summer" at the TU Berlin in 2000 focused on PET bottles that are used to form a cupola, the caps serving as connection with a foil. The assembly of about 3000 bottles created the form automatically. The simplicity of the idea is striking, as is the internal atmosphere with the light falling through the transparent material.[320]

320 Schittich, C. (Ed.): im Detail: Gebäudehüllen, 2001, p.62

- **Rhythms**

Architecture is subject to the environmental cycles: short term differences as well as longterm climate changes. Cultural cycles often relate to natural ones; rhythms of change are common. Architecture takes into account changes of natural and cultural environment. The building industry is particularly dependent on environmental conditions. Rhythms affect the opening and closure of buildings and energy consumption.

The durability of architecture also entails a certain rhythm of change and renewal. The difference between the Western and Eastern approaches to renewal is interesting: Japanese temples, for example, are regularly completely renewed, but are not regarded as new buildings.

- **Urban tissue**

Urban tissue exists as structure, as a pattern organising and ordering its elements, the individual houses. An urban structure can survive for centuries, if the needs of the inhabitants do not exceed its capabilities.

As soon as the individual buildings, representing "old" urban structure, do not "fit in" any longer - due to their scale or other conditions (need for infrastructure or different context) - they are prone to decline, and new urban structure is formed, for example old European town centres decay and difficult revival measures are experimented with. In such cases heritage protection plays an important role, but protection can, on the other hand, make renewal impossible.

Decreasing and shrinking urban areas pose a serious environmental problem. Once built environment has occupied an area, the changes in the natural environment are irreversible. Take abandoned coal mining areas: considerable effort as well as time is needed to re-naturalise such places.

The "normal" fluctuation of large numbers of people in urban areas presents big challenges for urban planning, the extreme case being the "death" of large urban areas left uninhabited.

Political decisions (e.g. the reunification of the two German states), bad economic situations, catastrophic events and wars (like those in many African countries, Afghanistan, Iraq, etc.) result in masses of people leaving their homes. The regional shrinking of the built environment due to mass migration is a new phenomenon in our time of continuous growth. The devastation of natural and cultural landscapes as a consequence of wars causes enormous damage.

Limitations in size

The comparison of sizes in nature and technology shows up the limitations.

There are inevitably design limitations for the sizes of buildings and urban structures. These limitations in architecture are due to human activity, the availability of light, air, and sun, requirements of access and supply and technical possibilities.

Human activity defines the size of the spaces for which they are designed. Usually the minimum amount of space for specific activities is provided, in order to make efficient use of energy and resources.

Light and vision are essential for the well-being of occupants or users. Requirements for light limit the density of cities and the depth (together with floor heights) of buildings. The connection of occupants with the world outside is an essential design parameter, which prevents built volumes from becoming too large. With new technical possibilities (e.g. light conducting fibres), the quality of artificial light gets closer to sunlight, but artificial vision of the outside has not yet been introduced.

In design for space, vision is an important factor for humans, and approaches to meet this urge (and the need to contact "home") through technological means are being explored, the sector being called "augmented reality". The artificial environments generated for this purpose pose new questions, which will be answered within the next few decades.

Real critical sizes for buildings and settlements are defined by structure, comfort, security and the need for infrastructure. Building regulations prescribe limits to sizes of and in buildings (e.g. without an elevator a building is limited in height to four floors), and restrictions are laid down in urban planning regulations as well. Vast urban areas would theoretically be restricted in size by their negative impact on the environment, and by pollution's negative impact on the inhabitants, but this has not prevented settlements from growing.

The smallest and the largest buildings are challenges that have been interesting for many centuries. Currently, there is another contest to build the highest skyscraper, with the Burj Dubai being expected to win with a final height of over 800m. The minimal shelter is the other end of the scale. The holiday house of Le Corbusier, "Le Cabanon" represents his idea of the minimal house for a single person, measuring 3.66m by 3.66m by 2.26m (according to the Modulor system that he developed), and is still an icon for minimised living.

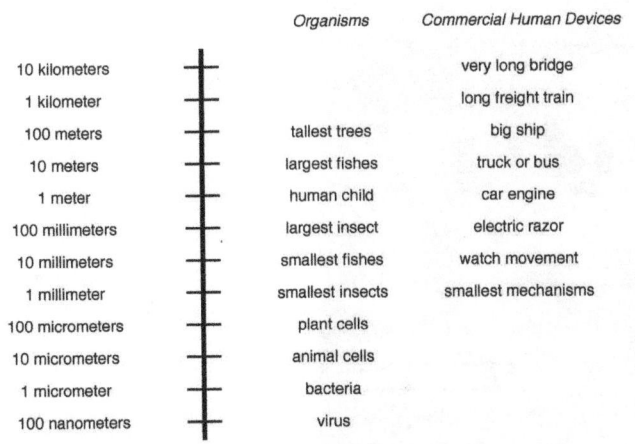

Fig.226 The size ranges of organisms and our mechanical devices, according to Vogel 1997.

The task to design high quality minimal living spaces is interesting for high-density areas and for very hostile environments, where recourse to the minimal is essential for survival. In a design for the environment of outer space, the minimal survival shelter is reduced to the space suit.

The stronger the pressure of population growth, the more futuristic the visions to cope with high-density are becoming. The "metabolism" movement that originated in Japan was based on the problem of living on an island with limited space, and has consequently developed visions of densely packed units, assembled on gigantic structures in Tokyo bay. Recently, the Netherlands, with their long tradition of high-density settlement have even thought about arranging parks and agricultural areas in multi-storey buildings, as demonstrated with the EXPO 2000 project in Hannover.

Fig.227 Burj Dubai under construction in October 2007, already rising 585.7m, planned to rise over 800m, designed by Adrian Smith and Skidmore, Owings and Merrill SOM 2006.

Fig.228 Le Cabanon, minimal house, Le Corbusier 1951.

New approaches | Architectural interpretation of life criteria

Fig.229 Taos pueblo settlement, New Mexico, 2004.

4.2.5 Order

"...The processes of life have led to the wonderful, three-dimensional patterns seen in organisms, hives, cities, and planetary life as a whole."[321]

Order is the most obvious criterion of life. Even if patterning due to self-organisation processes does occur in inanimate nature, patterns are considered characteristic for animate nature, and the transfer of patterns - form - is the most obvious possibility for the use of role models from nature. But, as described in chapter 4.1 on the general characteristics and the classical criteria of life, order in biology is more complex than that, and some of the phenomena involved are investigated for architecture in the following.

Integrity

Integrity in terms of wholeness is a basic characteristic of living organisms and differentiation from the environment is a precondition for life.
- **Integrity of buildings**

Buildings as architectonic structures usually appear as coherent units, distinguishable from one another. Even if single buildings belong to a higher-level organisation, e.g. housing development, town, city, they still are more or less independent units. Matters of property, use, maintenance and supply seem to make demarcation necessary. In some traditional settlements, for example the pueblo dwellings of the American Southwest and old Mediterranean villages and towns, we cannot discern individual buildings. These houses seem to have grown over time beside or above one another in an organic way.

Fig.230 Single family house in the "Fontana" settlement in Lower Austria, 2007.

The settlement itself is the unit, differentiated from the environment. The same phenomenon exists in informal settlement structures. This structural blurring may be caused by lack of material, scarcity of space, step-by-step addition and the envisioned transitional character of the dwellings.

The desired single-family house is an individual building, touching neighbouring houses as little as possible. The wish for an independent unit is very strong in many people, so they prefer even tiny single family units to any other option. Planned settlement structures of single-family houses appear more like a patterned landscape than an assembly of houses. Large developments have dimensions that are no longer comprehensible (e.g. Los Angeles).

321 Margulis, L. et al.: What Is Life?, 2000, p.4

Fig.231 Residential building in Shanghai, China, 2006.

- **Urban integrity**

Urban structures are distinct from their environment. Their appearance differs from other environments such as natural or agricultural environments. In hostile or in warlike environments the existence of the border has to be obvious, and must meet the demand for fortification.

The border of today's city is not linear - with increasing city size it spreads out at about the same ratio; spilling into the surrounding area.[322] Approaching a city, one can observe an increasing density of built structures. In the course of urban development, settlements have grown and merged into larger units. Recent urban structures cover vast areas, and in densely populated regions the borders of the original towns are completely blurred.

Urban structures of this size (the settlement area of Los Angeles has a diameter of 60km) are separated into regions, districts, or smaller units of functional or administrative connection. Often these borders are represented by the historical boundaries of older settlement units. Most borders consist of topographic features or traffic routes, but can also be arbitrary.

- **Identity - individuality**

The question of identity is relevant to the present situation of urban structures. Globalisation has led to an increasing uniformity in architectural style throughout the world, especially in cities. Even so, uniformity seldom reaches "sameness". Buildings still differ so much that it would be really difficult to find exactly the same appearance (e.g. English terraced houses, Chinese residential buildings made of precast concrete slabs). Differentiation does not necessarily lie in the building itself, but in its context and in its use. Even if identical buildings exist, their occupants render them individual.

Still, identification with a specific form of the built environment is no longer usual. The importance of regional characteristics which have evolved over centuries is decreasing, due also to the lesser importance of a particular region as a major influence. On the other hand, architectural identity is perceived more and more as an important part of cultural life. This finds an expression in numerous projects of so-called "proper" architecture, delivering landmarks in design which are geographically relatable because of their reputation. These are usually planned by famous architects and can provide an identity for their surroundings. People generally feel a need for individuality being expressed by architecture as well. This is usually still the case in traditional architecture and in single-family housing. With denser settlement forms identification is more difficult to achieve. Individual expression is concentrated on the interior, on furniture and on personalised outside elements, like balconies, entrance doors, windows.

---

322 Humpert, K.; Becker, S.; Brenner, K.: Formbildungsursachen regelmässiger Siedlungen in traditionellen Kulturen, table 60, in: Teichmann, K. et al.: Prozess und Form "Natürlicher Konstruktionen", 1996

Fig.232 The so-called "Hundertwasser-Haus", idea and concept by the artist Friedensreich Hundertwasser, planning by architects J. Krawina and P. Pelikan, Vienna 1985.

The need for individuality (and romanticism) was met in a famous housing project in Vienna, the so-called Hundertwasser-Haus, designed by artist Friedensreich Hundertwasser, opened in 1986. In spite of dissent among the architecture community, the house has become immensely popular and must not be missed on a sightseeing tour, but its reputation as an architectural project is merely due to the decoration on its surfaces.

- **Independence and autonomy**

By nature, urban structures have never been independent, but have needed supplying neighbourhoods to exist. Autonomous cities existed and still exist in the sense of self-governance, but economically, today's urban structures depend on international networks of supply of energy and resources, and thus are definitely not independent. Constant exchange is a precondition for the existence of urban life.

Visions of future settlements integrate the vision of independence (e.g. Soleri's arcologies, which were mentioned earlier, Cook's pinball machine, Hollein's aircraft carrier). Future settlements in space will have to be independent of the supply networks usually available on earth. Scenarios for settlements on Moon or Mars rely on harvesting resources (also for building purposes) and the establishment of production sustainable at least for the time of the mission.

As we know from former times and remote parts of the world, settlements and buildings can exist quite independently of global conditions, apart from the given site and its environmental conditions, and apart from a certain input of resources and energy for the inhabitants and maintenance purposes.

The integration of buildings in many kinds of supplying networks increases dependency nowadays. The breakdown of the networks would render today's buildings uninhabitable. Countermeasures are prescribed in many building regulations (e.g. emergency chimneys in case of an energy crisis, emergency staircases to replace lifts in power failures).

Energy harvesting technologies enable the autonomous operation of buildings. Integration of these efforts into the large supply networks is still difficult, as network operators do not yet allow autonomy (for as infrastructure networks rely on a high number of participants, house owners are forced to join the network).

- **Self-awareness**

Self-awareness goes beyond integrity and identity, meaning *"the conscious knowledge of one's own character, dealings, motives and desires"*[323]. If self-awareness relates to consciousness (which we still think is peculiarly human), the term is inapplicable to architecture.

Aspects of self-awareness are introduced within "smart" houses, as buildings contain information about the state that they inhabit. The knowledge about one's self requires the existence of a model as a background for the stored information and as a base for decisions. This result of research in psychology and cognition is transferred to computational models for the control of building operating systems.

323 New Oxford American Dictionary, program version 1.0.1, 2005

Fig.233 Interior view of Capsule element, Kurokawa, Expo Osaka 1970.

High order, low entropy

Of the many interpretations of order, *"arrangement, disposition"* and *"good arrangement; opposite to chaos"*[324] seem appropriate for this very important phenomenon.
"To create architecture is to put in order. Put what in order? Function and objects."[325] Le Corbusier

The term "architecture" itself is used in various disciplines, such as information technology, for an ordering system of any kind, but often a functional order is implied. In classic architecture of the built environment, order is about ordering elements in space, and thus ordering space itself. Abstract, formal or concrete, pragmatic ordering systems are present on all scales of design. Order can be derived from many sources: architectonic expression (e.g. form), the planned activity (e.g. size of circulation spaces), the practicable size of elements (e.g. construction, girders and floors: span versus effort and thus cost), the maximum processing size of elements (e.g. limited size of sheets of glass), the functional requirements (e.g. complex functional connections), infrastructure, aesthetic principles and many more possibilities. Ordering systems based on mathematical principles seem to be more objective than intuitive or artistic ordering.
Human activities require the suitable sizes of space in addition to, within or for the built structures. Planned order (concept), built order and functional order do not necessarily correspond, but order of concept should correspond to the order of functions.

Fig.234 Capsule Tower in the process of building.

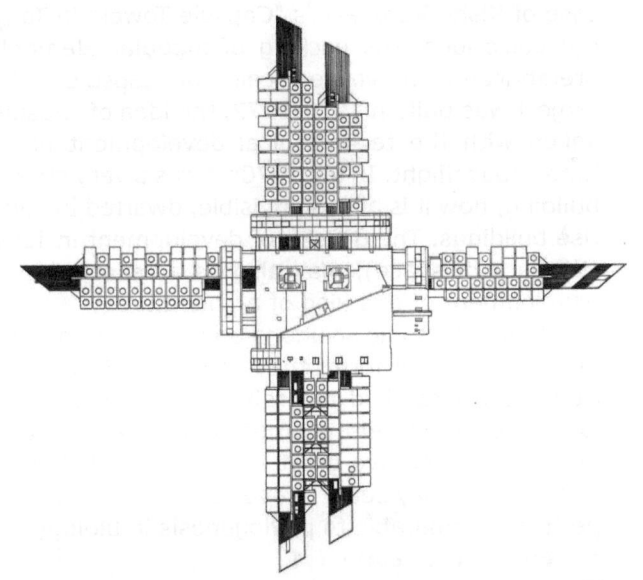

Fig.235 Elevations, Capsule Tower.

324 New Oxford American Dictionary, program version 1.0.1, 2005
325 Le Corbusier quote found on http://www.designfeast.com/thoughts/default.htm [12/2007]

New approaches | Architectural interpretation of life criteria

Scales of order

As in nature, where hierarchies of order exist over all scales of living matter (described in chapter 4.1. on the general characteristics of life), the built environment is also ordered at different scales.
Global scale is reached by the introduction of satellite navigation systems which are available for public use.

- **Regional planning and Urban planning 1:XX - 1:1000**

Order is determined according to the function of the space; space for specific activities such as transportation, parking, housing, and working is established.
On the city scale, order is not just provided on the 2D landscape, but also in terms of height (height limitations and zoning of height).

- **The building 1:500 - 1:20**

The whole of a building is subjected to some kind of ordering system. Underlying the completed project are the decisions made earlier to reach its goal, although they may no longer be visible. In the case of Kisho Kurokawa's "Capsule Tower" in Tokyo the basic idea was packing of modular elements, prefabricated minimised cells or capsules. The project was built in 1970-1972, the idea of capsules linked with the technological development of the time - spaceflight. In the 1970s it was a very striking building, now it is almost invisible, dwarfed by high-rise buildings. The density of development in Tokyo (the lack of space), may also be a reason for the development of this kind of architecture.
When architectonic solutions appear frequently but in slightly varied form, they are called typologies. Styles are less rigid but rely on a specific formal language or on sets of important aspects. Styles and typologies have great impact on architectonic expression. They seem to evolve over time, and are perhaps comparable to phylogenesis in biology.

- **Element 1:200 - 1:1**

The ordering system which defines the size of architectonic elements mainly relates to functional, process and material issues (e.g. cross section of a column, size and geometry of a window profile). Architectonic innovation is required to break the closed loop of use of specific elements and products available on the market: production of these elements in specific sizes and qualities; availability of elements in specific sizes and qualities; effect on design because of the availability of these elements. Design with elements not as yet produced or used in this way push production to its limit. For economic reasons this is easier with large-scale products.

The qualities of surfaces relate to architectonic expression (e.g. surface structure, colour, brightness, reflectivity...). This will change in future buildings with the integration of functional surfaces into architecture.

- **Detail 1:20 - 1:1**

The ordering system for the detailing of buildings is defined by the connection of the elements. Functional criteria have to be fulfilled (waterproofness, thermal insulation, structural requirements...). Processing matters are important for design solutions - the building sequence has to be taken into account.

- **Material and surface 1:1 - nano**

Order on the scale of material is established by material processing. The responsibility of architecture is in the choice of both material and surface finish, and in the stimulation of development through innovative application and materials research.
The ordering principles of all layers have to be taken into account and processed to form a coherent project. This is usually done by a step-by-step approach.
Not all wishes can be fulfilled (see the geometry of the Wittgenstein house for example - the abstract concept of coherence between the inside and outside ordering systems could not be reconciled due to material thicknesses[326]). The more abstract the main ordering principle of a project, the more difficult it usually is to implement the other inevitable orders in the same project.
Mistakes in order are very often the result of wrong interrelations of ordering parameters, or missing parameters. Ordering systems derived from specific activities/purposes are often imposed onto elements that should follow other principles (a classic example is the construction system of parking levels in basements being responsible for floor plans of flats in housing projects). Mistakes in the hierarchy of order happen when the relationship of importance between the design elements is unclear.
The information transfer in metric units does not communicate ideas of order directly, and this increases misinterpretations.

326 Turnovsky, J: Die Poetik eines Mauervorsprungs, Bauwelt Fundamente Bd. 77, 1987

Fig.236 Roof structure covering the great court of the British Museum in London, Foster and Partners with Buro Happold, 2000.

Qualities of order

- **What is ordered?**

Most obvious: material and elements in space are ordered by their assembly. At the same time, space is ordered as well, territories are defined. The use of space is ordered; function is one of the main parameters for design. The order of processes is defined in time: time related phenomena, e.g. activities like movement follow patterns of order. The process of planning and building is ordered.

Social order is strongly related to spatial order in architecture, and should also be taken into account.

- **Bottom up or top down**

Order can be created automatically through a process of arrangement of elements as in unplanned settlements (inherent order or emergent order), or imposed from the beginning as in urban masterplans. Planning is a top-down approach, whereas growth processes have a bottom-up approach to order. On an urban scale bottom-up approaches have led to the unplanned settlement structures we encounter worldwide. Characteristic patterns have emerged, which are the object of topological investigation. The control of growth, the establishment of a top-down order is difficult to achieve.

- **Systems of order**

Architecture is based on mathematical ordering systems. The definition of proportions relating the sizes of elements aims at the creation of aesthetic appearance. Formulas for the generation of harmonic proportions include the "golden section", the classic orders of building elements and Le Corbusier's "Modulor" system.

Other mathematical systems are based on important numbers or partitions. These mathematical or geometric ordering systems can have a connection to functional orders, but may also have merely symbolic meaning. Cultural and social ordering systems interact with the built environment. Ordering systems can be due to topographical conditions or orientation. Special parameters, like thermal comfort, can be used as ordering value. (One example of this is the zoning of floor plans according to thermal comfort, which relates to the function of the spaces, and saves energy.)

New ordering methods have emerged with computational means. Computer processed geometries are available as ordering systems. As an extreme example, the geometry of the roof of the Great court of the British Museum in London follows mathematical formulae.

"The project was designed by Foster and Partners, architects, and Buro Happold, engineers, and was fabricated and erected by Waagner Biro. The roof is constructed of a triangular grid of steel members welded to node pieces... The grid is triangulated for structural stiffness and so that it can be glazed with one flat panel of double glazing for each triangle of the structural grid."[327]

The structural engineer Chris Williams calculated the shape in close collaboration with the architects and Buro Happold. The form was found in a mixed scheme of sculptural, geometric and physical approaches, the latter referring to the physical processes of minimal surface, hanging chain and shell mechanisms.

327 Williams, C.J.K.: The analytic and numerical definition of the geometry of the British Museum Great Court roof, http://staff.bath.ac.uk/abscjkw/BritishMuseum/ ChrisDeakin2001.pdf [11/2007]

Fig.237 Great Court, the Reading room in the Centre.

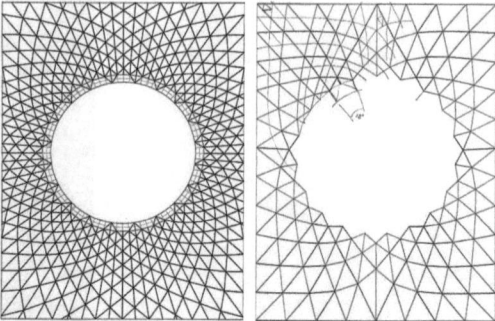

Fig.238 Evolution of the structural grid.

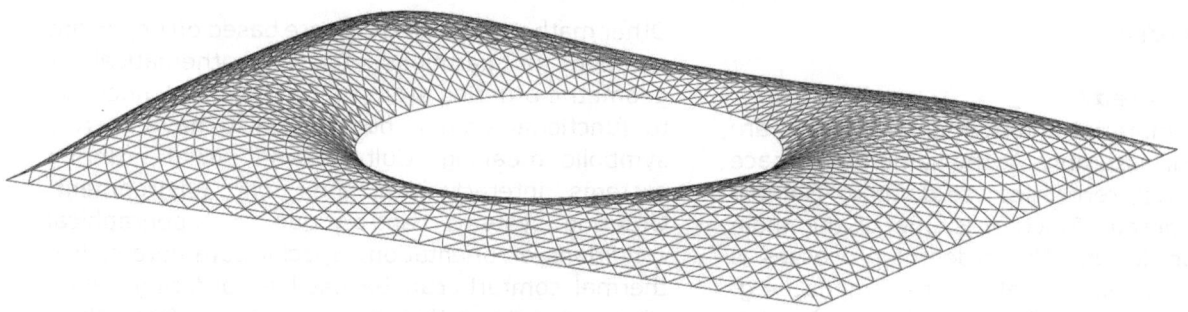

Fig.239 Glass roof geometry, isometric view of steel members.

Taking into account architectural, planning, structural and clearance requirements, the shape was defined analytically by weighting and summing functions.[328] *"The roof dropped 150mm and spread 90mm as it settled. In other words, the structure was built in one shape and then 'relaxed' into position."*[329]

The introduction of processes in planning and building is also done with computational means. "Morphogenetic" design strategies make use of genetic algorithms for design. With the introduction of time, order has lost its static concept, and is interpreted as a dynamic process.

One result is that the life cycle of architecture has increased in importance. The observed and planned time scale is extended beyond the individual building, including project development and large scale planning, and ideas about eventual demolition and recycling are also integrated.

- **Permanence**

The permanence of order in built environment is striking. Ordered systems like the urban structure of many medieval cities in Europe have survived for centuries. This is not entirely positive, as existing structures can also block urban development. The permanence of the built environment as a whole is still being underestimated.

Order on a landscape scale is provided through ownership of land. If the social and political situation remains constant, these systems and patterns can survive for centuries.

In biological systems, death is the precondition for further development. In architecture, the destruction of the existing is the precondition for the new.

328 Williams, C.J.K.: The definition of curved geometry for widespan enclosures http://staff.bath.ac.uk/abscjkw/OrganicForms/WideSpan.pdf [11/2007]

329 according to Johann Sischka from the implementing company Waagner Biro, in AD Architectural Design Vol 74 No 3, Emergence: Morphogenetic Design Strategies, 2004, p.76

Hierarchical levels

"Hierarchy is a system of ranking and organising things or people, where each element is subordinate to a single other element."[330]

Hierarchical levels in architecture do not refer exclusively to scale. As mentioned in "high order - low entropy", the rough levels of scale are: regional and urban structure, the building, the element, detail and material. In-between levels exist depending on the project's size.

Other notions of order have been presented above, and all these have to be integrated in the design of architecture for a coherent result. Aesthetics can be affected as well: the old classical orders demanded a hierarchy of architectural elements in order to achieve a specifically structured appearance, e.g. façade or internal space.

The easiest way to order things is to put them in line: hierarchical levels exist in biology as well as in architecture. The strictness of the notion "hierarchy" is matter of discussion in both disciplines: the discussion of complexity within the life sciences has also entered technology.

The assembly of elements creates patterns, and new characteristics emerge. As the architectural scale grows, new elements are introduced: emergent functional needs require circulation space, infrastructure and public buildings.

Aspects

As said before, the built environment exists on many scales and levels. In architecture, levels of order can concern matters of scale, space, construction, structure, material, function, speed, publicity, society, culture, aesthetics etc. The levels appear in a certain order, even though it is not always established on purpose, clearly hierarchical or introduced respectively in a historical setting. For planning, the categories, and thus the parameters for decision, have to be evaluated in order to find a balance. The relationship between the categories is specific for the project, and can also be specific for the company or the style. At best, all levels and categories relate reasonably to one another and form a project of high architectural quality. The influence of building regulations, conditions of implementation and unforeseen influences alter the values during the planning and building processes.

Within the categories on the next level, hierarchical order defines a gradation of elements (e.g. in structural matters: primary and secondary elements).

On the building scale, hierarchy of construction and of function is essential. The usability of the house depends on the quality in which human activities can take place. Sufficient space is most important but is limited by the high density of urban developments and the quest for profit. Hierarchy of public and private is a very important parameter for the configuration of space at all scales.

- **Main aspects**

Architecture can be subjected to a "big aims" function as a landmark, forming corporate identity or representation of any abstract idea. In that case, all other categories are subordinate.

This may be a challenging approach, but the final project is often unsuitable or random, e.g. in zoomorphic design, where the form of an animal is transferred to a building.

Diverse approaches to "Gestaltung" in architecture focus on different main aspects. Theories of design imply different main aspects being considered most important (Post modernism, Constructivism, Metabolism, Deconstructivism ...). The famous quote "form follows function"[331] expresses the relationship between two important aspects, and originates from the observation of nature.

- **Space and society**

In all scales of architecture, there is a hierarchy of accessibility, visibility and ownership that defines who is entitled where to do what. The interior and exterior of architecture creates a differentiation, which is used for technology (layering of a wall system or façade) and social order (relationship of spaces and functions).

The arrangement of public and private space depends on social and cultural conditions and relates strongly to the economic background. Social and built orders have mutual influence[332] and change over time. Social hierarchies are reflected in spatial order: individuals or groups of the population who are considered more important for some reason usually occupy the more esteemed space. This is usually the case on the family level as well as for societies. Space seems to be interpreted as another resource, and the relationship of spaces traces social relationships and possibilities.

---

330 New Oxford American Dictionary, program version 1.0.1, 2005

331 first published in Sullivan, Louis H.: The tall office building artistically considered. Lippincott's Magazine, March 1896. Sullivan used the phrase in the context of a description of design in nature and as instruction for architectural design.

332 as psychoanalyst Alexander Mitscherlich stated 1965 in "Die Unwirtlichkeit unserer Städte. Anstiftung zum Unfrieden."

The activities of society take place within public buildings. The kind, number and quality of these buildings shows the interest a society takes in the respective activity. Politics is responsible for the judging and regulation. Political power enables these usually large structures and the necessary infrastructure to be introduced to the environment.

Functional hierarchy

- **Functional hierarchy of buildings**

When putting up a functional order of buildings, which consciously represents a succession in time and importance, the basic functions needed for survival - housing and working - would form the base of the hierarchy. Public buildings represent a higher level in this functional hierarchy, in the sense that they are not introduced until a settlement has reached a certain size and density (e.g. school, hospital, opera, church, stadium...). If public buildings contain an infrastructural function, the need can easily be defined (station, power plant, waste water treatment plant...). Cultural spaces are the highest level of public building. They emblemise thriving and prosperity and require many resources for their establishment. Unfortunately the need for cultural space is not as easily measurable.

Zones where ownership is unclear or spatial or social assignment, do not "work" in terms of human activity, neither on the urban nor on the building scale. These areas can become difficult when their use is disputed, criminal activity is enhanced by inadequate surveillance, or they may simply be abandoned. Sometimes these zones (e.g. fallow urban areas, landscapes along borders or the Iron Curtain) have become re-naturalised over time, and accommodate valuable biodiversity.

- **Functional hierarchy inside buildings**

Functions, and thus activities within buildings, can be regarded as of more or less important. The notion of importance is in turn connected with the relationship between public/private and input/output.

Functions within buildings can include: representation, access, living, eating, working/producing, cooking, storing, cleaning, wasting, recycling. The sequence of activities should define order in space.

Multifunctionality does not necessarily contradict functional hierarchy. The usual approach is to have functional separation - so that an element, e.g. the layer of a wall, fulfils a specific task such as load bearing. Successful multifunctionality must combine functions that either have some similarity or complement each other.

Integration comes together with multifunctionality. Elements are multifunctional if they do not serve a sole purpose, but "integrate" other functions as well. Integration is a major issue in designing, and is often supposed to render components invisible: when technical devices, for instance, are "integrated" they are not discernable as separate units any longer.

All different aspects of a project have somehow to be integrated into one final design.

- **Speed as parameter**

Speed of activity is an important parameter for the order of space in built environment. Circulation and locomotion space is differentiated in relation to the speed of activity taking place. Elements relating to locomotion velocities fulfil circulation and transportation functions in buildings, cities and beyond. Motorised locomotion brings about additional functional necessities: even more space for manoeuvres, spaces for supply, storage, repair, maintenance and technology, as well as production and economic requirements. In cities, the speed difference between neighbouring spaces can be challenging and problematic.

Hierarchy of construction, structure and material

Within constructions, hierarchical levels and their interactions define the order in which load and forces are transmitted to elements. Apart from functional hierarchy, construction elements follow simple physical orders, e.g. the hierarchy of transmitting loads. Gravity enforces building from bottom to top. Load induced by dead weight and usage is usually transmitted from top to bottom. Different approaches, "taking loads for a walk" may be interesting for specific reasons, e.g. avoiding obstruction of ground floors or bridging large spans.

In a massive building the number of elements and levels is low. The roof construction is a load on the supporting walls, which are held by foundations. This is called the primary construction. In this case the bearing elements for the ceiling and other supplementary constructive elements constitute the secondary construction. Framed constructions have a higher number of elements and levels, through differentiation of functions and loads. In framed construction the loads are transmitted by linear elements. Secondary construction is needed to bridge the span between these elements. The closure of space is created by non-supporting elements.

The hierarchy of inside/outside is once more important for the layering of elements. Today's façades show a functional layering system, adapted to the environmental difference between inside and outside. The bigger the difference, the more important the layering, and the more effective the element as a whole.

Structure levels establish material hierarchy; material used for building is not structured in as many levels as living tissue. The reason why dead nature is used for building is that these materials are already pre-structured and thus already provide characteristics that can be exploited without further processing. The establishment of additional structure levels to materials would enhance the material's characteristics. This approach is currently only carried out in the design of lightweight materials, e.g. foam structuring of metals and plastics.

Hierarchy of process: building

The process of building requires the ordering of pieces to be put in place in a given timeframe. The sequence of things to be done is strongly connected to the hierarchy in construction. According to the size and complexity of the project, building can involve hundreds of companies and years of works. Whereas in former times it was carried out by architects, the control of the construction process - coordination, implementation and supervision - is now the domain of specialised companies. (This is one of the problems in the profession of architecture today: the separation of planning from implementation, and the widening rift between building architecture and design architecture.)

All activities can be roughly ordered as follows: foundation work, building shell, finishing and adding or installing equipment.

- **Process problems**

The sequence of activities carried out on a building site involves the presence of specific companies. With a new building, on average 50% of the work (measured in building cost) is done by the building firm, which is usually present on the building site without interruption. A range of different professions does the other half. The planning of building sequences sounds like a simple task (requiring information on when can materials be delivered, what is to be done, by how many people, therefore how long does it take and when can others use the space again etc.), but inevitably things go wrong, companies make mistakes, work packages take longer than expected etc.

So apart from plan, the quality of material and work, the art of implementing flexibility and options into a construction schedule determines if the project is on time.

If elements have to be introduced into a building too early or too late in the process, problems inevitably occur. It may be the case that elements are not implemented according to their respective grade of finishing because of their size (e.g. escalators as a finished product are put in place very early - protection measures are necessary), or mistakes in planning or production (e.g. work concerning the primary structure, the shell of the building, is carried out in the phase of finishing).

Challenges

- **Long term goals, higher level structures**

Higher-level elements have to be introduced retrospectively into existing urban structures after a regional order has been established. So the implementation of higher-level elements (highways, high speed tracks, public buildings, but also parks and "natural" elements) has a destructive impact on the existing order. The retrospective building of such a large-scale project provides the required functional service, yet the resistance from local inhabitants and users of the existing structures may be so strong that political power is needed to accomplish such a task.

- **Frame of reference changes**

In contrast to living organisms in a stable ecosystem, the frame of reference in our civilised world can differ and change enormously - think of the influences important for a village on a remote island, in contrast to an urban setting with international connections. Progress and development of refined communication technology change the frame of reference for architecture, sometimes rapidly. This can lead to discontinuous developments, involving for example the fast decline of existing settlements.

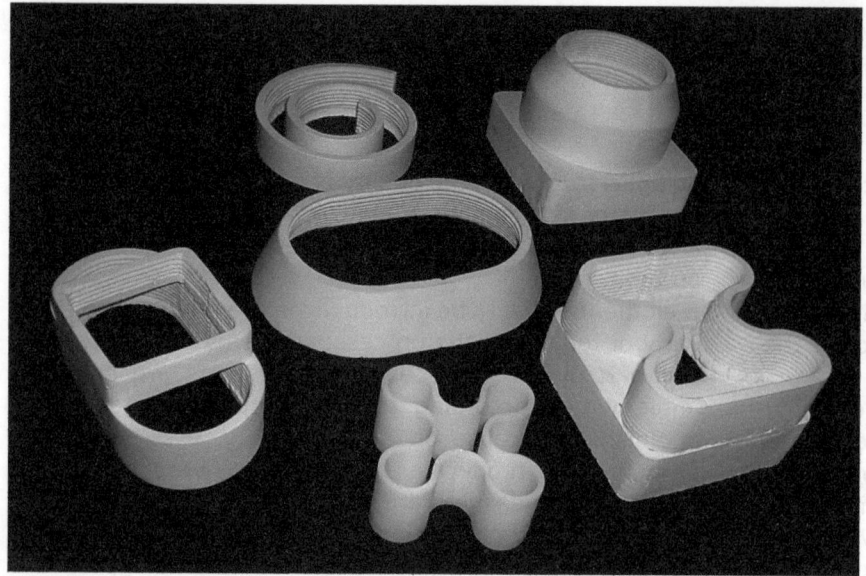

Fig.240 Concrete Crafting technology, 3D shapes, Khoshnevis 2004.

Fig.241 "Printing" process.

4.2.6 Propagation

In biology, time limitation of existence is connected to the phenomenon of propagation, and thus creates a continuity of adapted individual existence over time in a changing environment (see chapter 4.1. on the classical criteria of life).

To propagate means to *"breed specimen of (a plant, animal, etc.) by natural processes from the parent stock"* or to *"spread and promote (an idea, theory, knowledge, etc.) widely"*.[333]

- **Propagation of typologies and features**

Buildings do not propagate, at least not themselves. Propagation in the sense of self-replication does not exist in technology. In spite of scientists' efforts to generate self-replicating machines (e.g. the RepRap project at the University of Bath [334]) autopoiesis is still a phenomenon of living organisms.

If the notion of propagation is applicable to architecture at all, it can be applied to architectural features, or even typologies. This typological propagation is not a material process, but a process of distribution and passing-on of information, independent of the typology itself. It is referred to in architecture and art history as the theory of diffusion[335]. Knowledge of any architectural typology inevitably spreads in space and changes in time, thus referring to the process of evolution.

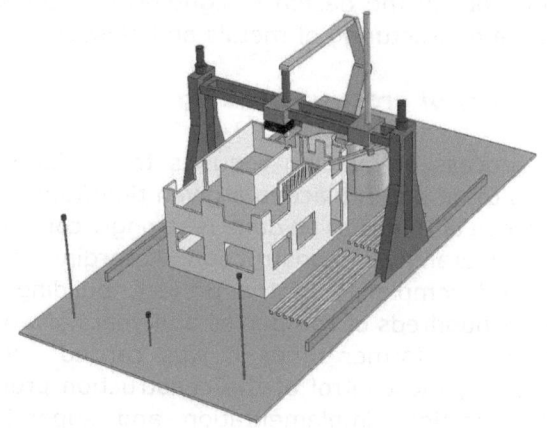

Fig.242 Scheme of construction of a conventional building using Contour Crafting technology.

---

333 New Oxford American Dictionary, program version 1.0.1, 2005
334 RepRap website http://reprap.org/ [11/2007]
335 Lehner, E.: Wege der architektonischen Evolution, 1998 pp.505

Fig.243 Printed dome.

- **Automated production as a first step towards propagation**

For decades, buildings have been at least partly erected by machines, but new methods of prototyping technologies and the basic idea of printing - the layered application of material - brings about a new quality of replication. The approaches towards automation of the building process are a first step in the direction of self-assembly of architecture.

Systems of self-building are experimented with by research groups. Digital fabrication techniques are common, spreading in all fields of production, therefore also in the building industry, mainly for technology and material fabrication. The term "fabber" has been coined for a 3D printer that produces all sorts of consumer goods.

Dr. Behrokh Khoshnevis of the University of Southern California invented the so-called Contour Crafting technology, a layered fabrication technology:
"automating the construction of whole structures as well as sub-components. Using this process, a single house or a colony of houses, each with possibly a different design, may be automatically constructed in a single run, embedded in each house all the conduits for electrical, plumbing and air-conditioning. The potential applications of this technology are far reaching including but not limited to applications in emergency, low-income, and commercial housing."[336]

The application includes future settlements in space. Materials for 3D print include concrete, ceramics and organic polymers.

Rupert Soar started his research at the Rapid Manufacturing group at the University of Loughborough, and currently works with the company Freeform Construction with a similar technology and a what he calls a "physiomimetic" approach, translating *"...natural construction processes that are the outcome of negotiation between accretion (additive) and destabilisation (subtractive) as a dialogue between agents of change"*.[337]

The group also performed research on termite mounds, aiming at the integration of adaptive characteristics into the new technology.[338]

The industrial application of these automated building methods will open the real world to design methods relying on digital techniques.

---

336 Khoshnevis, B.; Automated Construction by Contour Crafting - Related Robotics and Information, 2004, http://www.contourcrafting.org/ [11/2007]

337 Soar, Rupert: personal information, 2010

338 Soar, Rupert et al.: Beyond pre-fabrication, 2006, http://www.freeformconstruction.co.uk/ [11/2007]

Fig.244 Aldo van Eyck, Amsterdam's Municipal Orphanage, Amsterdam 1957-1960.

4.2.7 Growth

"Growth is the process of increasing in physical size and occupying space."[339]

Growth in biology is based on cell division, and growth concepts are discussed in two chapters of this book: 3.3 "Nature's design principles" and 4.1 "Life, biology" as one of the classical criteria of life. Growth in the built environment means physical increase of settlement surface, space, building size or building element. All growth depends on the availability of energy and matter.

Urban scale

In spite of the increasing consciousness of environmental problems, owing to the growth of built environment, the growth of urban areas is still considerable, and it has already led to the majority of the world's population living in urban environments. There are now vast areas of slums in many cities, especially the fast growing cities in the developing world. There are several different kinds of architectural growth in settlements.

- **Historic concepts, development of today's structures**

Settlement structures which are found today in areas that have been inhabited for centuries include "organic" grown structures which show signs of fractal geometry. Planned concepts include linear, square and circular layouts with radial extensions. The patterns are influenced by the need for protection, but further growth is hampered by earlier fortification structures. As well as fortification, the morphology of traditional settlements was influenced by parameters like available space and topography, economic conditions, allocation of land use and sociocultural conditions.[340]

- **Recent concepts, future development**

Growth nowadays occurs along infrastructural axes affecting large areas. Densification of existing settlements coincides with both vertical growth and uncontrolled urban sprawl: informal settlements, slums. Planned large urban developments (e.g. in China) try to counteract this, and strategies of so-called "smart growth" are experimented with on a 1:1 scale. "Smart growth" principles include compact building design and densification, in contrast to the preservation of existing open space and qualities provided by nature.[341]

Along with the scale of the settlement, there is a functional hierarchy of growth related to the use of buildings: housing, workplaces, industry/production, storage, infrastructure, and cultural facilities. Good urban planning provides the right relationship and spatial arrangement for the different areas in the equivalent developmental stages of the settlement.

---

339 New Oxford American Dictionary, program version 1.0.1, 2005

340 Mast, H. in Teichmann, K. et al.: Prozess und Form "Natürlicher Konstruktionen" 1996, Entwicklung großstädtischer Agglomerationen, table 81

341 "smart growth" principles are applied in some North American cities, http://www.epa.gov/smartgrowth/ [11/2007]

Fig.245 Eduard Böhtlingk "markies" 1986-1995.

- **Rules of growth**

Growth on an urban scale no longer occurs in a self-regulated way. In most countries, growth is nowadays regulated by development plans, which define land use and/or intensity. Growth within cities can either be limited by specific regulations, e.g. height restriction and site boundary (as in Vienna), or by the introduction of a factor exploiting the building site (as in New York).

Global growth of built environment changes the relationship between built structures and natural environment. With growing settlements and vanishing natural land area the contact between built environments and "natural" environment decreases in spite of the increasingly fissured border.

Strategies have been found to deliberately reintegrate nature into the cultural landscape, in order to cope with the human need to be in contact with the natural environment. Considering the speed of urban growth, there is no alternative. It is necessary to destroy existing built structures for the retrospective establishment of green corridors etc. As the functioning of natural ecosystems also depends on their size, corridors must enable the necessary exchange between regions of natural environment. Infrastructural measures for their connection (e.g. bridges and tunnels for animals) may appear exaggerated when observed from a regional point of view, but the consequence of the increasing fragmentation of natural ecosystems is an irreversible loss of biodiversity.

Growth of buildings

Typological growth of buildings, e.g. the development of skyscrapers, represents growth in the notion of phylogeny. Growth on the scale of individual buildings is interpreted as growth of available interior space and thus enlargement of the built structure. Space is enlarged by extending and adding of structures, laterally or in height. Volume increase by continuous extension can be achieved through unfolding and inflating. Sheltered space is extended by the two-dimensional extension of surfaces.

The possibilities of growth of buildings depend on their respective construction, form and geometry, openness and functional concept. The need for growth is a change in use, increase of activity and/or intensification, although this may be temporary.

- **Addition**

Growth by addition is the usual technical way to increase. A historical movement favouring growth by adding modular elements was the movement of Structuralism in architecture (a linguistic concept initially, then extended to anthropology) Aldo van Eyck's Municipal Orphanage, [342], is characterised by clearly defined modules, in a more or less flexible arrangement. For some reason the flexibility of those concepts has never been exploited.

- **Deployment**

Growth of space in the timescale of minutes and hours can easily be enabled by deployment. "Markies", the foldable caravan by Eduard Böhtlingk, 1986–1995, has become famous as a both classic and simple example of temporary space. Deployment of lateral parts increases the volume of a conventional caravan in a beautiful and also expressive way.[343]

342 Mollerup, P.: Collapsibles, 2001, pp.74
343 Detail Zeitschrift für Architektur+Baudetail, Mobiles Bauen, 1998, pp.1422

Fig.246 Stage for the "Seefestspiele" Lunz, Lower Austria, Hans Kupelwieser and Werkraum Wien, 2004.

The flower-shaped Venezuela Pavilion, Expo 2000, by Fruto Vivas Architects and Buro Happold, could open and close huge petals or leaves. Commonly only distinct parts of buildings can open or close.
The stage for the "Seefestspiele" (Festival on the lake) in Lunz am See enables the rhythmic and seasonal sheltering of auditorium space. The artistic concept was developed by Hans Kupelwieser. Development, architecture, structural engineering and management were carried out by the structural engineering office Werkraum Wien. Not only the roof, but also the stage is semi temporary: usually floating, it can be made to disappear below the surface. The roof for the auditorium can rotate and close the space, with the upper surface still providing terraced rows of seats. The movement is set off by a passive system: inflow and drain of a water counterweight minimises the effort for the movement.
A new notion of growth is introduced with the facility management of large buildings. The deactivation and activation of zones of buildings is comparable to the shrinking and increase of used space.
For the exploration of new environments, the packing and transporting of architectural space is essential. Tents provide this kind of temporary and moveable space, but are not suitable for particularly hostile environments such as polar or space environments.
The European Space Agency carries out and funds studies for bio-inspired space architecture. Within the frame of an Academia Study, suggestions for a deployable lunar base were designed, and will be further explained in the Case Study section in chapter 5.

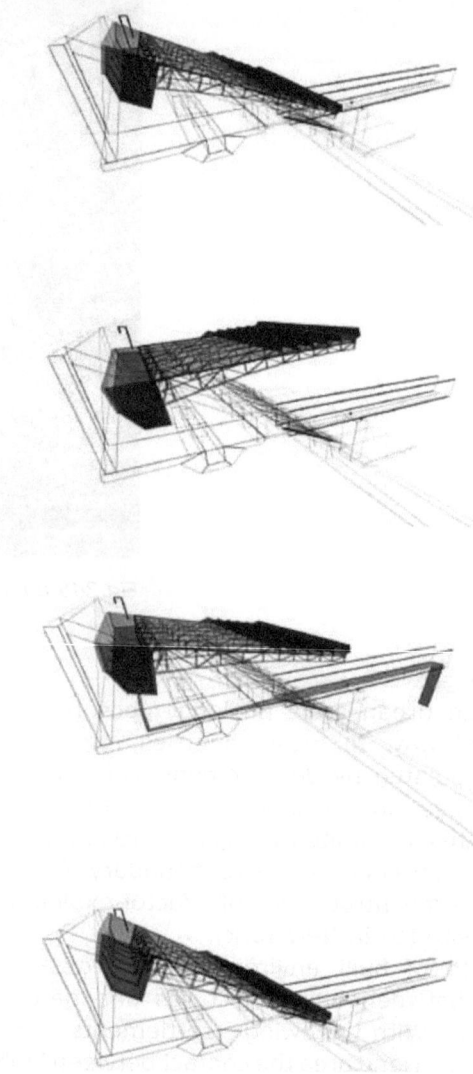

Fig.247 Rotation of the roof enabled by the water counterweight.

Fig.248 Willow runway of the Baubotanik group Stuttgart, 2005.

Fig.249 Overgrown handrail of the runway.

- **Growing elements, botanic architecture**

At the University of Stuttgart, living plants which continue growing are integrated as structural members in experimental projects of the "Baubotanik" group around Gerd de Bruyn. Willows were planted as the main construction for a runway and these have now grown over the metal parts of the project.

The fast growth of the willows will probably narrow the available space for users. Slowly growing plants would enhance the long-term usability and integrity of botanic architecture.

In Japan, the use of bamboo for living fences is an old tradition. Bamboo sprouts are bent and braided to form a vertical wall on the outside. The formation of shade roofs made of climbing plants can be found in many cultures. In Austria, vine is traditionally used for pergolas. In warmer climates, other climbing plants e.g. Bougainvilleas, form colourful shelters. Structural elements are needed for the support of the plants in this kind of space.

Fig.250 Sheltered space under a Bougainvillea in Laos, 2006.

Growth of material

When talking about growth of material, the timeframe of growth relative to the phase of the building is important. Growing building material with the idea of supplying it to builders has been always common (e.g. wood and other plant materials). But growth of material after being processed into a building is not.

Some building materials increase in volume to perform tasks like clamping and fastening, e.g. Polyurethane foam for the fixation and insulation of windows in walls. But usually, a building material that changes in volume is unwanted. Shrinking of material is a problem that has to be met by calculation and refined detailing.

The new production methods of architectural elements move towards natural role models. Instead of prefabrication and mass production, custom-made in situ production avoids transportation and storage efforts. Construction methods like extrusion and casting of material converge with natural processing of material. Biotechnology could also be exploited for the production of materials.

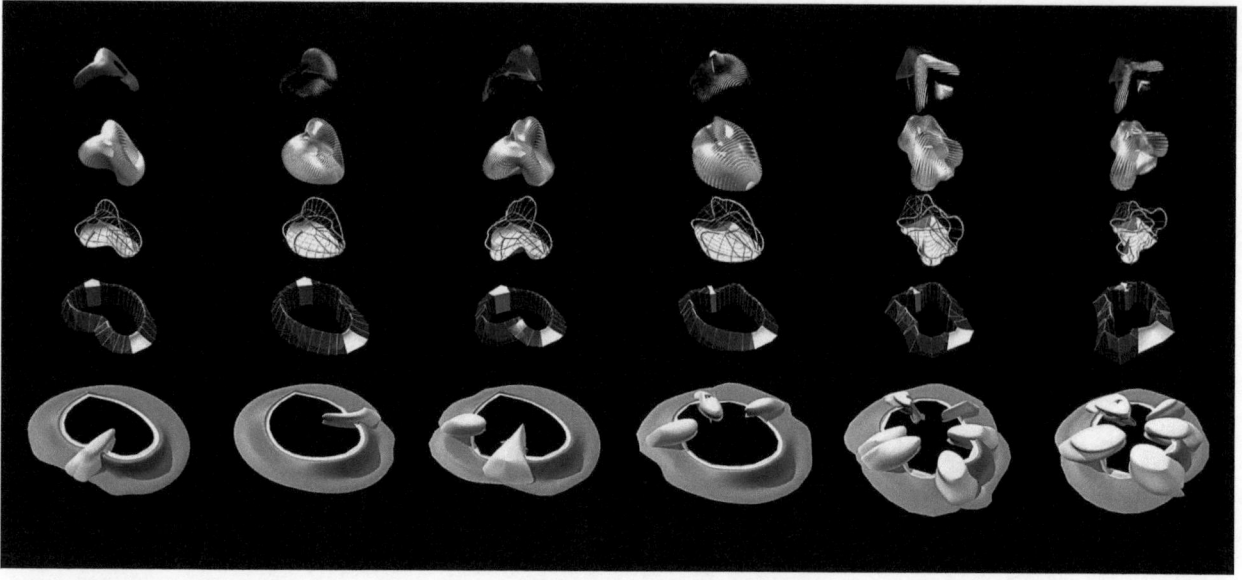

Fig.251 Embryological House, exploded axonometric of six variations showing: ground, clerestory glass, structural frame, cladding and photovoltaic shading, Greg Lynn, 1998.

Virtual growth

Apart from the concept of addition and deployment, growth on a building scale (in the sense of material or structural growth or the like) is transferred into a virtual environment. Morphogenetic design strategies work with virtual growth processes. With the implementation of the project into the real world the concept of change is lost.

Genetic Architecture, Greg Lynn, Animate Form 1999 - shows growth and development applied to the design process, in a virtual world.[344]

Biological growth is still used as a role model for computation. Functions found in nature can be transferred to growth processes in architecture.

Growth and destruction

Growth in an existing environment always entails a change in the environment, including destructive processes. Growth of the built environment results in the destruction of natural land or other built structures already on site. In particular this is always the case with densification.

Change is easily carried out in traditional building technologies, e.g. in adobe architecture. As the structure of settlements is that of an additive cluster, units can easily be added, removed and changed in size.

Adaptation to changed conditions can be achieved by resizing space, changing functions and reusing structures and spaces. With moveable architecture like tents and houseboats, the arrangement of buildings according to current conditions is even easier.

But if growth includes structural change of the settlement, e.g. by introducing larger elements, several smaller units have to make room for it. In democracies, complicated processes are required for large developments to be carried out in privately owned settlement structures. Migration is the precondition for large urban developments, and in autocratic countries this is often enforced.

Growth is always connected to instability and uncertainty, as is also the case in architecture. Even if it is "normal" that the built environment is changing, building sites influence all other ongoing processes in a negative way - streets are narrowed or closed due to cranes and other equipment, delivery traffic is considerable, dust and dirt on the building site pollute and disturb the environment. In general, processes of change are regarded as disturbing the "normal procedure".

---

344 Lynn, G.: Animate Form, 1999.

Fig.252 Overgrown entrance in the Ta Phrom temple, Angkor, Cambodia 2006.

4.2.8 Energy processing

"Energy is the property of matter and radiation that is manifest as a capacity to perform work".[345]

Energy can neither be generated nor destroyed. Basically, all types of energy can be transformed into another. In all energy exchanges, if no energy enters or leaves the system, the potential energy of the state will always be less than that of the initial state, but total energy remains constant. Heat energy increases constantly. Energy has the tendency to move from usable forms to non-usable forms, from higher concentrations to lower, and from ordered state to non-ordered state. Energy transformations reduce the amount of energy available to perform work.

Energy efficiency in biology is discussed earlier in the chapter on nature's design principles. Energy processing was discussed in the classical criteria of life section.

Just as in living organisms, architecture is open to energy exchange with the environment. The flow of energy in and out of a built environment is controlled in order to avoid unwanted side-effects. The methods of control depend on the kind of energy, and can be passive or active (the same can be said about material flow). Ecological design deals with energy efficiency in the built environment.

Forms of energy

Forms of energy affecting the built environment are: material bound energy, which - apart from fuels - is not available to perform work; kinetic energy; light; heat, which is less useful because its use in buildings is limited, and electricity, which is the most universally available and useful form of energy for the built environment.

Comparison between growth in nature and architecture

In contrast to growth in nature, building is usually carried out by the mere addition of elements and material. Other approaches have only been taken recently (e.g. the methods derived from printing mentioned above). Building in architecture can be compared to growth in nature, with respect to formation processes:

Growth happens in small units and is a distributed process of cell division, which increases exponentially. The elements produced are optimised for both effort and use. Interaction with internal and external influences is integrated into the growth process, and desirable for formation. The process takes place within a cycle of available energy and matter.

In contrast to biological growth, technological elements are mostly produced by subtraction of material, reducing the size of the element. The elements are relatively simple; refinement is prevented by economic efficiency and insufficient technology. Interactions with other processes are not desirable, as they interfere with the control of production. Supply of energy and matter for production purposes requires highly non-ecological methods, and has a negative impact on the environment.

Material bound energy

Building materials (as all other materials) contain so-called embodied energy. Chemically bound energy, which was needed for generating or producing the material, is one part of the embodied energy, which also includes energy used for processing and transporting the material. The terms "primary energy" and "grey energy" are used synonymously.

---

345 New Oxford American Dictionary, program version 1.0.1, 2005

Fig.253 Reflection mirror on the façade of the Hong Kong and Shanghai Bank, Hong Kong, Foster 1986.

There are many different methods for the calculation of embodied energy, and so different sources in literature include different values for the embodied energy of building materials.

Nonetheless, general advice is possible: heavy material should be sourced locally, as much more energy for transportation is needed than for lightweight material.

Aluminium, for example, contains a huge amount of chemically bound energy, consumed during processing, but transportation costs are relatively low.[346] The recycling potential of aluminium is excellent, so the material bound energy is not lost with the end of the building. Architects cannot usually influence the processing methods of materials, but they can influence economy and ecology by selecting materials that do not contain high amounts of embodied energy.

Light energy

Input of solar light, the main source of light in buildings, has to be controlled to avoid overheating and unwanted reflection and glare. Technical systems exist to enhance the flow of light deeper into buildings and thus enable larger buildings, which is more cost effective. Reflection and light conduction elements are common today in large-scale buildings, and is now also starting to be found in individually built houses.

Solar radiation carries a large range of frequencies of light energy as well as heat. This problem is met by new specialised types of glass, which are spectrally selective and can block ultra violet light (which has a damaging effect on interiors) as well as infrared (heat transmission). The production of glass fibre light ducts has also contributed to the separation of light input from heat entry.

346 Edwards, B.: Rough guide to sustainability, 2005, p.122

In spite of these developments the separation technique is not yet universally available, as the materials involved are still expensive.

Photovoltaic systems transform sunlight into electrical energy. Photovoltaic films and organic solar cells will soon be common as a functional surface on glass panes and building façades. "Energy harvesting" by exploiting solar energy is one of the keys to sustainable energy use in architecture.

Heat

Thermal energy is transferred by convection, conduction and radiation. Built structures soak up solar heat energy, and exchange thermal energy continuously with the environment.

Heating and cooling to maintain thermal balance are the largest factors in a building's energy consumption. According to the climate zone and building use, the comfortable internal environment differs from the conditions prevailing externally. To overcome the difference between inside and outside temperature, input of energy in form of material, electricity or hot/cool media (water, steam, air) has to be provided. Specific infrastructural elements are used to carry the heating or cooling media. Integrating low temperature heating systems in floor and wall areas is considered to be the best solution for constant temperature distribution and thus comfort. To prevent the equalisation of the temperature difference (attaining the "natural" balance), insulating material has to be used for the elements and layers that make up the boundary between inside and outside of the building.

Warm water is another essential resource needing energy. In developing countries the energy for cooking is still provided by the burning of any kind of available organic material. The huge volume of material used for this purpose corresponds to the equivalent amount of energy needed for this basic function, which contributes desertification in arid regions.

Superfluous, or "lost", heat of buildings is dispersed into the surrounding environment. On an urban scale, the thermal energy produced adds to the storage effect which occurs with any massive built element, resulting in the phenomenon of so-called urban heat islands.

The interrelation between light and thermal energy input into a building is solved with especially high architectural quality, but with considerable planning and execution effort in Renzo Piano's project in Houston, US, called The Menil Collection.

Fig.254 Panel production for Menil Museum, Houston, Texas, Piano 1986.

Fig.255 Panels attached to steel construction.

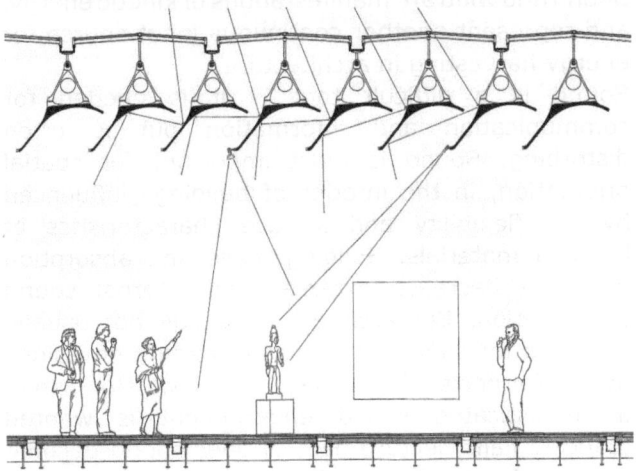

Fig.256 Detailed cross-section of the museum.

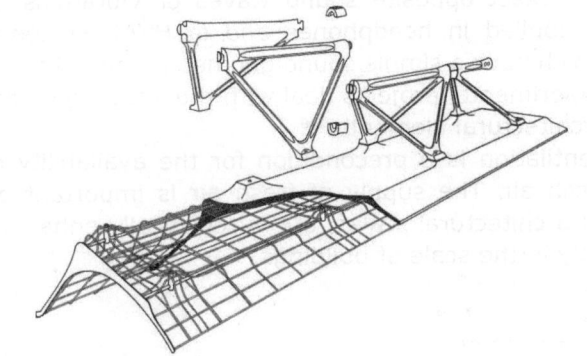

Fig.257 Steel construction and fixing of the panels.

Fig.258 Menil Museum in Houston, Texas.

- **Renzo Piano: The Menil Collection, Houston, Texas 1986**

Renzo Piano has been investigating new spatial structures and new materials for decades. He became famous together with Richard Rogers for their design of the Centre Pompidou in Paris in 1971-1977. The optimised constructive design of his buildings recalls constructions and forms from nature.

The Menil Collection was completed in 1986, accommodating a private art collection. The challenge and precondition was to design a presentation space with natural light conditions.

The project was developed in collaboration with Peter Rice from Arup Engineering. Piano decided to utilize reinforced concrete elements for the roof. First sketches showed a direct constructive integration of the parts into the primary roof construction, and then the load bearing function was separated from directing the light. The primary construction is made of spatial steel trusses that are shaped according to their load, and the precast concrete panels that control the distribution of light and stabilise thermal conditions in the gallery. The heat that enters the building through the glass roof is partly reflected by the horizontal surfaces of the panels. They minimise downwards radiation and thus protect the gallery space beyond from the layer of warm air that is created below the glass roof.

Fig.259 Interior space of the museum.

The concrete panels are responsible for the directing of natural light. The design was carried out in collaboration with lighting experts and the owner of the museum. The modelling of the forms was carried out by a combination of light tests, structural analysis and architectural shaping. Prototypes were produced of glass fibre reinforced composites. A conventional glass façade is the outermost layer.
During implementation, it turned out to be difficult to find companies for the production of these sophisticated elements. Finally the production had to be carried out by a UK company which was familiar with the production of refined precast elements. The production process had to be tested and confirmed with prototypes, because no finishing was possible. A spraying method was used to produce the panels according the defined criteria for construction, form, surface structure and colour.[347]
The particular quality of light and the contact of the gallery's interior spaces with weather condition outside causes a notion of daytime for the visitors to experience. The control of light and the prediction of interior conditions in this sophisticated way could only be done by modern computational methods. Implementation took place from 1982 to 1986, and 15,000 pieces of art are shown in turn.

Electricity

Electricity has become indispensable, as almost all of today's activities are connected to some kind of electric device. Operation of cities or a building's technical equipment, operation of all household appliances, work and leisure depend on the availability of electricity.
Electric energy is transformed into light, electromagnetic waves, heat and kinetic energy.
It is the most common form of energy expended in built environment.
Electricity is provided by supply networks, which until recently used to be national monopolies in Europe.

Kinetic energy

Kinetic energy is used for transportation and for actuation in control systems. Potential kinetic energy is stored in a building (see the destruction of the World Trade Centre towers - the sheer assembly of a building contains potential energy).
Sound and wind are manifestations of kinetic energy, and represent another continuous input source for energy harvesting in architecture.
Sound is a difficult topic - it is needed for communication and information but is often disturbing. Sound is most important for spatial orientation, in the interior of buildings influenced by the flexibility and surface characteristics of building materials. Building mass and absorption passively decrease external and internal sound transmission. But architecture is still not usually designed for sound. Only when extreme emissions occur (airports, highways, industry...) or when a sophisticated sound environment is wanted (concert halls, lecture halls...) does sound become an issue for the design of built environment. Active noise cancellation is a new method of controlling noise through eliminating sound waves by adding the exact opposite sound waves or vibrations. It is applied in headphones and in HVAC systems, which have a simple sound geometry, and in the US experimental projects deal with the integration into architectural elements.[348]
Ventilation is a precondition for the availability of fresh air. The supply of fresh air is important on all architectural scales, but is technically enhanced only in the scale of buildings.

---

347 Rice, P.: Peter Rice, 1994, pp.87

348 http://architecture.mit.edu/house_n/web/resources/tutorials/House_N%20Tutorial%20Active%20Noise%20Control.htm [11/2007]

Passive systems for natural ventilation rely on temperature differences created by differences in height, material etc., or the exploitation of wind energy.

Wind has recently been rediscovered as an energy source, wind power plants being subsidized by government bodies as alternative sources of energy. Wind energy can already be used on a regional scale, but it is not highly efficient (and technologically difficult as well - wind generators require fast moving elements and transformers, causing problems with sound and vibrations).

Strong aerial kinetic energy is felt more than heard. Feeling kinetic energy exerted on the human body by the elements (wind or water) is usually unpleasant.

Harvesting and transformation

On a global scale, energy is used for the production of building resources: materials. The industrial production of building elements and transport contribute to the energy embodied in buildings. Energy is "produced" by the transformation of energy from natural sources into usable forms. There are renewable and non-renewable sources of energy: fossil and nuclear fuels deposited in finite reservoirs are non-renewable. Furthermore nuclear, geothermal, solar, wind, hydro energy is all harvested from nature. "Alternative energy sources" are solar, geothermal, hydro, and wind energy. Decentralised harvesting of energy is more common in areas where supply chains do not exist or do not work properly (the wish for independence triggers decentralisation).

The "production", or more precisely the transformation, of energy takes place in power plants, and needs equipment. Facilities for the central production of energy are technically sophisticated, but at the same time part of basic infrastructure that we already regard as normal. Energy production facilities usually have a profound impact on the environment, as they either change the landscape considerably (hydroelectric power plants, regarded as one of the "cleanest" forms of energy damages landscape and ecosystems, as can be seen at the Three Gorges dam in China) or produce secondary impacts (e.g. toxic gases, nuclear waste). Storage and distribution facilities are necessary as well as buildings for the production of energy.

On a global scale, energy is also bound in matter, e.g. in traffic infrastructure. A continuous energy input is necessary to maintain our traffic network in the public space.

- **Energy transformation in buildings**

Depending on the source of supply of energy, transformation is most common from chemically bound energy to heat and electrical energy, and from electricity to heat, kinetic energy and light. In general, huge quantities of energy are lost in transformation due to low efficiency as - in contrast to many biological energy transformations - enormous amounts of heat is produced.

During a given timeframe, which corresponds to the state of the building, different transformations take place (planning, production, building, operation, dismantling, recycling): during the material production phase, chemical transformations occur; during the building phase, assembly prevails and kinetic energy is used and stored within the assembly; in the operational phase, electricity and thermal energy is used ("produced" on a large scale via kinetic/chemically bound or solar energy).

To sum up, energy processing on the building scale includes: embodied energy in the built structure - material and assembly, collection of solar energy and light, storage of energy (thermal energy within building mass and use of storage media, e.g. water), and input and output of thermal energy through different sources, heating and cooling, input of electricity for all sorts of uses and the use of kinetic energy for movement of parts and control systems.

- **Transport of energy**

The distance that energy travels is considerable. Supply networks have spread over almost all inhabited areas of the planet. Infrastructure for the transport of energy is important. Energy is transported via specific media embedded in large supply networks, controlled centrally.

Chemically bound energy is transported in pipelines or by ship, truck or train; electricity is easily transported over long distances in power cables.

Within buildings, it is again distinct infrastructural elements that transport energy. The thermal storage abilities of architectural elements and space are the only characteristics that appear independent of distinct infrastructure.

- **Control**

Energy in- and output in architecture is controlled whenever possible. Building energy management systems have been developed to control the flow of energy by processing information on energy usage and components that use energy, as well as environmental data. They represent a centralised method of control.

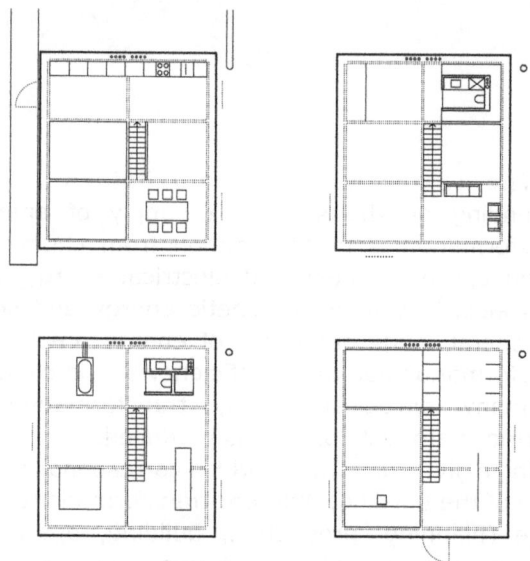

Fig.260 House R 128 floor plans.

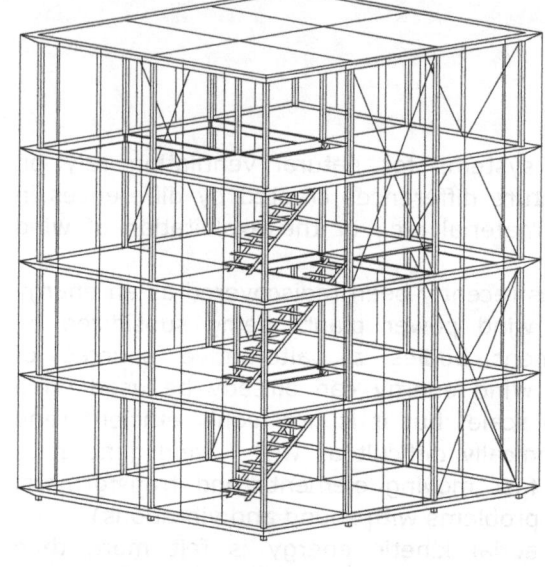

Fig.261 House R 128, Stuttgart, Sobek 2001, axonometric view of steel construction.

Sustainability in terms of energy supply

Alternative energy systems promote regional energy input from the environment together with transformation for intensive use. "Design with climate" makes use of energy conditions, e.g. uses mass to store thermal energy when big temperature differences exist. Some key points of advice for sustainable energy use are valid for any building task:
use passive systems first, use only what is needed, use as little energy as possible, use renewable sources and the most efficient way of exploitation.
Energy has started to be a category in measuring the quality of a building. In Austria, government funding of residential buildings already depends on the compliance with energy standards for building materials and operation. The energy management of buildings will in future be a required issue of planning, but the market price of architecture is still measured in price per m², and not in terms of energy. Different building types and uses have different needs in terms of energy. Ken Yeang suggested a categorisation in passive mode, full mode and mixed mode systems, that allows a reasonable guideline for the choice and implementation of technological systems.

- **Werner Sobek's R128**

Werner Sobek built the experimental "R 128" in 2001, in Stuttgart, as a totally transparent zero energy building. The energy concept is a very important issue in his design, just as are other components of sustainability such as comfort, functionality and, above all, recycleability. Sobek's ideas about aesthetics, construction, material and information technology culminate in the calm, reduced, smooth and transparent volume.

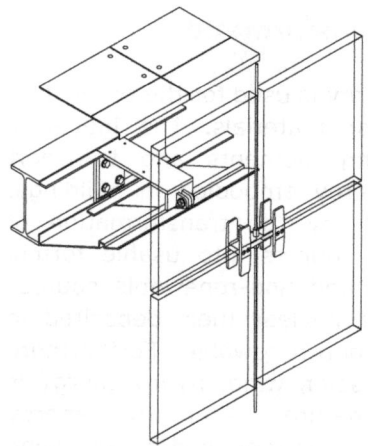

Fig.262 Façade detail.

The solar energy input suffices to heat or cool the air almost all year round. The problem of overheating is solved by an integrated water tank with a heat exchange system. Together with the production of electricity by photovoltaic elements on the roof, a mechanical ventilation system supplies fresh air. The whole energy supply is controlled and regulated by a complex computer system. A touch screen allows the user to overrule the system on every floor of the building. The house adapts to the environmental conditions by sensing, regulating and actuating its elements, characteristics that will be described in the next chapter. All technical elements are integrated into the corresponding layer of the building, so that the surfaces stay smooth and free from distinct control elements. The R 128 represents a high-tech approach to energy processing.[349]

349 ARCH+ Nr.157, Sobeks Sensor oder Wittgensteins Griff, 2001

- **Energy efficiency of traditional architecture**

Traditional architecture works with low energy use and low comfort, which may be enhanced by integrated passive systems. Material energy is normally low. Renewable materials are preferred, as the production takes place locally. Operating the building - cooking and heating - is done by burning material (chemically bound energy is transformed into heat). Traditional buildings are usually energy efficient, and can serve as a model to learn from. The thermal comfort of adobe buildings in hot and arid regions for example is astonishingly good. The aspects of quality in traditional buildings are a source of innovation for modern architecture. Scaling effects have to be taken into account for the transfer of principles. Traditional architecture is residential architecture for (possibly large) families. The solutions found are valid within this scale of the buildings size.

Influence of built environment

Buildings in general change the energy "landscape". Climate change is directly influenced by the built environment. Thermal energy storage induced by urban areas generates so-called heat islands, comparable with the global "continental" effect on regional climate. Wind conditions are changed as well. Blocked or enhanced air exchange due to buildings and lost vegetation cover also influences the regional climate. Natural ventilation has been a topic on an urban scale for centuries, the discussion being intensified by the increase of air pollution.

4.2.9 Reaction to the environment

Reaction to the environment implies sensing, signalling, regulation and actuation of any kind, and this is also valid for architecture. Reaction to the environment takes places in different scales and timeframes, in quick or slow reactions of the material, the individual building, the building typology or built environment in general. The process of reacting to environmental influences is fundamentally different according to the scale of observation. Reaction on an urban scale implies organisations or individuals as acting units, the laws being defined by social and cultural conditions. Reaction on the level of material may be a passive chemical or physical process. Reaction can have different manifestations, e.g. movement and locomotion, smartness, learning or innovation. Reaction can occur on the individual (building) and typological level, the latter being related to evolution.

In architectural applications, unforeseen building reactions to environmental influences are usually destructive (e.g. subsidence, vibration, deterioration...).

For controlled reactions, control systems are used and embedded in all kinds of technical systems. Sensing of a physical difference of some kind is the first step. A diagnosis evaluates the measurements, using a threshold, parametric relation, model-based comparison etc. As a result, actuation of any kind is initiated and the reaction activity is carried out.

Sensing: active and passive

The precondition for a reaction to the environment is the ability of sensing.

Knowledge-sensing is based on a knowledge base and complex decisions, e.g. the definition of the need for unfolding temporary space. Most reactions in technical systems are based on simple physical sensing.

Active sensing is implemented in all technical building equipment. Technical sensors can absorb information on temperature, humidity, air pressure, wind speed, light intensity, weight, movement, availability of chemicals etc.

Sensors are inactive most of the time (although some kinds of sensors consume energy for operation), and when activated transform information into an electrical signal. The signal has to be interpreted to give it meaning - parametric reaction generation, control circuits, or model-based diagnosis and decision follow.

Passive sensing does not require regulation processes; change of physical conditions directly initiates actuation. Passive sensing implies some kind of property change according to environmental (or internal) influence. Elastic deflection due to load, heating of building shell due to solar radiation are both physical influences that could trigger reactions.

Decentralised sensing requires the integration of sensory elements into architectural systems, no longer as add-ons but as integral parts. The disadvantage of passive systems is the invariability of reaction. Action depends on the physical parameter increasing or exceeding a threshold.

There is little possibility of retrospective intervention into a passive system. For example a pinecone opens with change of humidity - but the range of movement and reaction time cannot be changed. Hybrid systems combine the advantages of the passive energy efficient reactivity of systems that do not need sensing and actuating with an active adjustable system in case that the passive reaction is insufficient.

Active reaction to the environment is decided by human decision makers or by decision making technical devices. Active systems rely on some kind of information to assist in decision making, whereas passive systems act on the base of physical information. For architecture, the information of users and inhabitants is important, and the integration of user information is an important issue in the design of "intelligent" building systems, which will be discussed later.

Beyond the material and individual building, large-scale reactions occur on an urban scale, due to large-scale changes in the environment or local reactions to local environmental changes. Sensing and thus information can be physically present (e.g. noise of a new airport runway), so that the regional reaction takes place in the form of individual elements reacting (people moving away), and thus becoming part of a larger scale movement. In a controlled way, information would be available beforehand, and regulation could be implemented earlier. This requires another level of the planning and controlling body, which is represented by regional and urban planning departments. The integration of inhabitants into these large-scale changes of built environment is, however, not yet satisfying even in Europe.

Fig.263 Relocation of an Indonesian house, Waterson 1997.

Fig.264 Floating church of a boat settlements on the Tonle Sap lake in Cambodia, 2005.

Fig.265 Locomotion of a Californian home, 2004.

Reactions of architecture

Reactions on a building scale are planned and integrated into the architectural design. Buildings can "do" what they are planned for. The type of action is predefined by implementation.
Usually, individual elements sense influence, regulate and actuate a reaction. A building can react through: movement of building parts, opening and closing, thus controlling the access of living organisms, material or energy, change of properties of elements, change of properties of space and change of internal environment. Only rarely buildings can move as a whole.

Locomotion

- **Locomotion on the urban scale**

On an urban and global scale, migration is a reaction to different living conditions. The movement of parts of the population involves the abandoning of, and rebuilding of cultural environment. The growth of cities is part of an ongoing migration process which affects the whole planet.

- **Individual buildings and settlements**

At a building scale, the process of relocation requires preadaptation. Mobile buildings can change location only when they have been built to do so. Caravans and trailers are permanently mobile. Locomotion as a single event can be coped with by particular structural characteristics: either by enhancing structural integrity to allow transport, e.g. on a crane or a flat-bed truck, or easy deconstruction and reassembly.

Apart from enforced migration movements for political and economic reasons, it is cultural conditions that can foster the migration of people. The European impression of migrating differs from that in the US, where relocation according to economic conditions is considered normal.

Rhythms of changing environmental conditions, in particular the seasonal rhythms of availability of resources, have generated cultures of migration, which are embodied in architectural typologies, e.g. demountable structures like tents and yurts. Relocation is even easier on water. Boat settlements like those on the Tonle Sap lake in Cambodia are able to follow the rhythmic change of the lake's coastline.

Indonesian wooden structures are easily demountable and can be relocated, as described in her book by Roxana Waterson.[350]

350 Waterson, R.: The living house, 1997, title page

Fig.266 "Moonwalker", diploma project, Häuplik 2004.

In the area of the case study on Nias island, relocations of entire villages occurred, triggered by either social conflicts or problems with water supply.[351]

Settlement in unknown environments makes the ability to relocate, and thus to explore, very important indeed. Therefore many space architecture projects are designed for locomotion.

Movement

As in nature, movement allows adaptation to external influences. The movement of a component of a building is common and easy to achieve, e.g. sunshades, blinds, awnings, which exist in a variety of products and mechanisms. The movement of buildings as a whole is still exceptional.

- **Movement triggered by the sun**

The project "Heliotrop" reacts to the position of the sun by rotating as a whole. Rolf Disch designed the house at the beginning of the 1990s. The prototype of this project is located in Freiburg; the building has a circular floor plan, a cylindrical shape, and it is able to rotate with the sun, thus adapting to the different thermal properties of the outer shell. Heat exchange and photovoltaic systems on the roof make it a plus energy house.

351 according to oral information of Prof. Marshall in November 2007, the village of Bawögosali has been relocated. Similar histories are known about other South Nias villages.

Fig.267 "Heliotrop", Rolf Disch, 1990s.

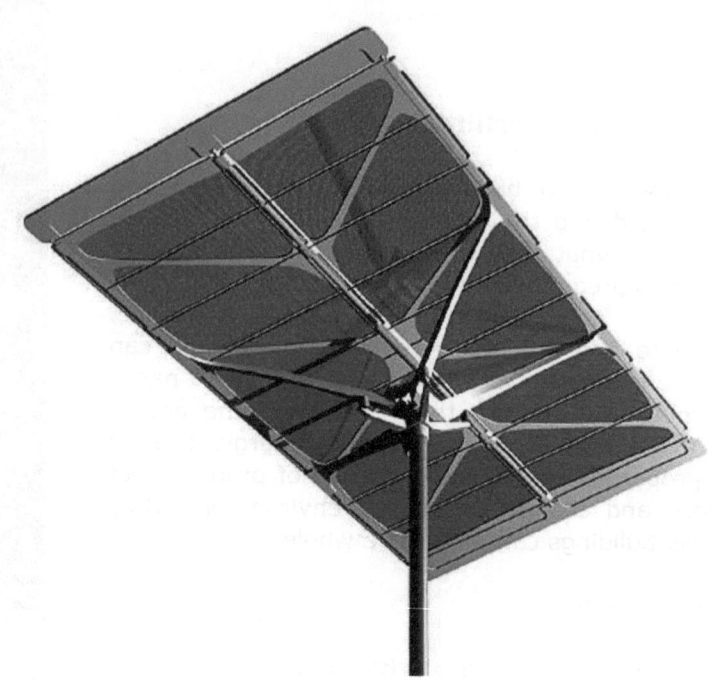

Fig.268 Cambridge Colonnade project, single "tree", Aarup and Vincent, 2003.

- **Movement triggered by wind**

Movement triggered by the wind was integrated in a project called the Cambridge Colonnade, carried out by Arup Associates and the biomimetics consultant Julian Vincent in 2003. The task was to create a sheltered but open pathway on a Campus along a canal. The proposed design uses wind to create a naturalistic atmosphere by allowing movement of the structure. Slender flexible columns carry elegantly shaped moveable roof elements with integrated photovoltaic panels, connected to each other by elastic springs. The force of the wind initiates the movement, which spreads wavelike over the whole structure, at the same time slowing the movement of the air. The aesthetics of the project reflects the aesthetics of the existing architecture. The project has not been implemented.[352]

Movement in a more abstract form, as movement of functions and activities, is also common in the built environment. Larger buildings in particular, e.g. office buildings for large organisations, need to adapt to constant internal movement. After many decades buildings are often no longer appropriate for their initial use, as patterns of activity can change fast. A solution for this problem is to create spaces neutral for their present use, but with maximum flexibility integrated. Yet there is a negative correlation between adaptation and flexibility in architecture: the better adapted to a specific use, the less flexible the building.

Fig.269 Cambridge Colonnade, elevation.

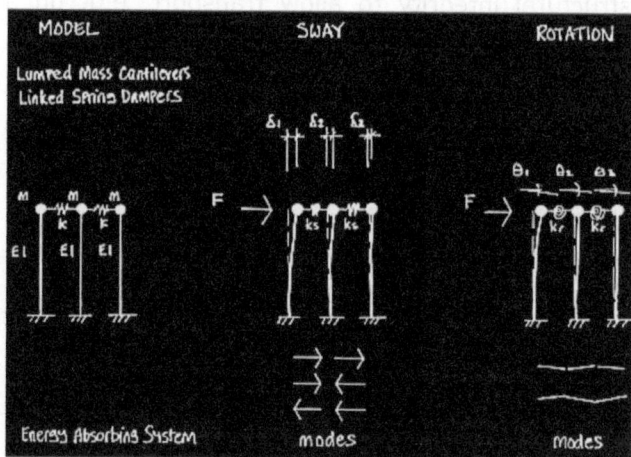

Fig.270 The range of possible modes of movement.

[352] Project presented by Julian Vincent at the Hannover fair 2004, Bionik Kongress

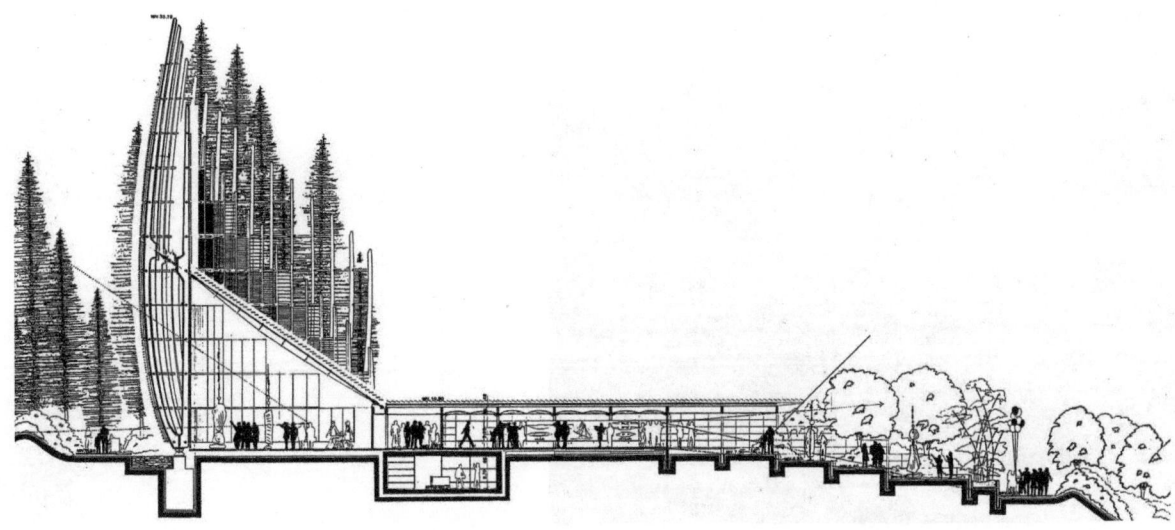

Fig.271 Centre Culturelle Tjibaou, Nouméa, New Caledonia, Piano 1998, cross-section.

Open and close

Movement of elements to open or close buildings has been known since the invention of doors and windows. The opening of a whole building, however, is exceptional.

Fruto Vivas Architects designed the Venezuela Pavilion of the Expo 2000 as a foldable flowerlike structure. 16 huge petals, steel structures covered by a membrane, could open and close the whole pavilion according to climatic conditions inside the building and external wind speed. Active sensing and hydraulic actuation was used.[353]

The consideration of opening and closing has to differentiate between the elements being permitted or forbidden to enter: access is allowed or denied.

- **Opening/closing for organisms**

In architecture, opening is either visual (windows permitting visibility) or actual, permitting access. Visual opening and closing is facilitated by specific elements and layering: openings in the wall, glass, curtains, blinds. Control of visibility is usually executed individually according to the human need for privacy or public view.

Opening for access is controlled by means of technology (the centuries-old system of locks and keys, or computerised access control, biometric scanners, together with automated actuation). Certain functions of buildings, e.g. in hotels, need dynamic solutions, where change can be implemented quickly. Building management systems provide the solution: they include access control, mostly for the internal distribution of access privileges. But it is precisely those buildings with high user frequency that still require human access control for security reasons.

- **Opening/closure for matter and energy**

In buildings, controlled input and output of matter - resources, water, gas, waste - is achieved by infrastructure. Input and output of thermal energy, sunlight, air, wind and energy is achieved by distributed but distinct openings. Ventilation is regulated by the user, and also in large-scale and low-energy buildings by technological means as part of a HVAC (heating ventilation air conditioning) system. Change of permeability is the equivalent strategy for reactive materials.

- **Opening and closure for wind energy on a building scale**

Renzo Piano used wind as an energy source for passive ventilation in his Jean Marie Tjibaou Cultural Centre, erected in 1998 in Nouméa, New Caledonia. The client was the Agence pour la Développement de la Culture Kanak. The aim of the Centre is the celebration and preservation of the indigenous culture of the Kanak people, who suffered under the French colonial rule. The site is a peninsula in a lagoon. Piano used curved wooden wind catchers (20m, 22m and 28m high) with a second layer façade to enable passive ventilation of the spaces behind. The cross-section shows the main order that the project follows: a slightly ascending 200m main axis, larger pavilion spaces to the left and conventional working spaces and exhibition halls to the right.

---

353 http://www.mitsubishi-automation.com/solutions/art18a.html#b [11/2007]

Fig.272 Centre Culturel in the background behind the characteristic Araucaria trees.

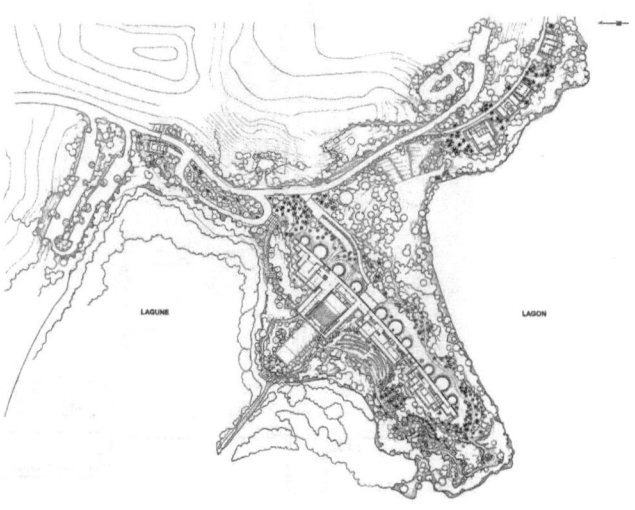

Fig.274 Layout of the whole project.

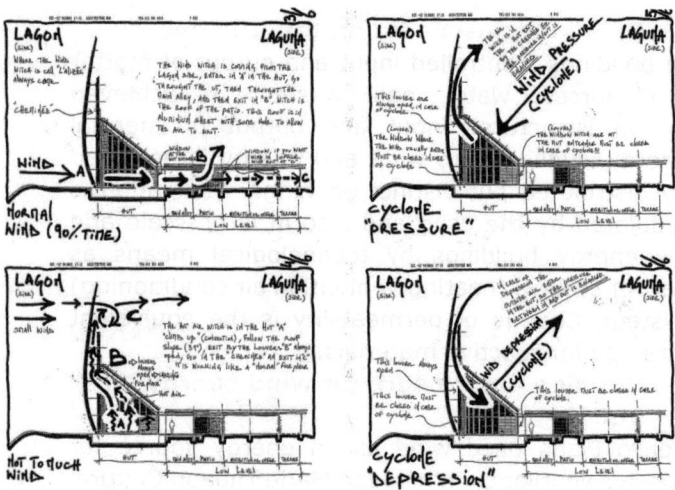

Fig.273 Ventilation during times of normal wind speed, and during cyclones.

The Arup Engineering Company calculated the ventilation concept. Glass shutters in the partitioning walls can adapt the building to different wind conditions.

Most of the time normal wind speed prevails, and the wind catchers act like chimneys exhausting the hot air. In the case of a cyclone the shutters to the adjacent spaces are closed and the upper louvres are opened to equalise pressure and minimise suction. Wind sensors and computers control the position of the glass louvres.[354]

The traditional "grand case" of the indigenous population is a circular structure up to 15m high, using the large ventilation space to provide a comfortable interior climate.[355] Possibly Piano's Centre Culturelle Tjibaou was also inspired by this traditional "natural construction".

- **Opening and closure on a material scale**

Reaction at a material level is illustrated by the experimental design of a Smart fabric inspired by stoma[356], the gas-exchange openings of plant leaves for ventilation according to humidity. Two layers of fabric are overlapped, so that the pores do not lie on top of each other. The distance between the layers is provided by a layer of a polymer which changes its volume with change in humidity. An increase of volume leads to a larger distance between the layers, which allows better airflow through the whole fabric system. The Smart fabric is a passive system for a ventilated skin.

354 Blaser, W.: Renzo Piano: Centre Kanak, 2001

355 Zöhrer, G.: Moaro - Die kanakische Architektur Neukaledoniens und ihre Stellung in Ozeanien, 2005

356 "smart fabric inspired by stoma", project by Michael Murauer for the design program "Bionik - natürliche Konstruktionen" taught by the author at the Vienna University of Technology.

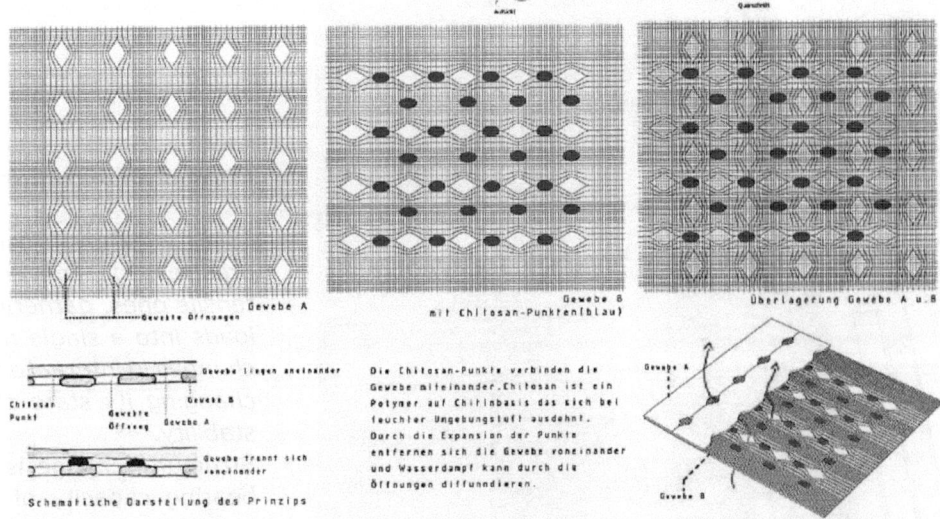

Fig.275 Michael Murauer, "smart fabric" - inspired by stoma: ventilation according to humidity, 2005.

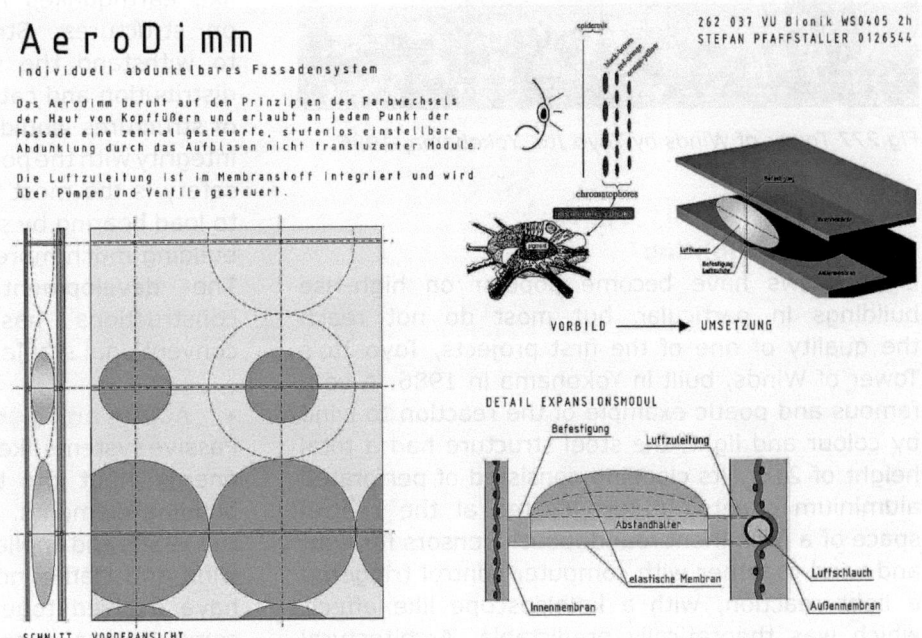

Fig.276 Stefan Pfaffstaller, "Aero Dimm" - pneumatic facade darkening, inspired by cephalopods, 2004.

Change of (material) properties

Change of properties concerns the transparency, visibility, colour, or any other adaptation of a material or a structure.

- **Transparency and visibility**

A common method to change the transparency of building façades is the addition of a semi- or non-transparent layer, e.g. an awning or a blind. Another method is so-called electrochromic glass, which react to a voltage by darkening.

The inspiration for the experimental design project "Aero Dimm"[357], a pneumatic façade with a colour change option, came from the colour change of cephalopod skin: between two layers of a façade, elastic membranes change volume by pneumatic pressure, creating dark parts on the surface. The ventilation tubes can be integrated into the inner membrane of the system. "Aero Dimm" could be used for façade darkening as well as colour change. The active regulation mechanism is not elaborated in the proposal.[358]

357 "Aero Dimm", project by Stefan Pfaffstaller for the design program "Bionik - natürliche Konstruktionen" taught by the author at the Vienna University of Technology.

358 Gruber, P.: The signs of life in architecture, SEB Glasgow 2007

Fig.277 Tower of Winds by Toyo Ito, Yokohama, 1986.

- **Coloured lighting**

Light shows have become popular on high-rise buildings in particular, but most do not reach the quality of one of the first projects, Toyo Ito's Tower of Winds, built in Yokohama in 1986. A very famous and poetic example of the reaction to wind by colour and light, the steel structure had a total height of 21m; its cladding consisted of perforated aluminium sheets. It was located at the central space of a prominent roundabout - sensors for light and wind together with computer control triggered a light reaction, with a kaleidoscope like effect, which was theoretically predictable. Architectural writer Charles Jencks called this an *"interactive architecture comparable with the landscape painting in Zen"*.[359]

- **Change of temperature**

All heating and cooling systems are included in this kind of change.

Other possibilities occurring in nature have not yet found analogues in architecture: change of surface properties (e.g. from hard to soft), change of density and weight (although these are used in systems for buoyancy and flight).

Change of structure

"But a truly responsive building would be prestressed, converting compressive loads into tensile ones, gathering all the residual compressive loads into a single mast. It would then respond to changes in internal and external forces by adaptively changing its state of prestress, giving lightweight stability."[360]

Structural change is useful for changing the load-bearing capacity of structures. Forces that affect buildings are not constant: certain influences - load, wind forces, weight of snow etc. - change the total load considerably. Catastrophic events like earthquakes have an even stronger impact on structures. Structures are usually designed to withstand the worst case of concurrent load distribution and catastrophic event. Efficient design of structures would balance the need for structural integrity with the possibility of repair: for inhabitants, safety is the most important. A dynamic approach to load bearing by structural adaptation could make building much more efficient in this respect.

The development of form-active membrane constructions has changed the building of conventional stable and rigid systems.

- **Active and passive adaptation**

Passive systems like dampers can transform external energy input into the kinetic energy of particular building elements. This strategy was developed in the 1950s and applied to high-rise structures to avoid wind and traffic induced vibration. Active systems have evolved together with the improvement of computational means for control. The advantage of active systems lies in the possibility of setting up the reactive behaviour, but the control system needs an energy input. Prestressed systems and controlled prestress have been experimented with for reinforcement of concrete structures. Change of stiffness and elasticity is particularly interesting in earthquake resistant design. Hybrid systems combine passive and active means of adaptability.[361]

359 Riley, T.; The Museum of Modern Art (Ed.): LightConstruction, 1995, p.166

360 Vincent, J.F.V.: Smart by nature, in Beukers, A. et al.: Lightness, 1998, p.46

361 Teuffel, P.: Entwerfen adaptiver Strukturen, diss. 2004, p.19

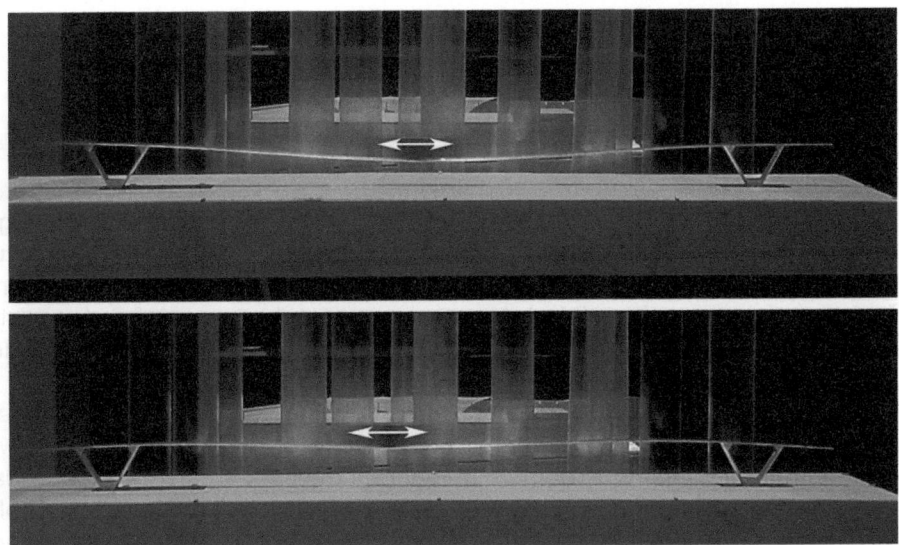

Fig.278 Adaptive bridge, exhibition model, comparison of passive load bearing and adaptive system, Patrick Teuffel 1994.

- **Load path management**

Werner Sobek is the head of the Institute for Lightweight Structures and Conceptual Design in Stuttgart. One focus of his research is the development of adaptive systems. He investigates the adaptation of structures by changing the stiffness or length of individual elements to manipulate the flow of forces. Patrick Teuffel has developed a design concept for adaptive systems, the so-called "load path management".

"A new possibility with adaptive structures is the replacement of mass through energy in using a 'virtual' stiffness due to actuation instead of physical stiffness."[362]

- **Pneumatic structures**

In pneumatic structures, pressure differences in structural elements change load-bearing capacity. In the pneumatic "Airtecture" building carried out by Festo, a company for pneumatic components, active pressure control according to wind loads, snow, and thermal expansion was applied to the structural elements. The "Festo muscles" were used as actuation components.[363]

Fig.279 Festo Airtecture building: structural elements consist of inflatable tubes with controlled pressure.

Fig.280 Fluidic muscles as actuation components.

362 Teuffel, P.: Entwerfen adaptiver Strukturen, diss. 2004, p.53
363 Schock, H.-J.: Segel, Folien und Membranen, 1997, pp.102

Change of space - growing and shrinking

Change of space as a reaction to environmental influences concerns urban settlements and buildings. Increase in space is related to movement and growth, and so the projects presented for "growth" are also valid for the chapter "reaction". Deployable structures can have diverse mechanisms of sensing, control and actuation. Quick or rhythmic change of environmental conditions triggers spatial reactions.

Self-repair

In nature, the concept of repair allows efficient use of matter and energy. Human technology tends towards perfectionism, but fortunately spending too much energy on negligible faults is prevented by reasons of economy. The number and frequency of faults and the necessity of repair is crucial for the total effort, the parameter which has to be optimised. A policy of design and implementation aimed at a specified fault tolerance reduces total effort.

Self-repair is a reaction preventing the element from total failure by some kind of adaptation. It is thought of as a passive process. Detection is due to a tangible physical parameter, which initiates repair directly. Self-repair is a self-organised process and the examples presented of crack-polymerisation and self-healing membranes are also models for self-repair.

Timescale of reaction

Timescales of sensing, processing and reaction in architecture are differentiated according to the scale of observation.

Urban settlements and structures exist for centuries, and processes of change can also take a long time. Long-term development in urban planning, e.g. a new underground line, takes decades. Several years/months are needed for the implementation of new buildings. Months/weeks suffice for change, renovation etc. Weeks/days are needed to carry out change in terms of temporary structures being put up or taken down. Days/hours are needed for thermal change etc., hours/minutes for repair and drying processes, to name but a few. Minutes/seconds are required for access, opening of elements, controlling flow of water and change of light and colour.

Information processing in nature and in technology is confronted with the speed-versus-accuracy dilemma: the faster information has to be processed, the less exact is the result.

In nature, diverse adaptations have evolved to meet this problem, e.g. in human vision the retina has a specialised focus region with more exact signal processing. Technical systems usually work with a predefined precision; fuzziness is already introduced in many control devices which allows faster and less exact reactions.

Fast reaction systems react to dynamic environmental influences, e.g. wind, traffic induced vibrations and earthquakes. Adaptive architecture as described above relies on physical signals being detected and processed fast enough for sufficiently fast actuation. So-called "performative" architecture carries out purposeful behaviour. Experimental projects work with remote or Internet based control. This kind of "interactive" architecture allows communication and is of interest for remote communication, e.g. in space design for long duration missions, where artificially enriched, so-called augmented environments are crucial for the psychological health of the crew. Real-time behaviour is envisioned for interactive architecture.

Phases

Reaction possibilities also relate to the different phases of architecture.

- **Planning**

In the planning phase, virtual modelling and reaction testing delivers valuable information about the characteristics of the structure and its future behaviour under environmental influences.

Reaction of buildings to the environment and environmental change is usually tested in more or less sophisticated but not exclusively virtual models, for: thermal conductivity and heating, reaction to wind loads, stability of construction, dynamic (time related) simulations - traffic, internal flow of users.

- **Operational phase**

Active sensing, processing and reacting is bound to the operational phase of a building, as these activities are integrated building components that start working when the "whole" of the system is established. The integration of reaction into other phases of architecture is even more difficult.

- **Introduction of stimulus-reaction processes into the design process**

This includes, for example, the interpretation of the design process as a continuous development flow according to selected environmental inputs, which is frozen at a specific point in time to be transformed into reality. This approach is used by evolutionary design.

- **Introduction of control systems into the building process**

Reactions in the building process could be carried out by passive systems, which already work although the structure is still unfinished. An integrated passive system is the unfolding process of a deployable structure controlled passively by inherent (smart) material properties, triggered by temperature or availability of oxygen or other factors. Growth concepts could also rely on passive systems.

The active system requires the establishment of a building entity, e.g. a building machine such as a 3D printer, or the integration of disperse building entities (e.g. many moving printers).

- **Disassembly**

Disassembly requires the same processes as building itself: some kind of destruction entity, or passive dismantling - integration of destruction processes into structure or material.

Smartness

Julian Vincent came up with this wide definition of smartness:
"Smartness can be a simple response which follows on directly and inevitably from the stimulus; or the outcome of an if-then construct in which a decision is made based on balancing the information from two or more inputs; or the ability to learn, which is probably the smartest thing of all, since learning can lead to a patterned model of the world (the brain is 'stored environment') allowing informed prediction."[364]

Smart or intelligent?

The term "smart materials" has come to be used as a collective noun including structures and materials which have some kind of responsive ability, showing adaptive behaviour. Smart and intelligent buildings and materials are referred to in the discussion of built environment. Various definitions exist for "intelligent building", but we seem to be far from "intelligence" as generally understood:
"Intelligence is the ability to acquire and apply knowledge and skills"[365]
- to find solutions to new questions, manage uncertainty. The term "smart" is softer, implying a lower kind of intelligence, or meaning mere expedient reactivity. Reacting sensibly and

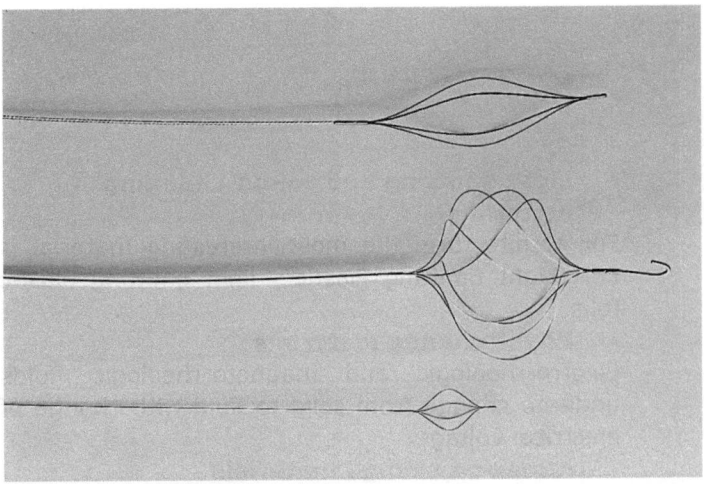

Fig.281 Nitinol, shape memory alloy.

successfully necessarily implies intelligence.

Smart materials

Smart materials have responsive ability and show adaptive behaviour. In most cases, they react with a change of electric voltage. They are not yet common in architecture, but will certainly be used more frequently in future. Some of the abilities mentioned above can be subsumed under "smart material behaviour". Many different classifications of smart materials are found in literature. A classification according to the kind of reaction is presented here, based on the research report on smart materials by the Institute for Lightweight Structures in Stuttgart, 1998.[366]

- **Shape change materials**

Thermostrictive materials like shape memory alloys, thermostrictive ceramics, thermostrictive polymers, piezoelectric materials and chemostrictive materials can change shape. A shape memory alloy (e.g. TiNi Nitinol) changes shape with temperature. It deforms to its martensitic condition at low temperature, and regains its original shape in its austenite condition at higher temperature.

Piezoelectric materials can transform mechanical strain into electric potential. Optical fibres use intensity, phase, frequency or polarization of modulation to measure strain, temperature, electrical/magnetic fields, pressure and other measurable quantities: they are excellent sensors. Thermostrictive and electro-active polymers have been experimented with as muscle-like actuators. Synthetic proteins extend at low temperatures and fold at increasing temperatures, so that they can perform work and bind water or other substances.

---

364 Vincent, J.F.V.: Smart by nature, in Beukers, A. et al.: Lightness, 1998, p.45

365 New Oxford American Dictionary, program version 1.0.1, 2005

366 Haase, W. et al.: Smart Materials, Recherche und Dokumentation, 1998

- **Light emitting and colour changing materials**

For architecture, the most interesting material is LEP, Light Emitting Plastic - light emitting plastic foils.

- **Phase change materials**

Electro-rheologic and magneto-rheologic fluids undergo change from solid to fluid with change of electrical voltage.

- **Adhesion change materials**

Adhesion change materials are used for functional surfaces.

- **Electron emitting materials, e.g. solar cells**
- **Energy storing materials**

Phase change materials, superconductors[367]

- **Materials and systems with added functionality**

Added functionality includes, among others, self-healing abilities and catalytic characteristics. The development of functional surfaces materials is connected with progress in nanotechnology. Self-repair and self-healing are interesting features in the design of materials for architectural application. Self-healing of pneumatic structures is already being investigated[368]; self-repairing waterproof layers for flat roofs have not yet been produced. Some few smart materials, e.g. self-cleaning surfaces and sensors, have found entry into building by integration into industrial products, but most are still waiting for a possible application in architecture.

Learning and innovation

So-called expert systems can learn and solve new problems. Learning is needed to cope with a changing environment and unforeseen situations. Working with changing environments is standard in architecture, but is not yet applied to buildings. Change can affect environmental input - climate, sound, resources; technical input - new infrastructures develop; ground and settlement in/on it; use/inhabitation/purpose of the built structure that the individual building is embedded in. Catastrophic events and other unforeseen developments add to the list.

Learning systems integrated into control and regulation systems change the "behaviour" of buildings.

---

[367] Classification according to Haase, Walter et al.: Smart Materials, Recherche und Dokumentation, 1998

[368] Speck, Thomas et al; Self-repairing membranes for pneumatic structures in Proceedings of the Fifth Plant Biomechanics Conference, 2006

Flexibility can to some extent allow retrospective adaptation (e.g. change of use/inhabitation - neutral floor plans). Flexibility in construction interpreted as increased load bearing capacity provides multipurpose opportunity (e.g. roof extension, additional cladding, security...) but is less efficient.

Innovation

On a typological and technological level, progress is made by innovation. In architecture, innovation is not only sought for as an enhancement, but also as a status symbol, the psychological and social need for differentiation being an important trigger. Innovation in architecture happens on both ends of the scale of financial potency, but not in all fields. It occurs in projects where pressure to produce novel solutions is great. There are several reasons to look for innovation:

- High density - pressure of population growth (Japanese metabolism, Ken Yeang's developments for urban areas in Malaysia)
- Ecological and economic conditions - lack or scarcity of resources (As the case study about Nias will show, the development of very subtle wooden constructions was enhanced by the area's remoteness, whereas today the lack of timber leads to an increase of concrete structures)
- High "importance" of the building, being a role-model or a status symbol (top public buildings, e.g. stadiums, global headquarters of organisations, and those countless examples of status symbol of wealthy or eminent people e.g. Bill Gates)
- New challenges, new environments, e.g. underwater developments, arctic developments, space architecture
- Catastrophes and failures
- Individual differentiation, inventiveness

Difficulties

- **Single solutions - no mass market**

Innovation in architecture as such is not easy to enhance, as designs are single solutions. This is different in other disciplines with mass production. For this reason, spin-off solutions for problems of material, implementation technology or even design issues are introduced into architecture from other disciplines - the defence industry, mechanical engineering, the car industry, aerospace industry, industrial design etc., concerning material, implementation technology even design issues.

Fig.282 Millennium footbridge in London, by Foster Assoc. and Arup, 2000.

- **Architecture is a holistic long-term solution**

Architecture is a long-term solution: no change is expected for many years. The long timeframe enhances a conservative approach. Architecture affects the entirety of life, not just a limited timespan like as motor-car or furniture.

- **Safety issues**

The safety of people who work and live with architecture has to be guaranteed. For this reason, and to clarify responsibilities, architecture is a highly regulated field. Safety issues are met by building regulations, and production and building companies are strictly responsible for safety, at least in the Western world. Regulations exist on any level of design and make innovation a difficult art, which requires much effort and energy. Innovation on the building level has to comply with all regional building regulations. Innovation on the level of material and products has to undergo certification testing to meet requirements of stability, durability, fire protection, health etc. The building industry is therefore notoriously conservative.

Enhancing innovation in architecture and urban design needs more investment into research of the built environment. The disconnection of the experiment from application by means of simulation is already common and virtual models are used to visualise future developments in architecture, urban planning and construction engineering. For many reasons, modelling is still insufficient to predict all implications of our future buildings.

An embarrassing and enlightening example is the Millennium footbridge in London, designed by Foster Assoc. and calculated by Arup engineers, which had to be closed three days after the opening due to heavy swaying. The phenomenon responsible for the movements was not well known until then: synchronous lateral excitation, or people-excited lateral vibrations. It was known before that with every step of a pedestrian a slight outward horizontal force affects the ground, but in this case stabilising and synchronising effects increased the movement to an unknown extent. Months after the failed opening in June 2000, the "Wobbly bridge" was retrofitted with dampers.[369]

This incident has made it clear that we still don't know enough to simulate reality. The conclusion is that working with analogue models is still important. In architecture, the creation of regulation free space for real life simulations - e.g. building exhibitions - would enhance development. Experimental projects all over the world are well known, visited and studied by architects, students, and the public - see the Expo 2000 buildings in Hannover.

For the future, a hybrid effort seems reasonable: the still inevitable use of real models combined with virtual representations.

---

[369] http://www2.eng.cam.ac.uk/~den/ICSV9_06.htm [11/2007]

Fig.283 Interactive art work Aegis Hyposurface, dECOi 2001, Birmingham.

Fig.284 Detailed view of Aegis Hyposurface.

Fig.285 MUSCLE, ONL 2003, interior space.

The reactive building

A truly reactive building would integrate many forms of environmental influence and carry out many forms of reaction. Newest architectural designs in virtual reality, e.g. in Second Life, experiment with all sorts of reactivity. In real life, few projects aiming at reactivity have been carried out.

Mark Goulthorpes (dECOi) and an interdisciplinary group of researchers designed the so-called "Aegis Hyposurface" in 2001 as an interactive art-work for the foyer of the Birmingham Hippodrome Theatre.

"The piece is a facetted metallic surface that has potential to deform physically in response to electronic stimuli from the environment (movement, sound, light, etc). Driven by a bed of 896 pneumatic pistons, the dynamic 'terrains' are generated as real-time calculations."[370]

Spatial reaction is achieved at least in an experimental wall project.

Kaas Oosterhuis and his ONL office investigated what they called the "building body".

Oosterhuis has published on topics like "hyperbodies" and "swarm architecture" inspired by a biological paradigm. His "MUSCLE" project presents a progressed form of reaction, involving space and structure. Nonetheless, it is still proto-architecture, an experimental project designed for an exhibition. *"The MUSCLE is a pressurized soft volume wrapped in a mesh of tensile Festo muscles, which can change their own length... The balanced pressure-tension combination bends and tapers in all directions... The public connects to the MUSCLE by sensors, and by input through sliders on the computer screen. The sensors are attached to the reference points of the construction. Coming closer to the sensors triggers a reaction of the MUSCLE as a whole... But meanwhile ONL has programmed the MUSCLE to have a will of its own. The MUSCLE may not want to go there, and may try to crawl back. Then a true interaction starts, and the outcome of the transaction process may be unpredictable. The MUSCLE is the prototype for an environment that is slightly out of control."[371]*

370 http://www.sial.rmit.edu.au/Projects/Aegis_Hyposurface.php [11/2007]

371 http://www.oosterhuis.nl/quickstart/index.php?id=347 [11/2007]

Fig.286 MUSCLE, ONL 2003.

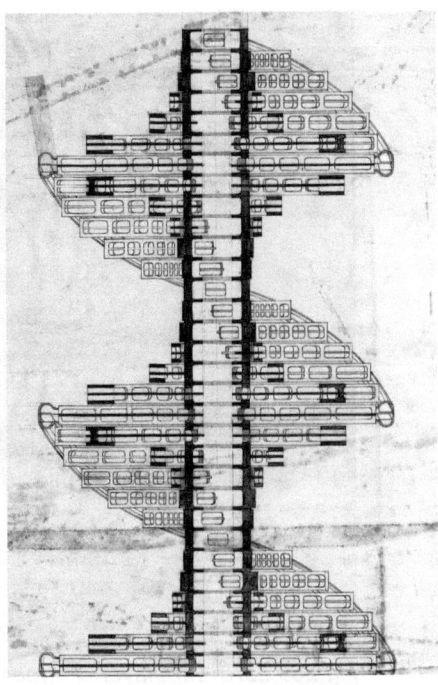
Fig.287 Kurokawa project, section.

4.2.10 Homoeostasis and metabolism

"Homoeostasis is the tendency toward a relative stable equilibrium between interdependent elements, esp. as maintained by physiological processes."[372]

"[Metabolism is] the chemical processes that occur within a living organism in order to maintain life. Two kinds of metabolism are often distinguished: constructive metabolism, the synthesis of the proteins, carbohydrates, and fats that form tissue and store energy, and destructive metabolism, the breakdown of complex substances and the consequent production of energy and waste matter."[373]

Within a living entity, dynamic equilibrium of input and output of matter and energy is achieved, as described earlier in 4.1.3. on the classical criteria of life.
For architecture, homoeostasis and metabolism can be interpreted as the ability to establish an internal system of dynamic equilibrium, by harnessing, transforming and emitting matter and energy to the environment.
Metabolism and symbiosis as stated by Kisho Kurokawa in the 1960s was meant to implement change, exchange and constant renewal into architecture, using natural role models (tree, spine, trunk, DNA etc.) for the organisation of elements.

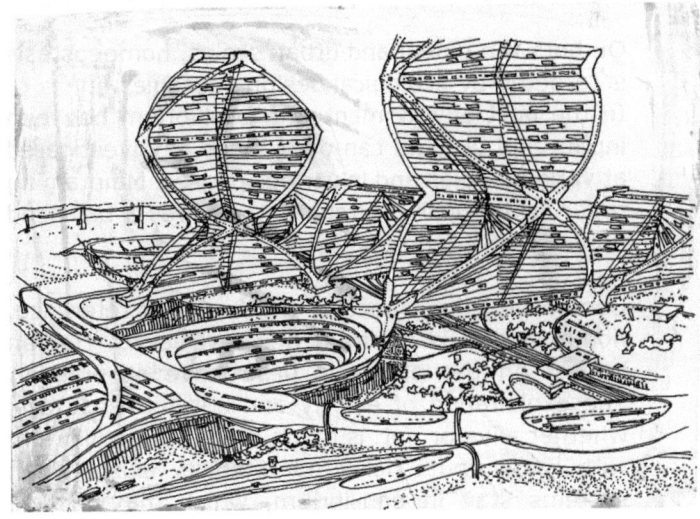
Fig.288 Kisho Kurokawa metabolist project, 1960s, shaped like DNA.

[372] New Oxford American Dictionary, program version 1.0.1, 2005

[373] New Oxford American Dictionary, program version 1.0.1, 2005

New approaches | Architectural interpretation of life criteria

Fig.289 Earth with layer of atmosphere and moon seen from the Space Shuttle Discovery.

Scales

On both the global and urban scales, homoeostasis is targeted by ecological design strategies.

In the built environment, the equilibrium between inputs and outputs can once more be investigated at various scales and in various phases. Maintaining order in the built environment requires a balanced flow of matter and energy.

- **Global scale - ecology**

Considering the planet as the frame of reference, homoeostasis is far from being achieved. Human impact on the planet is destructive at present. The exploitation of energy and natural resources, whether alive or not, is by no means in equilibrium with the outputs delivered. Stability exists when systems stay in equilibrium, which may include continuous change: the global ecosystem has always been changing, but recently the speed of change has become alarming.

Human activity is no longer limited to local level. Networks of resources have spread all over the globe and the provenance of products is not transparent to customers. In general staying regional would involve a "back to the roots" approach, abandoning important achievements of civilisation. "Sustainability" should aim at a balanced state on a global scale. "Globalisation" entails diversification in industry and economy - its advantages may outweigh increased transportation in terms of cost, but may not in terms of resource and energy use. The latter should be the determining factor deciding the question of whether it is better to "go global" or "stay regional".

Weighing possible implications on a global scale is a difficult task, and errors are waiting to happen, but nevertheless this is all we can do to know more about what we are actually doing.

Ecological design approaches try to identify and evaluate impact inputs and outputs - Ken Yeang suggests biodiversity as the final measurement of impact.

In terms of biodiversity the sheer existence of a building is a threat to nature, if we envision primeval forest covering the same site. If we talk about cultural landscape, the comparison between before and after the construction of the building is still a challenge for architecture to integrate as many living species as possible, apart from humans.

As already mentioned previously, on the global scale the share of the built environment is substantial, considering the consumption of resources and energy on the one hand, and the output of pollution on the other.

"Building use: 50% energy, 42% water, 50% material, 48% agricultural land loss
Building related pollution: 42% air quality, 50% global warming gases, 40% drinking water, 20% landfill waste, 50% CFCs/HCFCs, and 50% indirect responsibility for coral reef destruction"[374]

As the share of the built environment is considerable, architecture ought to be a highly influential discipline. The awareness of that influence should make energy and environmentally friendly design more worthwile goals.

374 Edwards, B.:AD Architectural Design 71, Green Architecture, 2001, p.108 fig.1

Fig.290 Biosphere2, Arizona desert in 2004.

- **Urban scale**

On an urban scale, the flow of matter and energy affects large areas, and the impact of densely populated areas on the natural environment is destructive. Economic control loops deliver the supply of goods according to demand - information is manifold, no centralised control ensures supply. Centralised control as realised in Communism is not dynamic enough to meet the needs of people.

Supply networks deliver energy. Control of energy delivery is not centralised, but central (government) bodies may become active if things get "out of control". Altogether, the sum of activities and interactions is complex and difficult to predict on a regional level.

- **Building scale - all phases of existence**

Just like processing, inputs and outputs can be measured on building scale. An internal environment is established for the operation phase, as the "milieu", in which humans prefer to live. The bigger the difference of parameters it requires compared to the external environment, the more energy is needed to maintain it.

Ordered according to their processing of matter and energy, the two ends of the scale of architectural categories are: space design as the high-tech approach to establish environment for human survival in an extremely hostile environment, versus traditional architecture as the low-tech approach to provide shelter with as little effort as possible, in a regionally selected environment which already delivers better conditions.

- **Autonomy**

The independence of buildings from their environment was tested in Biosphere2, in the Arizona desert. Biosphere2 is an assembly of buildings (living, library, infrastructure, "biomes" - greenhouses), which was created to provide shelter and livelihood for eight persons for a scheduled time of two years from 1991 to 1993 within an artificial and autonomous ecosystem. The experiment delivered important scientific results, showing the difficulties of predicting the behaviour of ecosystems:

"The major atmospheric observation of the two-year closure was the decline of oxygen. This was definitively traced to a two-step process of oxygen loss to soil organic matter producing CO_2, plus the CO_2 being captured by structural concrete to form calcium carbonate."[375]

The building materials were carefully selected, but it had not been anticipated that the structural concrete would capture such amounts of oxygen.

---

375 http://www.biospheres.com/ [11/2007]

Milieu

Milieu is used in life sciences for a set of parameters, which refer to the existence of specific organisms. In social sciences, milieu characterises a social environment which is important for specific people. Milieu in built environment denote the sum of architectural influences that determine the life of people. Within buildings, it encompasses some important parameters for comfort in internal environments:
- Temperature
- Humidity
- Ventilation and air movement
- Light
- Electromagnetic radiation

Some of the parameters cannot be regarded as independent; to give an example, the temperature felt by a human being depends on air movement, temperature and humidity. Human experience delivers a general sensory output of comfort, which takes into account more than one parameter. For the evaluation of comfort, many parameters have to be related to each other.

- **Influences**

Not all existing influences on the built environment can be used or even exploited. Most influences must be at least partly averted, and determine built environment through defence and control mechanisms. There are continuous, stable influences and discontinuous influences, often following a "natural" rhythm of a statistical kind (50 year flood, rain per m^2 or year). Input to the built environment contains both matter and energy. Matter/materials are: air, water - precipitation and water for use, land, minerals, fossil fuels and agricultural products.

- **Processing and storage**

Energy processing was already discussed earlier, and represents a considerable part of metabolic activity in the built environment. Processing of material for the building industry and built environment in general happens industrially. In buildings, occupants' and users' activities process material by means of infrastructure.

Storage facilities are needed for material of any kind and these take up large tracts of the built environment. Storage space is also needed for waste processing.

- **Output**

Output of buildings is material waste, ash, heat (losses by transmission, hot exhaust of gases), gases and sewage. Such output is not used as thoroughly as one might expect. Much energy is lost in the form of heat; the recycling of waste from building activity has only recently become a serious market factor (e.g. use of demolished concrete for road substructure, recycling of plastic for insulation material...)

Stability of the internal environment is guaranteed by continuous input, and by active control of the flows of matter and energy.

- **Hierarchies**

Standard architecture is and has to be embedded into large regional networks of supply and waste management. A hierarchy of metabolic activities is established, especially in densely populated urban areas. Traditional and informal architecture shows a comparably small flow of matter and energy, usually limited to the house and its immediate surroundings. Ecological building tends towards a re-individualisation of metabolic functions, the "production" of energy and waste management. This is not a step backwards, considering the efficiency of local action in biological systems.

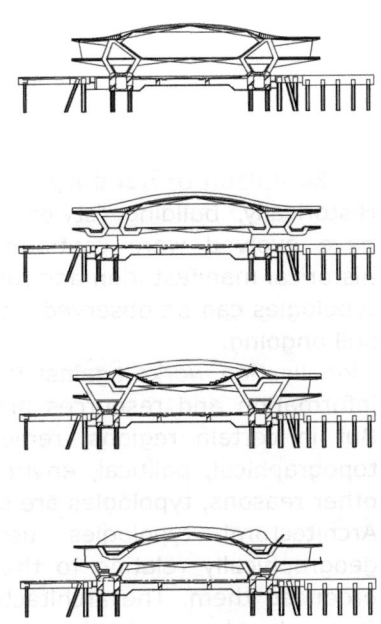

Fig.291 Section series.

Fig.292 Yokohama terminal, upper layer of topographic space, FOA, 2002.

4.2.11 Evolution and natural selection

"[Evolution is] the process by which different kinds of living organisms are thought to have developed and diversified from earlier forms during the history of the earth."
"The gradual development of something, esp. from a simple to a more complex form"[376]

Evolution includes all processes which have led life from its beginning to today's diverse manifestations. Evolution and natural selection are discussed in chapter 4.1.5, and approaches for transfer to technology are shown in chapter 2.2 on the fields of research in biomimetics.

Transferred to architecture, "evolution" has at least three different notions:
- Evolution of architecture
- Evolution in architecture
- Evolutionary architecture

Evolution of architecture

In the course of the evolution of humans and animals, architecture has evolved as a characteristic expression of the species. Animal buildings have evolved over millions of years. Existing typologies are bound to the respective species. Variation of built structures is limited to features needed for local adaptation (spider's web, termite nest).

The human species has always needed shelter from environment, but "Gestaltung" of shelter cannot be traced back to its beginnings, and a historical interrelation between species and building typology cannot be stated. The phenomenon of forming the environment to the respective species needs is known in the animal kingdom as well. Apart from the many temporary buildings like birds' nests, the buildings of many insect species form an interior environment the survival of which the individuals depends (ants, termites, bees etc.). The buildings of beavers change the environment considerably. But these achievements are small compared to the way human species has adapted the environment to its needs. By the impact of human activity on the environment, evolution of the human species is projected towards the outside. This may be a first step in the evolution of a higher level of organisation towards Lynn Margulis' biosphere as living entity.

The evolution of the built environment started with the assembly of shelters or buildings. With growing settlements, urban structures have evolved along the development of society's complex structures and cultural achievements. Examples of urbanisation are known from all over the world.

---

376 New Oxford American Dictionary, program version 10.0.1, 2005

Evolution in architecture, subject approach

Evolution in architecture is interpreted as an analogous term for the process of adaptation to internal and environmental conditions, in the sense of phylogenesis by empirical trial and error, variation, evaluation and survival. The development of architectural typologies occurs by evolution in terms of slow gradual adaptation.

Adaptation can happen continuously, and can be observed well in traditional architecture: old building typologies, having developed over long periods through gradual enhancement in a trial and error manner through innovation measures, which slowly enhance the building, or parts of it. Variability is provided by differentiation. Selection happens according to many influencing parameters, resulting in a general evaluation, which decides the survival of the typology.

Discontinuous evolution in architecture is bound to momentous inventions, related to progress in technology.

This kind of "macroevolution" has delivered characteristics that we consider "normal" features of today's architecture:
- Closure of internal space
- Separation of load bearing and covering

"...the steel frame pioneered by Louis Sullivan and the Chicago School from the 1880s meant that internal structure and cladding could be regarded as separate."[377]

- Separation of openings for access and vision (and also today's infrastructure)
- Vertical piling of spaces
- Integration of functional compartments (building infrastructure)
- Metabolic phenomena in internal environment

It cannot be determined when in history these characteristics were introduced. Apart from convergent developments historical tradition is not good enough to deliver allocations in time. In general, the velocity of evolutionary developments is different in technology, and is now accelerated in virtual reality.

- **Evolution of building typologies**

Historically, building "styles" and typologies that have evolved were continued by tradition. The historical manifestation and further development of typologies can be observed worldwide, the process still ongoing.

Globalisation works against this differentiation, as information and resources are spread worldwide. But in certain regions, remote and isolated for topographical, political, environmental, cultural or other reasons, typologies are still being formed.

Architectural typologies used to be spread geographically relating to the cultural areas that produced them. The architecture of organisations (e.g. churches, schools, companies) differs on grounds of function and identification, and is found dispersed about the built environment.

- **Evolution of new architectural typologies: skyscraper and topographical architectures**

Exceeding height limitations has always been interesting in architecture. Social differentiation is manifested in height differentiation, as height provides a better viewpoint, signifies superiority and importance. High temples and towers were built for religious and political reasons.

The newest typology we are confronted with in present architecture, the skyscraper, has developed over the last century. Skyscrapers evolved at the end of the 19th century due to some major influences:
- Progress in technology: use of steel as building material and development of elevators
- Densification of urban settlements due to economic growth in the US (Chicago, New York)

The limits of this typology have not yet been reached. With new materials and technologies seemingly unsurpassable height limits are still exceeded: From the 1980s until recently the height records were held by Eastern Asia (Petronas Towers, Taipei 101) and will move to Dubai with the opening of Burj Dubai in 2008.

Topographic architecture, the introduction of artificial landscapes, is another new achievement that has appeared within the last decades. Especially in urban environments and for infrastructural purposes, vast areas are covered with spatial structures, integrating added functionality for the public. The event landscape, as proposed by Reiser+Umemoto,[378] or built by FOA - Foreign office architects in Yokohama 2002, may be a new and spreading typology.

---

377 Aldersey-Williams, H.: Zoomorphic, 2003, p.15

378 Competition entry for New York in International House, London (Ed.): AD Architectural Design Nr. 68, 1998, p.88

- **Influencing factors for evolutionary adaptation in architecture**

Hard and soft factors influence evolutionary adaptation in architecture. Hard factors are material, population density, ecological and environmental issues like topography, available material and available energy source. Soft factors are knowledge, technology, society and culture.

Architecture tends to integrate all available factors, just as the long British tradition and availability of knowledge and technology for shipbuilding and sailing has influenced high-tech architecture.

- **Individual "evolution"**

Evolution of individual buildings would mean immediate adaptation to changing conditions, and is more related to "reaction". But gradual adaptation to changing internal and environmental conditions can happen over a long timescale.

Medieval architecture in Europe is still being used, and has been adapted to new requirements countless times. Change meaning change of structure, use, material and surface has always been carried out.

Innovation occurring by this process in a single building can enter architectural evolution on a typological scale when communicated.

The ability to adapt architecture immediately to a changing environment would render typological considerations unnecessary.

As in nature, convergent characteristics have also developed in architecture. For specific technical problems, similar solutions have been found in distant regions of the globe, which cannot be explained by theories of diffusion, e.g. the use of symbolic decoration of the exposed gable peaks on roofs.

Evolutionary architecture

Life Science discoveries have led to technical mimicking of evolution.

In architecture, computational evolutionary processes are also introduced at an increasing rate. The use of genetic algorithms makes sense where situations are complex and no prediction models exist. Evolution is imitated by artificially generating recombination, variation and natural selection, using parts of the biological process for the design process.

Eugene Tsui claims basic principles of his "evolutionary architecture", but aims at an ecological and spiritual approach. Tsui is on the forefront of architects using evolutionary principles as reference for formal organic translations.[379]

But it is already the common architectural design process that shows similarities to evolution and natural selection: design of variations alternating with phases of reflection, evaluation and selection. Transfer of phases, or parts of the phases, to a virtual development process is done by algorithmic design and morphogenetic design procedures.

- **Use of genetic algorithms**

The use of any process and development principle in architectural design inevitably leads to a dissociation of the design process from the designer, to some extent to self-generating architecture. The development is determined by rules set by the designer, but the end product is not or only partly predictable. Methods of this kind need high effort in the development phase, and became possible with computer simulations.

Genetic algorithms process populations of abstract representations to optimisation problems, to evolve towards a solution. It is a search technique, used in computing when no exact solution can be defined and implemented in many technological devices and control systems.

A group around John Holland investigated genetic algorithms in order to abstract and explain the adaptive processes of natural systems, and to design artificial systems software that retains the important mechanisms of natural systems.[380]

---

[379] Tsui, E.: Evolutionary Architecture, Nature as a basis for design, 1999

[380] Goldberg D.E.: Genetic Algorithms in Search, Optimization, and Machine Learning, 1989 p.1

Fig.293 "paraSITE" project, ONL architects, 1996.

Based on the work of John Holland[381], architect John Frazer published "An Evolutionary Architecture" in 1995. *"An Evolutionary Architecture investigates fundamental form-generating processes in architecture, paralleling a wider scientific search for a theory of morphogenesis in the natural world. It proposes the model of nature as the generating force for architectural form."*[382]

The idea of using evolution as a tool involves the application of biological terms phenotype, genotype and genes to architecture. The genes of architecture are interpreted differently.

Frazer uses the notion of ideas as individuals, subordinate to evolution and introduces the *"population of ideas"* and *"...rich genetic pool of ideas"*.[383] Bill Hillier interprets genotype in urban structures as a "deep" urban pattern underlying the visible phenotype. His theory of "space syntax" makes the genotype visible, and can be used as a tool for urban design.[384]

Kaas Oosterhuis also writes about "genes" of architecture, describing the project paraSITE:

"The form-genes of this paraSITE consist of two intuitively drawn closed curves which are related to each other by choosing values using standard commands of the 3D Studio program. Beside the form-genes there are the social genes that describe their way of implementation in society, and of course the intelligence-genes which predict the responsiveness of the volume and the smartness of the skin of the parasite"[385] Oosterhuis interprets "genes" as the determining aspects of the project. Morphogenetic design projects using computational methods for form-generation at present produce proto-architectures. Greg Lynn represents the contemporary "hybrid space" architecture. He uses animation studies for the design of variations of forms. His models enforce the association of organic forms. The "Embryological House" became famous, not only for the design process raising (still) unanswered questions of selection[386], but also as a symbolic project which anticipates the end of mass production in favour of architectural mass customisation. The exactitude of digital programming in connection with automated production techniques is expected to construct buildings of increased complexity and individuality. Lynn says the houses are *"nodules [showing] the budding and elaboration of linked but variant geometries"*[387], the term embryological hinting to the un-developed or developing[388], the interpretation of animation as developmental or evolutionary process.

381 Holland, H.J.: Emergence, from Chaos to Order, 2002
382 Frazer, J.: An Evolutionary Architecture, 1995, p.9
383 Frazer, J.: An Evolutionary Architecture, 1995, p.100
384 Hillier, B.; Specifically architectural theory, 1993
385 Oosterhuis, K.: The genes of architecture, 1994 published in Oosterhuis, K. et al.: Sculpture City, 1994
386 Abel, C.: Virtual evolution - a memetic critique of genetic algorithms in design, 2006
387 Lynn in Feuerstein, G.: Biomorphic architecture, 2002 p.76
388 Feuerstein, G.: Biomorphic architecture, 2002 p.76

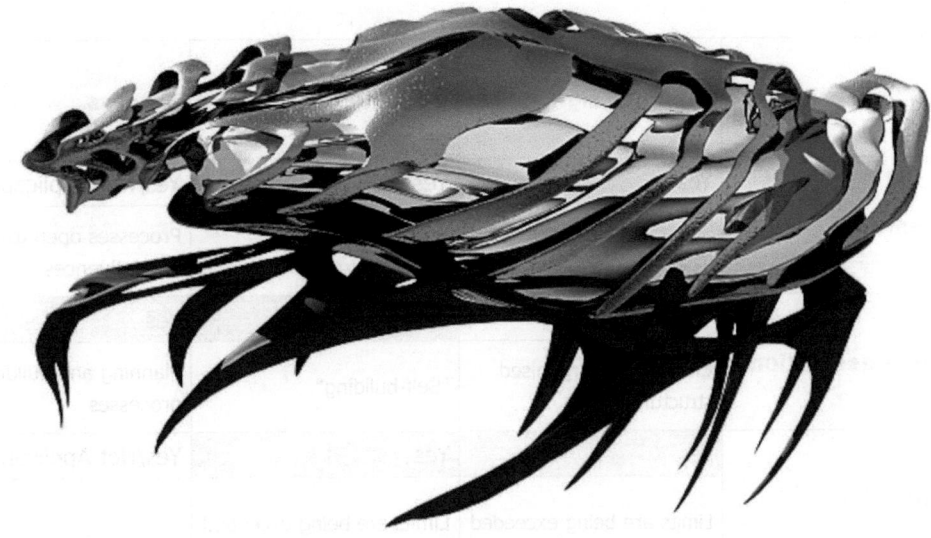

Fig.294 Marcos Novak, "AlloBio".

Marcos Novak, investigating actual, virtual and mutant intelligent environments states:
"One of the fundamental scientific insights of this century has been the realization that simulation can function as a kind of reverse empiricism, the empiricism of the possible. Learning from the disciplines that attend to emergence and morphogenesis, architects must create generative models for possible architectures. Architects aspiring to place their constructs within the nonspace of cyberspace will have to learn to think in terms of genetic engines of artificial life. Some of the products of these engines will only be tenable in cyberspace, but many others may prove to be valid contributions to the physical world."[389]

Evolutionary processes are just one of life's criteria, and cannot be implemented in another field without consideration of other life criteria. The envisioned future architecture of the same people who are on the forefront of propagating evolutionary architecture will also involve other criteria of life, e.g. reaction and adaptation abilities.

There are numerous other examples of architects starting to use these design strategies.

- **Memes, optimisation and frame of reference**

Interpreting virtual evolution in architecture as an optimisation process - what are the characteristics, and what are the criteria for optimisation? How do tangible characteristics and aspects relate and influence each other?

Which aspect can be taken for optimisation in an artificially evolutionary way, and does the subsequent integration of this aspect have emergent consequences that cannot be predicted?

Critics argue that the "ideas" used by evolutionary design resemble the so-called memes, discrete units of cultural information, if anything. The social and cultural mechanisms for the dispersal and propagation of memes are fundamentally different from the biological process of evolution and natural selection.[390]

Optimisation in architecture depends on the parameters taken into account for a decision. A good architectural solution is efficient at considering the particular frames of reference which define the importance and value of design parameters.

Optimisation does not imply the existence of a best solution, or a best design. It can only be judged by a specific framed viewpoint, which relates to, but is not exclusively defined by, geographic area, population, social and cultural influences, economy, etc. Moreover, it is a set of criteria that are important for the project. The context is crucial for any evaluation.

But no matter which frames of reference to which an architectural project was designed to relate, the judgement of quality follows criteria of use, experience by the user and impact of the built structure to the environment.

---

389 Novak, M.: Transmitting Architecture: The Transphysical City

390 Abel, C.: Virtual evolution - a memetic critique of genetic algorithms in design, 2006

	Urban Design	Building	Process	Material and Structure
Openness	Yes	Yes	Yes/Not Applicable	Yes
	Cities as open systems	Buildings as open systems	Processes open to input and influences	Permeability
Self-organisation	Yes	Yes/No	Yes	Yes
	Cities as self-organised structures	"Self-building"	Planning and building processes	Material processing and structuring
Limitation	Yes	Yes	Yes/Not Applicable	Yes
	Limits are being exceeded	Limits are being exceeded		Limited in space and time
Information processing	Yes	Yes	Yes	Yes/No
			Processing is based on information	Not enough inherent information
Order	Yes	Yes	Yes	Yes
Propagation	No	Yes/No	Yes	No
	More growth than propagation	Propagation of typologies	Cyberspace	
Growth	Yes	Yes	Yes	Yes/No
	Urban growth	Extension and increase in size	Growth processes used for design	Sustainable materials, but no growth in applied state
Energy processing	Yes	Yes	Yes	Yes/No
				Production, but no harvesting
Reaction	Yes	Yes/No	Yes	Yes
		No intelligent reaction yet		Smart Materials
Homoeostasis and Metabolism	Yes	Yes	No	No
Evolution and Natural Selection	Yes	Yes/No	Yes	Yes
	Evolution as development	Yes on typological level		Evolution as development

Fig.295 Translation diagram, architectural interpretation of criteria of life, light grey fields: important issue in current architectural discussion, dark grey fields: as yet unexplored, a strategic search for innovation should follow.

4.3 COMMENTS AND HITHERTO UNEXPLORED FIELDS

In spite of the observed increase of life criteria in modern architecture the emergence of these criteria does not occur in a specific order, and there is no natural sequence of life's criteria emerging in built environment. Some of the criteria are basic inherent characteristics (order, limitation), whereas others cannot yet be found in reality (propagation in the notion of self replication). There seems to be no timeline of appearance of the criteria in the history of architecture. Reaction to the environment in a passive way is also a very old concept: nomadic architecture anticipated today's mobile architecture.

Energy transformation and homeostasis are recent achievements, while evolution processes as a new option in design is the youngest phenomenon.

There are no projects showing all criteria of life. The translation diagram on the previous page presents a rough overview of the applicability of the terms to architectural interpretation at different scales, together with examples for each interpretation. Light grey fields are important in current architecture discussion and experimental designs; dark grey fields are still unexplored and should be followed by a strategic search for new architectural vision.

- **Activation of elements**

The combination of the criteria of life delivers even more interesting visions. The activation of architectural elements requires energy processing, information, reaction and metabolic characteristics and seems to be the most interesting approach that is currently taken towards a living architecture.

For instance, the connection of information and energy processing, of metabolic features and reaction could deliver a completely different strategy for enclosure and openness. Experimental projects like the Water Curtains of MIT and the Blur Building of Diller + Scofidio[391] realise a more ephemeral approach to enclosure and definition of space, activation of architectural elements being a precondition.

- **Intelligent management of information and material**

Strategies for minimising information processing include the use of structures that do not need much information, and the multiplication of existing structural units, with fractal geometry being the role model. With analogue measures, effort and error rate in planning and implementing in architecture and technology could be reduced.

This does not imply a simplification, but an increase of complexity by the exploitation of more structural levels. Simultaneous information processing would enhance the whole course of planning. A general integrated environmental model should be available for planning purposes. Access to information is often used as an instrument of power, but open access has in many disciplines proved to enhance development. The growth of information depends on information processing by tradition, passing-on of knowledge. Even if architectural plans usually represent one, specific, solution for a specific site, the plans of buildings should be available after completion (except when that conflicts with security). In this way development could continue much more easily.

Information on the scale of material is connected to matters of order and hierarchy. Introduction of new structure levels in building materials would generate more efficient materials - in Julian Vincent's words: *"In nature, shape is cheaper than material."*[392]

- **Limitation in space and time**

The most efficient strategy is either a coherent life cycle approach to limitation in time or strategic renewal of building parts. The life cycle approach defines the life expectancy of the building as a whole as the basis for choosing particular materials and structures. This avoids the use of expensive long lasting materials in temporary structures, which would then have to be recycled.

The strategic renewal approach is based on long life expectancy as well. Construction and primary elements are designed for long-term survival and specific elements have to be replaced regularly (e.g. roof covering in contrast to structural components) This approach requires architectural details to be designed for dismantling.

Similar physical limitations in space are valid for biological systems as well as architecture: space and surface for exchange of resources and access defines volume/surface ratios of buildings, leading to specific architectural and urban patterns. Gravity and environmental forces limit structural design. In general, the limitation of energy and material leads to intelligent and efficient use; design and innovation are still enhanced by shortage.

- **Order**

Physical laws, growth processes and survival determine order in biology, whereas order in architecture is determined by environmental conditions, the user's patterns of activity - e.g. movement -, and the technological means to plan and build.

---

391 http://www.thecityreview.com/arcnowt.jpg [11/2007]

392 Vincent, J.F.W. in: Beukers, A. et al: Lightness, 1998, p.44

Fig.296 Blur Building Swiss Expo 2002, Diller + Scofidio, artificially created cloud.

There are fundamental differences between order in nature and order in architecture, but both are based on the same mathematical and physical principles.
In architecture, time related aspects are becoming more and more frequently implemented. Bottom-up design methods are experimented with, integrating kinds of self-organisation. Their introduction into planning and building processes is difficult, but could create better efficiency.
Finding the right frame for decisions is essential for good design. The definition of the key part of the whole, formulation of problems and solutions is also a matter of hierarchical levels. Bottom up - top down balance in design is a matter of scale and frame: some effects cannot be observed locally and therefore decisions cannot be drawn on local levels. For these phenomena top-down approaches are necessary.

- **Survival**

Architecture projects suffer from a problem well known to organisms: they have to work, and experiments endanger the survival or the security of the users. But in contrast to organisms that exist in real time we have the means to establish a safe "playground" where experimental architecture could be tested without putting anybody at risk, apart from the virtual playground we already use.

- **Propagation**

Propagation in architecture has not yet come beyond the passing on of information.

- **Growth**

Growth on a material scale, as surface increase, is not yet used or investigated in architecture, as architects still only dream of a really elastic but structural material for the skin of a building. Architecture grows by addition, in contrast to growth by cell division in biological organisms.

Urban growth has not yet reached its limits. Protecting nature reserves from urban growth is essential for the future, as we do not know to what extent landscapes can become "cultural" and still provide sustainability, in other words: how much natural environment has to be left for the biosphere to survive. The building of green corridors and active integration of habitats for other species into human architecture will be tasks for the future, the design of conservatories being perhaps the first step.
Growth related to users' needs is another topic not yet dealt with. The dynamic activity of people rarely results in a dynamic architecture, at least not on a building scale. The need for change is commonly met by a change of place: moving to another flat, another part of town, even migrating to another country. Today's architecture does yet not grow with user needs. A comparison of the overall efficiency of the two strategies in terms of energy and matter would be another interesting study.

- **Energy processing**

The biggest difference between living systems and buildings in the use of energy is that instead of the excessive energy use in technology, information is used for problem solving in nature.[393]
We are working at mimicking some of nature's technologies to harvest and store energy (artificial photosynthesis etc.). Energy efficiency is a major topic in architecture, and increasing activation of building elements will increase the need for decentralised energy harvesting in order to reach the goal of zero consumption from central sources. We have achieved some success in harvesting, but - to cite an example - chemical storage of solar energy is still a challenge.

[393] Vincent, J.F.V. et al.: Biomimetics - its practice and theory, J.R.S., 2006, p.8

- **Reaction to the environment**

The use and further development of intelligent building systems including sensory devices, diagnostic models, actuation and control measures is a standard in today's architecture.

Reaction of building as a whole has not yet been invented, not even at a conceptual stage. Houses do not move, they do not get out of the way; they do not take an active role in every-day life. Apart from structural flexibility and actuation, the ability of overall reaction implies a kind of building mood, a common parameter as expression of an overall status or overall quality and as a decision making tool (just as the "sentient building concept" proposes for sets of parameters[394]).

Emotion as a complex reaction could be artificially introduced as emergent quality.

Control of building reaction is carried out by active means, as the newest developments in architecture show. Investigation of swarms and other phenomena created by multiple agents or influences hint at a deeper understanding of biological models and aims at emergent behaviour of systems.

Astonishingly, features that "care" about users are not experienced as being convenient, but as being difficult to deal with because of communication and control difficulties. As a consequence, features involving the sense of touch are considered more scary than nice to have, even if it sounds poetic to have a responding environment (Can you lean against a wall, and it will respond? If you fall - can the floor become soft to hold you? Could you tell the walls of your room to become transparent, or become permeable to wind? Could you form a comfortable seat out of the shell surrounding you?...).

Kas Oosterhuis thinks beyond responsiveness and visualises active participation of architecture: *"True hyperbodies are pro-active bodies. True hyperbodies actively propose actions, they act before they are triggered to do so. Hyperbodies display something like a will of their own."*[395]

- **Homoeostasis and metabolism**

Undoubtedly metabolic processes in architecture are on the increase. Buildings without processing of matter and energy in operational phases are no longer imaginable. Pneumatic structures even depend on metabolic flow for their own structural integrity - this is a new dimension in development.

- **Evolution and natural selection**

The implementation of evolution and natural selection in technology (in the sense of using the process as a tool) is a very recent achievement. Technology and culture can be interpreted as evolutionary achievements as well.

Adaptation and differentiation as the outcome of evolutionary design processes are used to enhance the quality of the built environment. Other characteristics which are being explored by life sciences (e.g. preadaptation, co evolution) exceed the present interpretation and will be important for future transfers. Combining the advantages of the process of evolution and the advantages of creative design seems a reasonable hybrid strategy for future design.

- **Conclusion**

The role models in nature deliver ideas of what could be pursued in the future.

"Intelligence" of a building would involve new and unforeseen reactions - we are still waiting for buildings showing this ability.

There are many reasons why we do not yet have many more buildings boasting the qualities of living organisms. Apart from technical problems and the generally conservative building industry, the question of power and control over the system and the environment is a crucial topic. Implementation of characteristics of life is easier in temporary structures and buildings of experimental character. Many renowned architects experiment at this 1:1 level, using their building tasks to develop specific ideas and push ahead industry.

The superposition of biological paradigms onto architecture is not equivalent to the claim of architecture being alive. The discussion of the criteria of life in architecture clarifies terminology, and thus improves understanding and connection between the fields of biology and architecture. Taking this into account, an innovative development will inevitably will take the direction towards more sustainability, ecological compatibility and quality.

In the future, the findings of life sciences research should be made more easily accessible. Open access policies will support the spread of information and enhance development. Architects can deliver considerable research into traditional architecture and natural constructions, but basic information in life sciences has to be delivered by other disciplines. Working in interdisciplinary projects is a precondition for knowledge transfer and should be supported to increase the exchange of information in all fields.

---

394 Mahdavi, A.: Sentient buildings, from concept to implementation, 2005

395 Oosterhuis, K.: Hyperbodies, 2003, p.55

Architecture	Organism
Idea, project development Planning	Evolutionary development of species
Production	Conception, creation of egg cell, mitosis
Implementation, building	Embryonic growth Birth - aliveness Growth
Operation	Fully grown state, normal life, propagation
Damage, abrasion	Injury, illness
Repair, renovation	Recovery (self-healing), Medical treatment
Change	Metamorphosis
Vacancy, abandonment	- no analogy in nature
Decay, dismantling	Death
Recycling	Recycling

Fig.297 Comparison of the lifecycles of architecture and organisms.

4.4 A LIVING ARCHITECTURE

4.4.1 Life in architecture, or when is architecture considered alive?

By its nature, architecture is not alive. It is part of inanimate nature, and is subjected to the same physical principles and processes. Even if some of life's criteria exist in individual projects, technology is still far from creating artificial life.

In spite of that, architecture is said to have a lifecycle. In the above diagram the lifecycle of a building and that of an individual organism are compared.

The processes leading to the different stages are fundamentally different, but the comparison shows parallel stages. The existence of settlements and cities is independent of the individual unit to some extent. They are renewable structures which also evolve. Today's phenomena include storage of gadgets rather than activity, space created for cars rather than for people, development of uncontrolled urban sprawl and "death" of large urban areas.

The usage of architecture is the significant parameter for "aliveness". Artificial life is a tempting issue in the discussion of life sciences and architecture. In other fields of technical application (e.g. robotics) this may well be an important topic.

In spite of the presence of some already existing criteria of life in architecture, the whole range of these criteria has not been found in one single architectural project. Applying the classical life sciences theory we can therefore say definitely that architecture is not alive, and we can assume that it will not be alive in the near future. The question of if architecture could be alive is by no means trivial: futuristic architectural visions include features of aliveness. Considering Science Fiction architectures and the attempts towards intelligent buildings, we can state that aliveness is not appreciated by users, who see their power to control their environment endangered. Aliveness of technical systems gives people an uneasy feeling. And, if architecture were alive, we would have to live with death and failures as well as successes.

On the other hand the expression of "architecture being alive" is commonly used for life in architecture, assuming architecture being used and valued highly by the occupants. Life in architecture is a sign of high quality, and can be stated by:
- Occupant satisfaction
- Use of space, frequency of activity (also constant over time - as against shopping zones that are "dead" in the evening)
- Integration of architecture in the social and cultural lives of people
- Exchange of matter and energy with its environment
- Slow increase of entropy - good maintenance: energy/material input by maintenance measures to stop normal decay
- Added value for the environment, design of the environment

The expressions "aliveness of architecture" and "life in architecture" have different meanings. The discussion of life of, and life in, architecture, can deliver the means to talk about the quality of architecture beyond flows of energy and material. There is no single measurable parameter which indicates architectural quality, but the values mentioned above are investigated when we are sufficiently interested in the quality of our built environment.

4.4.2 Artificial life in architecture, the artificial designer designs itself

For good reason, automation has not yet been introduced into architectural design to the highest possible degree, given the computational means available. Architectural design renders visions as reality. For automated design processes, the generation of aims and visions that a project should fulfil would also have to be automated in order to reach reasonable results. This would require the automated system to know about the current status of our built environment, the potentials of action and the vision of the future.

We still do not manage to build individual technical systems that - due to their artificial intelligence - can "survive" in the complexity of their environment, so we seem to be a far cry from creating built environments by artificial systems. If we don't yet manage to create artificial life forms, how could we create an artificial creator?

Simulations are already integrated into the planning process. The hybridisation of automated processes and human design seems to be the most reasonable prospect at the moment.

Automation is implemented for other good reasons:
- Coping with the complexity of a given task
- Fast processing of large amounts of data
- Automation as an argument for the seemingly objective results of a design process
- Optimisation of particular parameters
- To enhance innovation
- To achieve the unexpected

Automation processes increase the distance of the designer from the project (insertion of a design tool) and can result in reduced responsibility for the outcome, but on the other hand, the increased use of simulation of the real world should help to ensure that the best design for the given building task can be produced.

Experiments in virtual space require good models of built environment and all the processes involved. In general, we do not yet have models of our environment sufficiently sophisticated to make decisions.

Better information will hopefully soon be available for public access (Google Earth and 3D data of cities are already available). As soon as models are sufficiently reliable, experiments can take place in a virtual environment.

4.4.3 Creation of environment

Consciousness of ecological backgrounds is no doubt increased by the investigation of nature, but the decision to go actively for sustainable design has to be taken independently of biomimetic approaches. Architecture should assume more than a passive auxiliary role for the creation of our future environment. Investigating changes of environmental conditions through architecture delivers the knowledge needed to assign architecture an active role in the creation of the cultural landscape.

5 CASE STUDIES

In the following, selected examples of a biomimetic approach to architectural research and design carried out by the author will be presented. Together, the case studies show the diversity of possible applications of natural phenomena in architecture.

- **Adaptation and evolution of traditional architecture on Nias Island**

The study of traditional architecture on Nias Island is a research project, based on a field trip in 2005. In the context of biomimetics, it represents the approach of research of the quality of traditional architectural typologies, which have "evolved" over a long timescale. The study includes the characteristics of adaptation to environmental conditions that may be of further use. [396]

- **Transformation architecture**

Transformation architecture is interpreted here as purposeful translation of principles from nature to architecture. The application fields are experimental, one of the innovative fields being space architecture. Various design projects apply natural phenomena. [397]

- **Lunar Exploration Architecture**

The study "Lunar Exploration Architecture" was carried out in 2006 by an interdisciplinary design team at the Vienna University of Technology. Biomimetics was used to find innovative designs for a deployable structure for a Lunar base. [398]

- **Biomimetic design proposals**

Visionary concepts based on a biomimetic approach show starting points for the development of highly innovative products and architectural solutions. The projects were created within the frame of a biomimetics course led by the author at the Vienna University of Technology.

5.1 ADAPTATION AND EVOLUTION OF TRADITIONAL ARCHITECTURE ON NIAS ISLAND

Traditional architecture has evolved over a long timescale. The qualities of architectural traditions can be rediscovered for modern interpretation. Traditional architecture which has survived until today is well adapted to environmental cycles and therefore interesting to investigate. As long as traditional architecture exists, it can be used as a source of information for sustainable and well-adapted building techniques. The architecture on Nias seems to be an excellent example, but as many other traditional typologies it is on the verge of disappearing, together with the societies and cultures that have been relevant for their formation as one of the main conditions.

In the context of biomimetics and natural structures, adaptation is used as a filter for the observation of development or evolution of building typologies and particular elements.

5.1.1 Introduction

A first trip to Nias Island took place in the course of a field trip to North Sumatra in 2003. [399] The study of traditional architecture on Nias Island was initiated by the author and Dr. Ulrike Herbig after the tsunami and the big earthquake that had devastated the island of Nias and a large part of North Sumatra in 2005. The interdisciplinary research project aims at documenting the unique architecture and culture. The field trip in summer 2005 showed that although heavy damage had occurred, the traditional architecture had proved outstanding resilience to the disasters. Apart from architecture, the group especially documented changes in culture and society by means of interviews and film. [400]

The author's approach to Nias architecture is sustained by the fascination for efficient and aesthetic building construction, the key features being adaptation to the local conditions, in particular earthquake resistant design and the evolution and differentiation of house typologies on the island.

---

[396] An introduction to the topic of Nias architecture and first results of the research project were published in Gruber, P.; Herbig, U.: Settlements and Housing on Nias island, Adaptation and Development, 2006

[397] Gruber, P.; Imhof, B.: Transformation: Structure/space studies in Bionics and Space Design, in: Acta Astronautica Volume 60, 2006, p.561-570

[398] An excerpt of the study was published in Gruber, P. et al.: Deployable structures for a human lunar base in Acta Astronautica Volume 61, 2007, p.484-495

[399] Herbig, U. et al.: Documentary Film "Architecture and Culture of Northern Sumatra and Nias", 2005

[400] The group consisted of Dr. Ulrike Herbig (surveyor) and the author as project leaders, Markus Schweinzger and Victoria Herbig (camera), and the students Mia Mechler, Tanja Gombotz and Hubert Ackerl. A first visit in June 2005 was made by Dipl. Ing. Johannes Melbinger, who later abandoned collaboration. The project was supported by the local Capuchin missionary Johannes Hämmerle, who has been working and doing ethnological research on the island for more than forty years.

Fig.298 The most impressive structures: the King's house "Omo Sebua" in Bawomataluo, with megaliths in the foreground, in August 2005.

The research project

The project aims at the investigation of the advantageous qualities of the traditional architecture. The documentation and understanding of these qualities will help in the process of future design and implementation into modern architecture.

The fieldwork was carried out within five weeks in August/September 2005, four months after the earthquake. We took the following steps: collection of information in the capital Gunung Sitoli at the Pusaka Nias Museum and with international aid organisations, finding of locations of the villages or houses, rough assessment of damages and comparison with existing documentation. We followed this by selection of specific objects, building documentation by drawing and measuring, interviews with the owners, photographic documentation and filming. Measuring was facilitated by Laser distance meters and a Total station (instrument for surveying that includes theodolite, angle and distance measurement).

A rough survey of damage was done in eleven villages in the south of Nias, where the most traditional villages still exist. Wood samples were taken back to Vienna for testing. After tests at the Institute for Building Construction and Technology at the Vienna University of Technology in 2006, the processing of the data of the overall project is still ongoing; colleagues from other disciplines working on photogrammetric interpretation and a documentary film. The data of the chosen houses have been processed, and 2D plans and 3D models drawn.

These models were used for visualisation purposes and to create the virtual models for structural analysis carried out for the South Nias house typology. The next step would be to include in a future project micro-tremor analysis of the documented buildings, which could evaluate the analysis done so far. The hypothesis about the effectiveness of the houses in terms of earthquake resistance - the most interesting adaptive feature - was created together with collaborating engineers, on the basis of the fieldwork investigation.

Situation and environmental conditions

Nias is a small island 120km to the west of Sumatra. One of the many names of the island is "Tanö Niha", the island of men. It is about 150 km long and 50 km wide and together with smaller islands to the south and north covers a total area of 5,625 km². The topography of the main island of Nias is characterised by large rivers, valleys and mountains up to 887 metres high, similar to the west coast of Sumatra. But in contrast to Sumatra, there are no volcanoes and thus no fertile volcanic soil. The climate is tropic, warm with a humidity of 80-90% and an average of 250 days of rain per year with frequent storms and heavy rains.

Fig.299 Panorama view of Sorake beach in South Nias, a well known surf-spot. The reef in the foreground was uplifted by the earthquake in 2005.

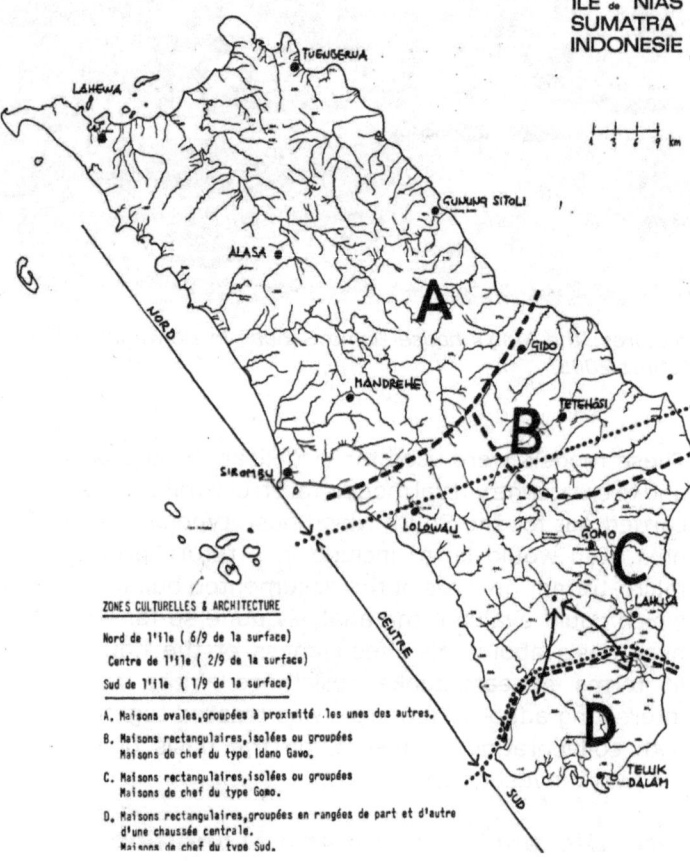

Fig.300 Zones with different architectural typologies, Viaro 1984.

In the 19th century Nias was still densely covered with primeval forest rich in species, which has survived only in small inaccessible mountain regions. Wherever cultivation is possible, extensive rubber and patchouli plantations cover the island; their export is the main source of income. Along the rivers and the coastal plains bananas, coconuts, cacao, and rice are harvested. Animal husbandry - breeding of pigs and chicken - is also important. Pigs are needed as means of payment for the many feasts and rituals, as required by the "Adat", the traditional law.

The sea is traditionally regarded as dangerous, so only small scale fishing is common today, and it is not a significant factor in traditional economy and culture. Surfing became very popular in the 1980s, and has led to small-scale tourism in the south.

Nias lies on the fracture zone of the so-called Eurasian and Indo-Australian tectonic plates. Another indigenous name, "hulo solaya-laya" is translated as "the dancing island" and refers to the frequent earthquakes.[401] As Nias seems to be a geologically young island, ground conditions are very unstable and differ regionally.

On the whole, the infrastructure on Nias is rather poor, but was improved by international help after the catastrophes. Only the main roads from the capital are paved. Heavy rains, landslides and large rivers make it difficult to maintain the traffic routes. The road along the east coast is relatively new and in good condition due to reconstruction efforts. Smaller inland routes are still gravel roads or mere footpaths connecting the villages. Electricity is only available along the major roads and water supply is still a problem in most regions. Because of the destruction of natural plant coverage, good building timber is scarce.

401 Hämmerle, J.M.: Nias - eine eigene Welt, 1999, p.15

Population

Due to the rough topography the population is concentrated in coastal areas and fertile valleys, with about 65,000 people live in the region of the capital Gunung Sitoli on the east coast. The largest town in the south is the port Teluk Dalam. Other big towns are isolated, only accessible by mountainous roads or small boats, e.g. Lahewa in the north and Sirombu on the west coast. Sirombu was particularly badly hit by the tsunami. The traditional villages, of which few have survived until today, were built inland. The coastal settlements were founded in the 18th and 19th centuries as bases for trade with the Netherlands. They have grown significantly and accommodate the majority of Nias population nowadays. Although many people left the island after the catastrophic events, the number of inhabitants is still rising. Population growth is causing social, economical and ecological problems:

1900 150,000 estimated
1961 314,829
1996 633,630[402]
2003 760,000[403]

The history of settlements in North Sumatra and Nias is unclear, but historians assume that migration started 1,500-2,000 B.C. Settlers came from Southeast Asia and imported a culture cultivating rice, processing metal and erecting megaliths.[404] As natural resources on Nias have always been limited people depend on economic exchange with neighbouring Sumatra, and trade was and still is of high significance. It therefore seems natural that an Arab trader, who mentioned an island off Sumatra in 851 wrote the first known record of Nias. It was then known not only for its abundance of pigs, but also as a source of light-skinned slaves. The people of South Nias were especially dreaded for their martial behaviour, head hunting and enslaving other tribes.[405]
Nowadays South Nias villagers perform their dances and their famous stone jump for tourists.
The Netherlands colonised Nias in the second part of the 19th century and missionaries followed, so that

Fig.301 Stone jump in Hiliamaetaniha.

Fig.302 Dancers in Hiliamaetaniha.

Fig.303 Stone figures at the Museum Pusaka Nias in Gunung Sitoli.

the majority of the population today are Christians.[406] Written records and research of the hitherto oral history of the island began at that time. The slave trade ceased, and the island was charted in 1822, under the English governor Sir Thomas Raffles.[407] After a short period under Japanese rule, Indonesia claimed independence in 1945 under Sukarno, later Suharto. By the time of the field trip, Nias belonged to the district of North Sumatra, and was divided into Kecamatan Nias and Nias Selatan.

---

402 Hämmerle, J.M.: Nias - eine eigene Welt, 1999, p.18

403 Figure mentioned by UNHABITAT Dercon, B.: Housing Damage and Reconstruction Needs in Nias and Nias Selatan. Evaluation of the Aceh MapFrame 3 Data, 2006

404 Wiryomartono, A.: Cosmological - and spatiotemporal meanings of a traditional dwelling in south Nias, Indonesia, Dissertation, 1989, p.19

405 Hämmerle, J.M.: Nias - eine eigene Welt, 1999, pp.15

406 A large number of churches is scattered over the whole island, whose typological analysis would be worth another study.

407 Viaro, A.M.: Urbanisme et architecture traditionels du sud de l'île de Nias, 1980, p.4

Fig.304 Meeting place in Bawomataluo, feast, Schröder 1917.

Fig.305 Nobles in South Nias, Schröder 1917.

Society and culture

The significant changes in society and culture within the last century have already blurred the individuality and achievements of Nias culture, and will continue to transform it to a modern society. Remnants of tradition have survived in a few remote places, with photographs and written documentation of oral history. In the 19th and 20th centuries explorers, colonial civil servants and missionaries documented the unique culture and its transformation.[408] Pater Johannes Hämmerle, who has been working on Nias as a Capuchin missionary since 1971, has collected valuable evidence on cultural heritage, both oral tradition and artefacts, and also founded the Pusaka Nias Museum in Gunung Sitoli.

The structure of Nias society as it exists in a rudimentary form today, is stratified - a hierarchical class society with a village chief, nobles, common people and slaves.[409] The differentiation is becoming more and more obsolete but society is still far from egalitarian.

The traditional villages were inhabited by a single clan, and coalitions and frequent wars were fought between neighbouring settlements.

The structure and activities of old Nias society, and its mythological cosmic order, is reflected in the traditional settlements.[410] The chief's house, the so-called "Omo Sebua", signified the ruler's importance by its central position, size and artistic finish.

"[The chief's] house is the most complete binary structure possible within the customary laws of the village. It is understood that the form of the house as well as all of the elements contained within it are part of the fertile land of the ruler and are reification of the total cosmos."[411]

The "Niha" as Nias inhabitants call themselves agree that the origin of their culture lies in an area called Gomo, in Central Nias.[412] Differences in social organisation and culture, best visible in language, art and architecture, exist between North, Central and South Nias.[413] The west of Nias has recently been referred to as a fourth cultural zone by Hämmerle.[414]

The territories were isolated by topography and developed independently, yielding particular settlement and house typologies.

---

408 Documentation of the travels of the Austrian Brenner-Felsach were recently published by Mittersakschmöller, R.: Eine Reise nach Nias, 1998.

409 Feldman, J.A.: The Architecture of Nias, 1977, p.43

410 Nias culture was extensively investigated by ethnologists, art historians and architects. Related to architecture, the most important studies were published by Feldman J.A. 1977, Viaro A. 1980 and Wiryomartono B. 1989.

411 Feldman, J.A.: The Architecture of Nias, 1977, excerpt of abstract

412 Feldman, J.A.: The Architecture of Nias, 1977 pp.28

413 Feldman, J.A.: The Architecture of Nias, 1977, pp.40

414 Hämmerle, J.M.: Nias - eine eigene Welt, 1999, p.22

Fig.306 Omo Sebua in Hilinawalö Mazingö, 2005.

Qualities of the traditional architecture

Architecture in Nias has evolved over a long period, in a constant process of empirical effort for improvement. The information implemented in the developed typologies is the source of innovation we want to use to develop further solutions, in a process of "quality mining". The outstanding resilience of the houses to earthquakes might be the reason for the survival of these building types.

- **Form and aesthetics**

The diagonal bracings of the substructure, the tripartite structure of the houses in general, the elaborate and cantilevering façades and the impressive roof are all very strong architectural forms which are easily recognizable, allowing identification. All have been recorded in the form of signs, abstract drawings and models.

- **Spatial quality**

Feldman and Wriyomartono, among others, investigated the relationship of space to the activities and mythologies of the Niha. Space is handled very carefully as a means of social order. The use of space is strongly and hierarchically defined, and characterised by a specific grade of privacy or publicity. This hierarchy of space defines the activities which may take place, and enables the inhabitants to live very closely together, on village as well as house scale. The quality of the interior spaces inside the houses is very high: orientation is provided by the large window openings, which provide just enough daylight to avoid the heat, and to make living comfortable. Order is provided by different floor levels and by light separation walls.

The elaborate designs (and sometimes ornamentations) of construction elements and walls in the front room give a very calm and stable impression. According to tradition, the dark volume of the roof is not separated from the living floor by a ceiling: it provides an endless dark heaven.

- **Excellent craftsmanship**

Craftsmanship was excellent in Nias. But only few "tuka nomo" (i.e. masters of house-building) are left, and traditions and techniques are on the verge of being lost. The technology of woodworking has parallels in megalithic culture: similar ornamentation has been found on stone and on wood. The detailing of the architecture displays craftsmanship in woodworking at its best, as found in other countries like Japan, Russia and Austria.

- **Climatic condition**

Due to the good ventilation, the large volume of air in the roof and the ability of wood to adapt to very high humidity, the climatic condition inside the house is excellent without active air conditioning. The construction of the house as a whole also provides enough covered outside space, so that any kind of activity is possible even in times of strong rain.

Fig.307 Interior view of house in Bawömataluo.

Fig.308 Carvings in the Omo Sebua in Bawömataluo.

- **Sustainability**

In the past, all material used for building could be found on the island. Lacking any data on environmental conditions of the past, sustainable material use in building cannot be proved. If we project the estimated global building related material use of 50%[415] onto Nias, building must have contributed to the disappearance of many kinds of trees on the island, and the disappearance of primeval forest in general. Nowadays there is a lack of timber for building, and in particular the North Nias houses with their substructure of Manawa Danö hard wood can no longer be built.

- **Earthquake resistance**

The construction of the houses seems to withstand even very strong earthquakes. The security of the inhabitants has been de facto proved, as according to our survey nobody was killed in the collapse of a traditional house during the big earthquake in March 2005. Earthquake resistance is the most interesting feature for research and future design.

---

415 Edwards, B.; International House, London (Ed.): AD Architectural Design Nr.71, 2001, p.108 fig.1

Fig.309 "Pancake" type of collapse, Central Nias.

Fig.310 Collapsed building in North Nias.

5.1.2 Earthquake March 2005

According to geologists the big earthquake of March 28, 2005 reached a magnitude of 8.7, severe to violent peak ground acceleration and severe to violent ground velocity in the Nias region, the centre of the island being hit harder.[416]

Geological contour maps provide information on the movements in the area of Nias, showing immense vertical uplift along two ridges in Nias and Simeulue:[417] parts of the west coast of Nias were lifted 2 metres. Enormous deformation and ground liquefaction took place, destroying bridges and streets, infrastructure and buildings.

The worst damage done by the earthquake was in the populated coastal areas, especially in the capital Gunung Sitoli, intensified by bad alluvial soils and cheap concrete structures.

In the whole island of Nias, 758 people died, 705 were badly hurt, 781 hurt and 84,388 persons internally displaced, accommodated in camps or with host families.[418] About 50% of all public buildings and many roads and bridges were destroyed. Out of 122,652 housing units 71,000 were damaged or destroyed.[419]

The generally poor economic situation of the island was boosted by the recovery efforts of the government and international aid. But this upswing came together with high inflation and a serious shortage of raw materials, both hampering reconstruction until today.

Damage recorded in August 2005

Our own survey shows the damage to traditional Nias architecture, especially in the South Nias villages. In most of them, less than 50% of all houses are still Omo Hada (traditional in construction), the others being replaced by wooden Malayan-style houses or simple concrete bungalows. Large modern village extensions change the picture. In the villages we visited, between 5% and 30% of the total number of houses damaged were traditional ones.[420] In other words about one third to half of the remaining traditional houses were damaged by the quake. In most cases, damage occurred to older buildings and those badly updated. The main reason for damage to traditional houses is bad maintenance, and the devastating earthquake was the trigger event for collapse.

There seems to be a correlation between the economic situation, which depends on the situation of the village, and the maintenance of the traditional architecture. The more remote, the poorer the villages tend to be - people do not have the resources for repair and maintenance of the houses any longer. The lack of building material aggravates the situation. In Bawomataluo the largest traditional village, with a connection to a major road, only minor damage occurred.

In North Nias, the traditional houses have suffered at least minor damage by the earthquake. As the number of North Nias houses per village is limited to a handful, no statistical comparison was done. In Central Nias, there is a large variety of traditional buildings, and no specific order in the structure of the settlement. Due to the short timeframe of the field trip to Central Nias, we were unable to carry out a proper statistical comparison here. But it was evident that in general, damage to the central area of Nias around the village of Gomo was heavy.

---

416 USGS US Geological Survey: shake map 28th of March 2005.

417 Briggs et.al.: Deformation and Slip Along the Sunda Megathrust in the Great 2005 Nias-Simeulue Earthquake, in Science vol. 311, p. 1897-1901, 31 March 2006

418 IOM International Organisation for Migration: Post Disaster Damage Assessment on Nias and Simeulue Island, June 20, 2005

419 BRR numbers revised by UN-Habitat to: 39,022 not damaged, 75,635 less than 50% damaged, 7,995 more than 50% damaged, (published 23 Jan, 2006, in Housing Damage and Reconstruction Needs in Nias and Nias Selatan.)

420 figures according to field trip Gruber/Herbig 2005

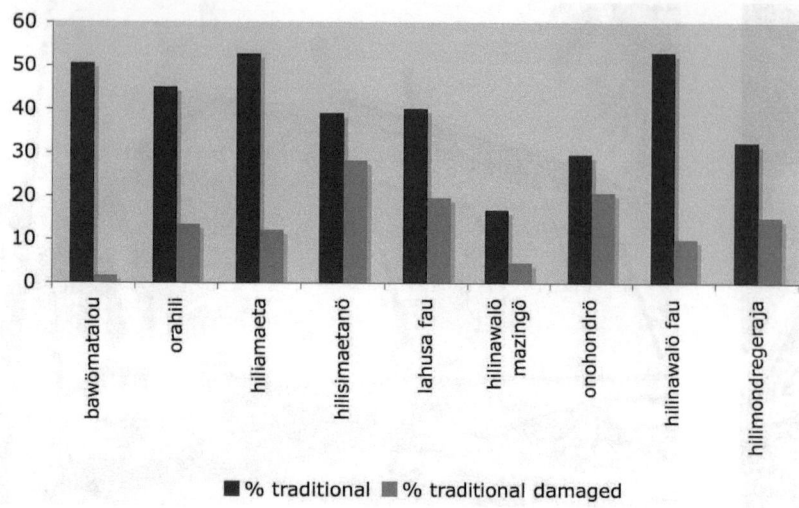

Fig.311 Statistics of damage to traditional houses in South Nias.

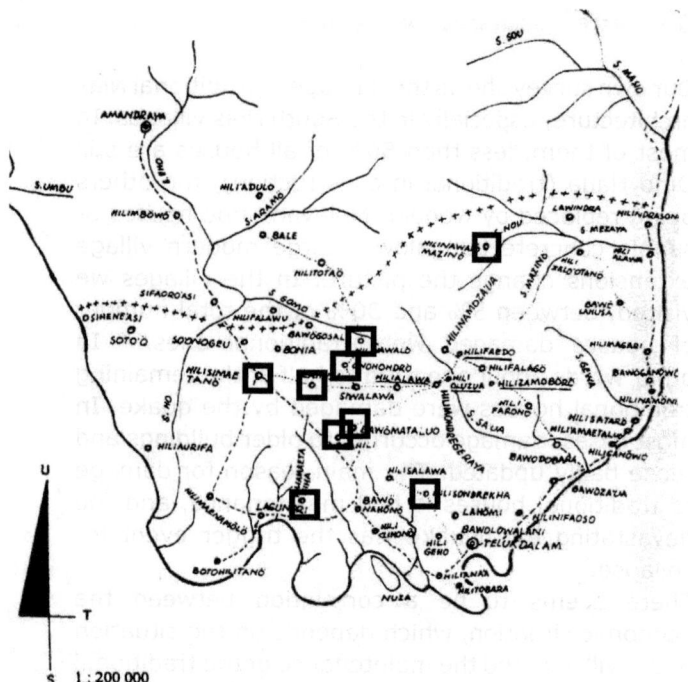

Fig.312 Map of South Nias, villages visited marked with black squares.

The percentage of damaged traditional houses in South Nias is appalling, considering the already low percentage of traditional houses in the few villages which can still be called "traditional". The statistics show that an average of 38 percent of the traditional buildings suffered damage by the earthquake. Considering the poor economic situation it is likely that many of the damaged houses will be torn down rather than repaired.

The assessment was carried out four months after the catastrophic event, so we can also assume that completely damaged buildings had already been torn down, and that collapsed houses had been removed by that time.

Variation in damage level

We observed big differences in the level of damage, for both modern and traditional architecture. This corresponds with observations made by scientists of the Centre for Disaster Mitigation in Bandung. The following factors are responsible for the variation: amplification of peak ground acceleration due to local ground conditions, ground liquefaction, differences in vulnerability of the buildings and differences in vulnerability of their foundations.

As the ground conditions vary enormously, scientists suggested a microzonation hazard map for North Sumatra and Nias.[421] This map, an important basis for further decision making, is not yet available, and by the time the field study was carried out no efforts had yet been made towards a joint approach of all aid agencies to fund a survey.

421 Centre for Disaster Mitigation in Bandung: Survey of Geotechnical Engineering Aspects of the December 2004, Great Sumatra Earthquake and Indian Ocean Tsunami and the March 2005 Nias–Simeulue Earthquake, 2006

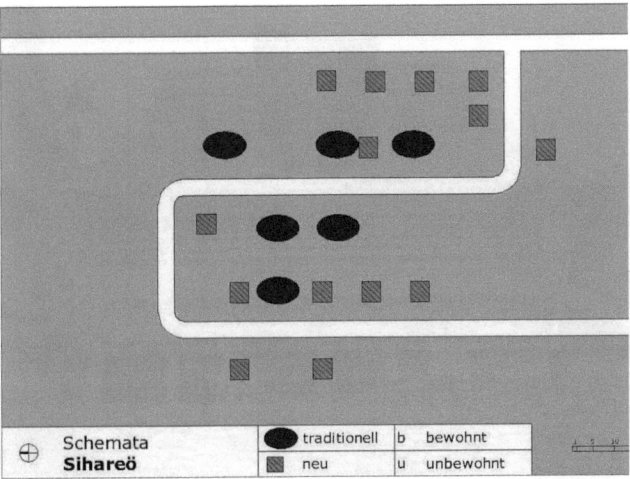

Fig.313 Sihare'ö Siwahili, North Nias, Müller 2005.

Fig.314 Settlement in North Nias, Sihare'ö Siwahili, 2003.

5.1.3 Settlement structures

All traditional settlements in Nias are located along connecting streets and pathways, and some of them can still only be reached on foot. The pathways inside or close to settlements are carefully paved. Hills and slopes are terraced. Especially in South Nias, impressive staircases with flanking symbolic animals and elaborate drainage systems were built. The processing of stone is well known and also used for megaliths, huge stone figures and other symbolic forms placed in front of houses. The manifestation of this megalithic culture varies according to cultural zone. The existence of fortified and paved settlements in a rural society is unusual, and partly explained by rampant warfare. The slave trade brought wealth to South Nias, facilitating this impressive architecture. The location of the settlements was very carefully chosen. Important parameters were the availability of water, the possibility of quick drainage, the stability of the ground, the vicinity of agricultural areas and the vicinity or distance to other settlements.

Correlation between house typology and settlement structure

There is a necessary correlation of both settlement structure and building typology with society in this area. The spatial order reflects social and cultural conditions and needs. In former times, there used to be a king's or chief's house (Omo Sebua) in every village or region. The four remaining structures in South Nias could be visited during the field trip.
In the following, the interdependency of building and settlement structure will be discussed - representing an example of mutual influence between the two patterns, which refers to hierarchical levels in design of built environment, and the internal adaptation of architectural typology.

- **North Nias - disperse settlements**

In North Nias, settlements are found along the hilltop roads. They consist of groups of independent oval houses, oriented with their longitudinal sides parallel to the street, but single cottages are also found. The houses do not touch, and are often at some distance from each other. In former times, the villages were fortified with bamboo fences or earth walls. Traditionally, megaliths are placed in front of the houses, symbolizing the connection between the living and the dead, as well as reflecting the social status of the house owner.
The houses were entered from the village square, through a trap door underneath the house.
A front staircase or a front porch has replaced this entrance or a front porch as this defensive feature is no longer needed.
The oval floorplan of the North Nias house is highly autonomous, which makes extensions and annexes a difficult design task. The open part of the façades usually takes more than a quarter of the total length of the house. The entrance annex is visible from the street, situated at the start of the long axis. The kitchen and bathroom annexes which became necessary after the Dutch prohibited fireplaces within the houses are rather unstructured hut-like extensions in wood and/or concrete on the side of the house facing away from the street.

Fig.315 Settlement in Central Nias, dispersed settlement, Gomo area, 2005.

Fig.316 House in Central Nias, Gomo area.

- **Central Nias - dispersed settlements with single or combined houses**

In Central Nias, settlements are not located on hilltops but found in valleys along the big rivers cutting through the mountains. Here the settlements consist of a less rigid assembly of individual or coupled houses along streets or small plazas. The orientation of the houses is with their ridge parallel to the public space.

Although the settlement history of Nias is said to have its roots in Central Nias, nowadays the architecture of this region appears as a hybrid of northern and southern styles. As in the villages of North Nias the settlements are clusters of single buildings. But in contrast to the north, the houses can be combined and are situated with their eaves facing the village square. This orientation and the rectangular floor plan are also found in the South Nias villages, but the precise definition of space does not exist here.

The space in front of the houses is paved with stones and used for drying agricultural products or laundry. Stairs and steps define spatial relationships. Decoration and ornamental art characterise the architecture in Central Nias. Carvings, patterns and animal representations on the facades serve as protection for the house and its inhabitants. Symbols give information on the state of fertility of the family, for example the number of women living in the house.

The Central Nias house is the most variable type of building, so the existence of a distinct typology is more than questionable. The floorplan is rectangular or T-shaped, with lateral eave-like extensions. The construction allows lateral extensions, as well as combinations of two or more houses. The façade can be open on three sides, the lateral façades constructed in a way similar to the front.

Fig.317 Corner situation in Hilimondregeraya with South Nias house types.

Interestingly, the combination of two Middle Nias houses exists in two variations: sharing a common entrance-space, or standing close together and thus having two separate entrances at their sides (in two big houses the separating walls were removed and the spaces connected). The combination of both variations creates a row house structure like that of South Nias villages.

The annexes for bathroom and kitchen are easy to combine with the rear of the house. In spite of this advantage, the architectural quality of the annexes is no better than in other regions.

Fig.318 Street view of Bawomataluo, traditional houses in different sizes and colours, 2005.

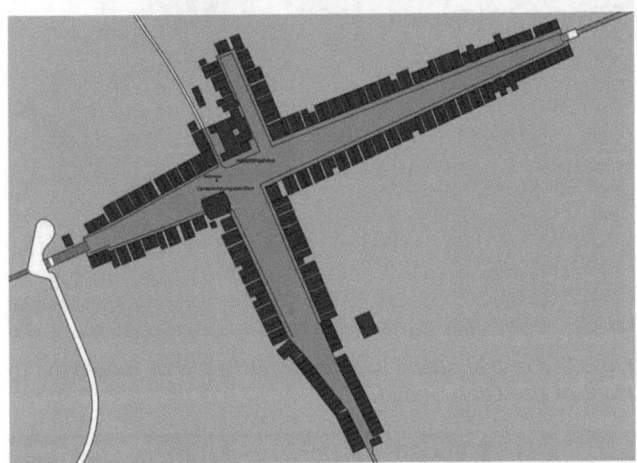

Fig.319 Map of Bawomataluo, Müller 2005.

Fig.320 Orahili, view from Bawomataluo, 2003.

- **South Nias - dense settlements with row houses**

Villages in South Nias are situated on hills and are named after their location. In former times of frequent warfare and headhunting raids, an outer palisade of sharpened bamboo stakes with a deep ditch behind fortified the village. The most impressive South Nias settlements are planned highly ordered structures with a definite centre, where the Omo Sebua is located. The primary element is the central axis extending from a distinguished "navel" of the village. The orientation of this main axis is independent of the cardinal points, following the main mountain ridges. As a result, many traditional South Nias villages around Teluk Dalam have a Northeast-Southwest orientation.

The paved street's width gradually increases from the narrower entrance staircases to the central meeting space in front of the chief's house. The entrances and centres of the villages are always accentuated, with megaliths placed in a particular zone and meeting places in the centre.

Several hundred dwellings are arranged on either side of the main street in big villages. Covered entrance terraces are shared by the pairs of adjacent households who live in coupled houses.

The street pattern is in some villages is extended to L or T shapes, and public baths are located in the vicinity of the villages. The South Nias type is a classical row house similar to the European tradition, with one distinct front façade and structural sidewalls. The floorplan is rectangular. As the sites are narrow, between 3.20 and 5.20 metres, the floor plans extend at right angles to the public space. Two adjacent houses share a narrow entrance, which is an independent platform and staircase, giving access both to the two houses and to the space behind the row of houses. Informal doors allow connection of adjacent houses to provide escape routes. The kitchen and bathroom annexes often double the overall length.

In spite of the long extension being at right angles to the street, the ridge is parallel - demanding a very high roof.

- **General remarks**

The relationship between building typology and settlement structure is very close and coherent.

It seems as if they have been designed for each other, and perhaps they have evolved in mutual influence. Only very rarely did we observe that the houses did not fit the settlement structure (e.g. in Hilimondregeraya on the hill where the former king's house was standing, and in Bawöhili close to Sifalagö Gomo, where Central Nias houses stood densely packed between modern houses)

Modern annexes have been added to all traditional houses containing kitchen, bathroom/toilet and stables. Privacy determines their distance from the main house and the public kitchen space. In Central Nias, the back yard was also used for fishponds. In general, order decreases with distance from the public street and the centre. We did not identify any storage buildings. In former times, rice was stored in wooden boxes on entrance platforms.

In contrast to other architecture typologies there is no polarisation of the opposite sides of the streets. (The Toba Batak in Sumatra have living houses and sopos, storage buildings, on either side of the public street, facing each other.) Zoning of height defines hierarchy and privacy, in the paved public space as well as inside the houses. Bath and meeting places used to be common institutions, with elaborate stone furniture.

Fig.321 Centre of Bawomataluo, Bale and megaliths in the foreground.

Fig.322 Central space in Hilinawalö Fau with megaliths in front of the Omo Sebua.

Fig.323 Bath place in Hilimondregeraya.

Fig.324 View out of the Omo Sebua in Hilinwalö Mazingö, with central walkway and space in front of the houses used for temporary shelters and for drying products and clothes.

Society and space in South Nias

Society and space have a strong relationship in South Nias, and seem to be mutually adapted.
Every building has a fixed position in social life, connected to many rituals and feasts. The architectural elements are named in Nias language, and have a particular place within the cosmological order. Feldman, Viaro, Ziegler, Wiryomartono, Beatty and others have all investigated this correlation in detail.
The settlements have distinct entrances and centres, the distance from centre and height correlating with the social importance of a family.
The chief's house, the so-called Omo Sebua, was usually the largest structure, located at the centre of the village. In some villages more than one Omo Sebua existed.

- **Centre**

The Omo Sebua in Bawomataluo (with a zinc roof) is located at the centre of the T-shaped village with a zone of megaliths in front. Apart from the one in Bawomataluo, the Omo Sebua are placed on higher platforms with staircases, which enhances their impressive appearance. In Hilinawalö Mazingö the Omo Sebua is located most impressively at the end of the inclining main street axis.
At the time of the field trip, only four Omo Sebua still existed in the South of Nias, and according to oral information one of those partly collapsed in 2008. The loss of these structures is due to the change in society, now headed by an elected political chief instead of an inherited king.

The former chief's families have lost their wealth and cannot afford to maintain the huge houses any longer. The loss of the Omo Sebua seems to be followed by a rapid decline of the traditional architecture in the village. A new function and ownership structure would have to be found to keep the structures alive.[422] Formerly, meeting houses existed as a special typology of covered open pavilions. Apart from one remaining building in Bawomataluo, which resembles a "bale"[423], the typology of meeting houses is already lost. However, modern meeting houses are common and situated in the centres of the villages. In many villages, stone benches serve as meeting places in front of the Omo Sebua. The meeting place in Hilinawalö Fau shows a particularly elaborate design.

---

[422] This is hampered by the present owners, the remoteness of villages and the island in general, and the ignorance of international organisations towards non-massive architectural heritage. The Omo Sebua of Hilinawalö Mazingö is on the UNESCO list of the 100 most endangered sites, but this does not bring means nor any consideration of change. The website Niasonline reports the Omo Sebua in Hilinawalö Mazingö having been reconstructed in 2006 by the British Turnstone Tsunami Fund. http://niasonline.net/eng/2008/01/07/nias-not-just-stone-jumping-and-surfing/

[423] in his talk at the Vienna University of Technology in October 2006, Jerome Feldman pointed out the similarity between the Bale in the pacific region and the Bale in Bawomataluo.

Fig.325 Street view in Hiliamaetaniha, covered walkway in front of the houses (mbele mbele), drains (elea), megaltih zone (öli batu), working space (ewali), central walkway (dodo lala).

- **Hierarchy of space: transition from private to public**

The most private space in a traditional village is the rear of the house. Guests are welcomed in the living room at the front, which is itself separated into different levels. Covered entrance terraces between two coupled houses are shared by pairs of adjacent households. Neighbouring houses are also connected with doors to provide escape routes, which were needed in the past. The space below the house is just used to store material. The front space below the roof overhang along the street is a semi-public space, used for working, socialising and for transition. A drainage gutter defines the border.

The next area towards the street is reserved for displaying the house's megaliths. This zone is called "wall of stones" ("öli batu") and indicates the rank of the householders. The megaliths are a kind of petrified model of the social hierarchy and feasts of merit. The stones are classified by gender, and come in a variety of forms, including menhirs, benches and circular seats.

The space between the öli batu and the public walkway in the middle of the street belongs to the individual house and has to be maintained by the owner. It is used for drying agricultural products and laundry. After the catastrophic earthquake caused heavy damage to the houses, this space was used for temporary shelters.

Fig.326 Interior view of North Nias house in Sihare´ö Siwahili.

Fig.327 Roof space of North Nias house in Sihare´ö Siwahili.

Fig.328 North Nias house in Tumöri, the roof covered with plastic.

5.1.4 Typologies of the "Omo Hada"

The surviving houses can be categorised into distinct typologies according to the cultural regions[424]. Divergent typologies are found in literature, but seem to have disappeared. Alternative house forms developed where traditional houses had been torn down.

- North Nias houses have an oval floor plan, slanting walls all around, rows of vertical pillars and diagonal bracings (X form) in the substructure, and a huge hat-like roof. Two central pillars lead from the first floor to the ridgepole.
- The hybrid typology of Central Nias houses has not yet been fully examined. Central Nias houses have rectangular or cross-shaped floor plans, often slanting sidewalls, slanting front façade, and V-shaped diagonal bracings in the substructure. Very different layouts of the floor plan are possible.
- South Nias houses are always coupled, have a narrow row house-like rectangular floor plan, straight load bearing sidewalls, a slanting front façade, v-shaped diagonal bracing in the front façade and a very high roof. They have a public room in front and sleeping rooms at the back. The front room is lit and ventilated by an opening which stretches over the whole street façade, secured by a wooden grid.

Constructive elements of the cantilevered front façade create different floor levels in the interior, used as benches for sleeping or sitting and for storage purposes. The standard typology of the South Nias row house is a rectangular-shaped elevated house construction, oriented with the eaves towards the street. The substructure is made of four rows of strong pillars ("Ehomo"), reaching from ground to first level. Diagonal posts ("Driwa") support them as in North-Nias houses. But in contrast to North Nias typology, here the V-shaped columns are situated at the very front, acting as support and as representative elements. All the house posts rest on foundation stones, on the one hand to prevent them from rotting and on the other to make the construction as a whole more flexible. The space created beneath the house is used for storage and as an animal shed.

Three houses of different typologies have been documented in detail, and will be described in the following chapters.[425] The individuality of the houses was recorded for calculation purposes, as static analysis can not be carried out on a typological level. In spite of the diversity, there are common characteristics which connect all Omo Hada.

---

424 Feldman, J.A.: The Architecture of Nias, 1977, pp.40

425 The recordings were carried out by Gruber, Herbig, Ackerl, Gombotz and Mechler in August 2005.

Fig.329 Roof space of Omo Sebua in Onohondroe, light entry shows heavy damage to the roof cover.

Common features

The "Omo Hada", the traditional house, is built according to traditional law. This "Adat" regulates all activities of life, and therefore all steps and rituals related to the building process. Entirely made of wood and other plant material, a traditional house in Nias provides living space for one family. Apart from the few remaining Omo Sebua (the big houses of the kings are entered from below through trap doors), today's houses are entered from the sides, by means of ladders or staircases sometimes covered by annex buildings. Kitchen, toilet and mandi, the freshwater basin, are placed in shapeless annexes at the rear. All traditional houses in Nias have large lattice openings in the front façades and window flaps in the roof: this latter kind of opening is peculiar to the island of Nias and cannot be found elsewhere in the Indonesian Archipelago. The common characteristics and differences in construction are described in the following.

- Tripartite structure

The most obvious feature of the Omo Hada is tripartite structure, the vertical zoning into three distinct levels, which relates to the mythological interpretation of the cosmos.

- Separation from ground

All houses rest on a grid of vertical and oblique posts placed on slabs of stones. The three-dimensional substructure offers great resistance and elasticity because it does not settle into the ground. The separation of the house from the ground is an important concept for traditional earthquake-resistant construction and resistance to ground moisture and insects.

- Height difference for spatial organisation

The first floor - the living floor - is separated into public, private and transitional spaces either by wooden walls or changes in floor-height. This element to organise space is most elaborate in South Nias houses and is also used on the village level.[426]

- Box as living floor

The living floor is a very stable box-like structure. Even if the substructure collapses, the box persists.[427] The lightness of the roof ensures safety for the inhabitants.

- No furnishing

The houses are barely furnished; the inhabitants' belongings are stored in chests. The most important piece of "furniture" is a long plank below the louvres, which the tenants use as a bench. In South and Central Nias, this element is a structural component of the façade.

- Lattice openings and window flaps

Large openings running along the front façades provide good ventilation and a superb overview of the neighbourhood, ensuring good control of contact between inhabitants and outsiders.

- Steeply pitched roofs

The steeply pitched roofs are a notable feature of Nias houses. Average South Nias houses reach a height of more than 10m. Many houses are still roofed with palm leaves although the use of corrugated metal is becoming more popular.

The high multi-storey construction rests on two main pillars in the North and Middle Nias types, and on sidewalls in South Nias.

The roof construction is a light 3D-frame structure, minimising material. Large overhangs protect the wooden connections from rain and provide additional outside space, which is used for working. The impressive Omo Sebua reach a height of around 20m.

Visualisation

The abstract 3D models that are presented in the following make the common features and the differences between the typologies particularly visible. In the next chapter, three detailed architectural surveys are presented.

---

[426] Wiryomartono, A.: Cosmological - and spatiotemporal meanings of a traditional dwelling in South Nias, Indonesia,1989, pp.250

[427] According to interviews with the village chiefs, in the 11 South Nias villages we visited nobody was killed during the 2005 earthquake due to the collapse of a traditional house.

Fig.330 North Nias house in Tumöri, entrance side.

Fig.333 North Nias house in Tumöri, street view closed side.

Fig.331 Central Nias hous in Hilianaa, entrance side.

Fig.334 Central Nias hous in Hilianaa, front side.

Fig.332 South Nias house in Hilinawalö Fau, front.

Fig.335 South Nias rowhouse in Hilinawalö Fau, entrance side with modern extensions on the rear.

North Nias house in Sihare'ö Siwahili

The North Nias house in Sihare'ö Siwahili belongs to Razi Zebua, who lives there on her own. Minor damage had been caused by the earthquake, but had been already been repaired when we visited the house. Other houses in the same village were badly damaged and torn down. The house of Razi Zebua is in a very good state, the wood still being strong and unaffected.

Fig.338 North Nias house of Ibu Razi Zebua in Sihare'ö Siwahili.

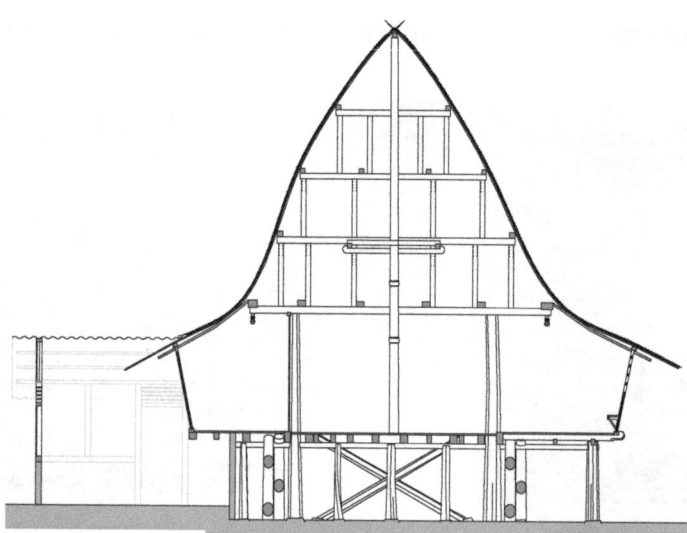

Fig.336 Cross-section 1-1.

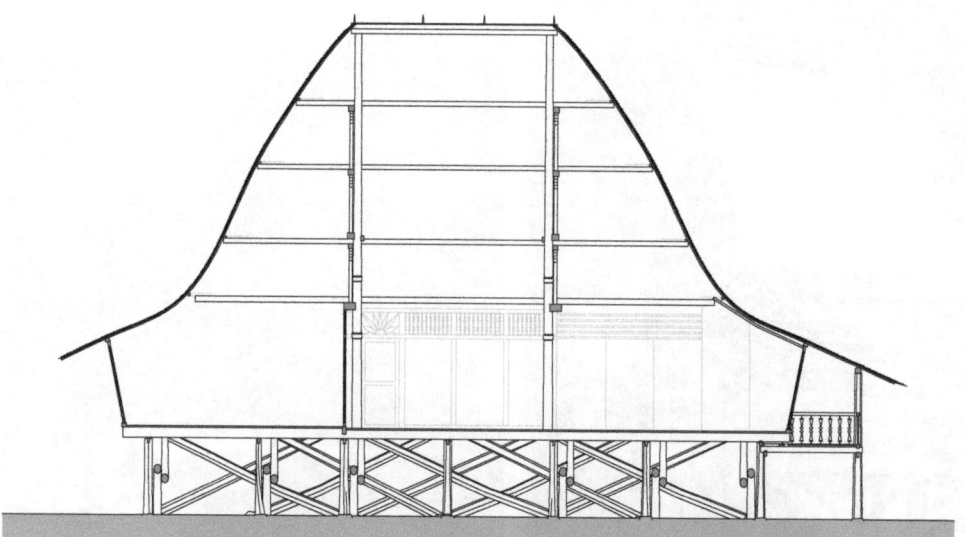

Fig.337 Cross-section 2-2.

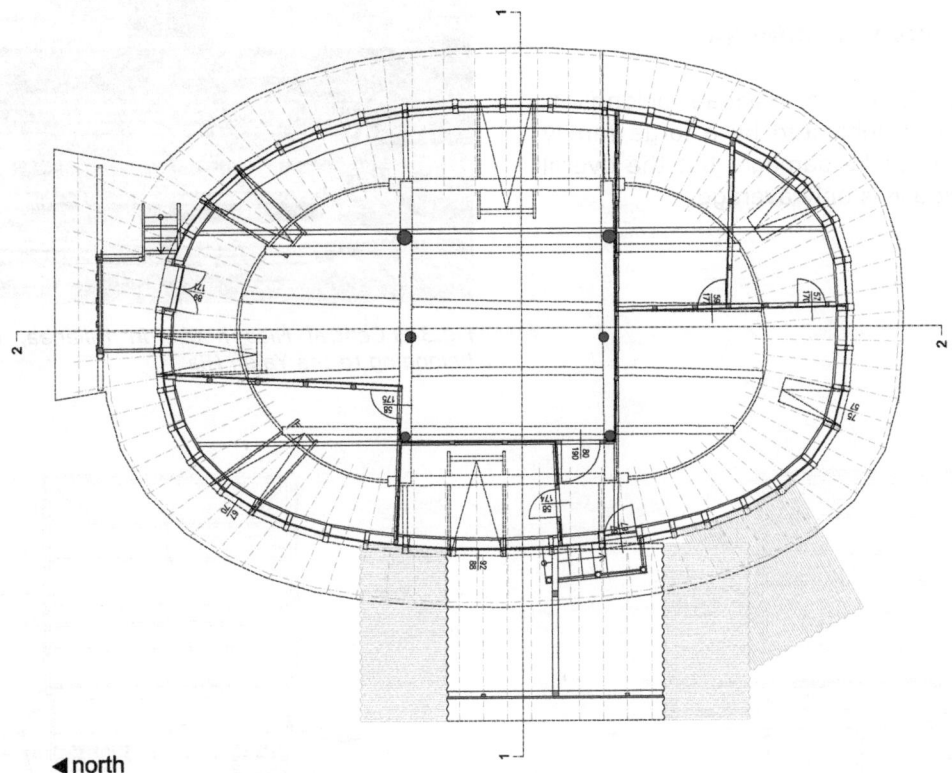

◀ north

Fig.339 Living floor.

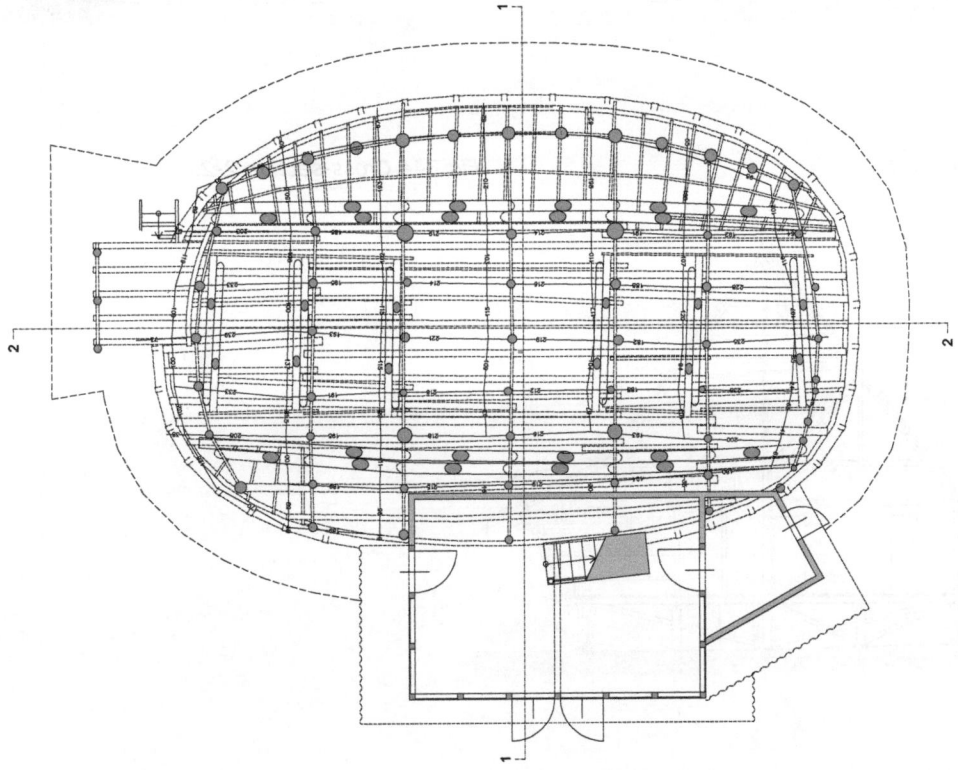

Fig.340 Substructure.

Central Nias house in Hilianaa

The Central Nias house in Hilianaa belongs to Ina Yakin Hia and is inhabited by a large family. There was no earthquake damage, but the overall condition of the house is only average.

Fig.343 Central Nias house in Hilianaa, Gomo region, belonging to Ina Yakin Hia.

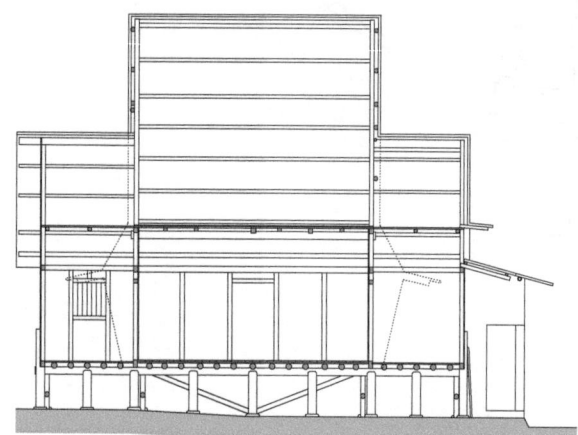

Fig.341 Cross-section 1-1.

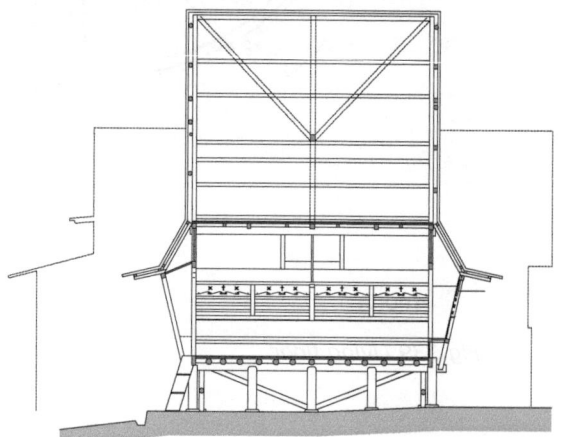

Fig.344 Cross-section 2-2.

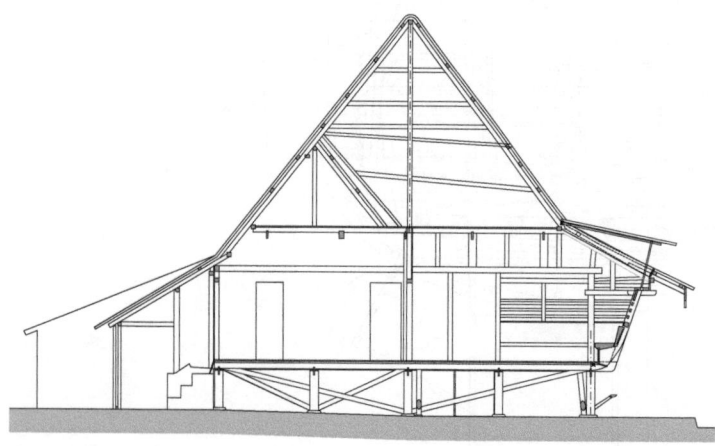

Fig.342 Cross-section 3-3.

Fig.345 Decoration of the façade.

Fig.346 Carving on the backrest of the bench.

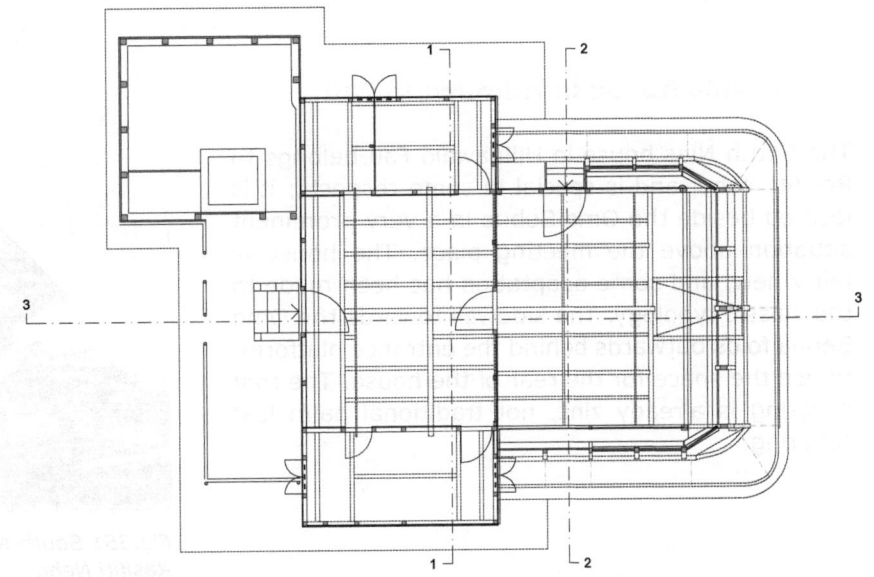
Fig.347 Living floor.

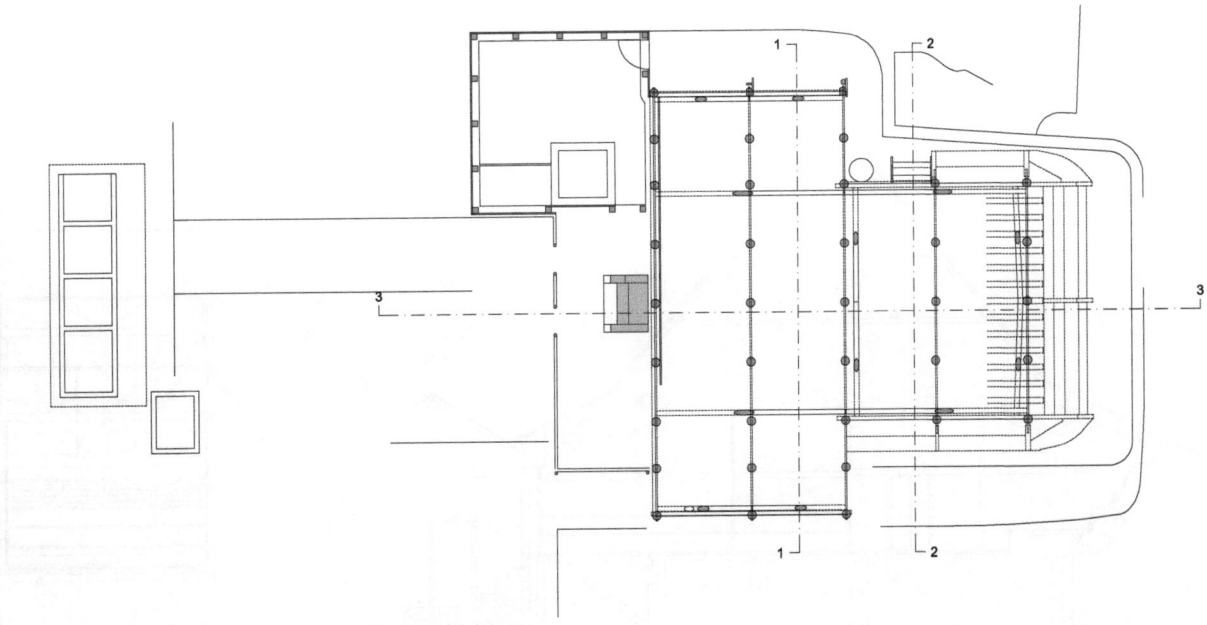
Fig.348 Substructure and backyard, with stables for pig.

Fig.349 Panorama view of Hilinawalö Fau, centre of village to the right of the photograph.

South Nias house in Hilinawalö Fau

The South Nias house in Hilinawalö Fau belongs to Rasifiti Nehe, and is special in some respects: it is located beside the Omo Sebua in a very prominent situation above the meeting place. The house is fairly new, and some adaptation has been made to the classic typology. The sidewall towards the Omo Sebua folds outwards behind the entrance platform, to use the space for the rear of the house. The roof covering is already zinc, not traditional palm leaf covering.

Fig.351 South Nias house in Hilinawalö Fau, belonging to Rasifiti Nehe.

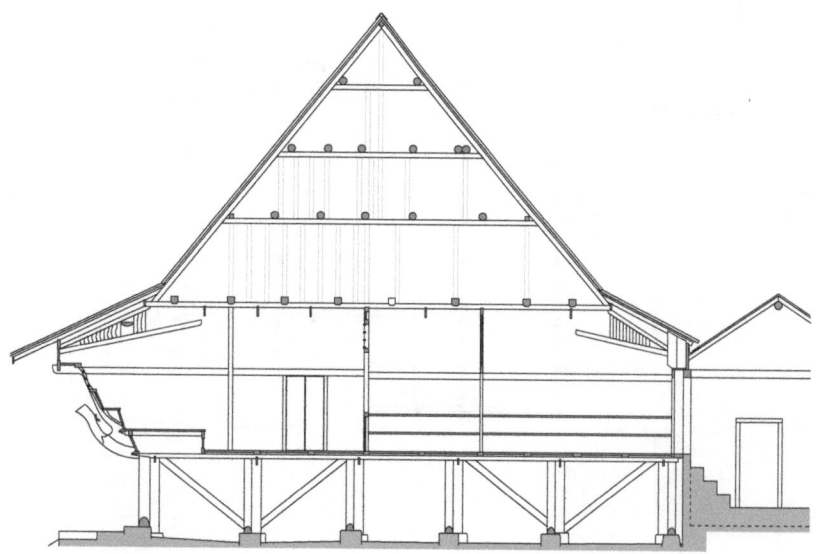

Fig.350 Cross-section 1-1.

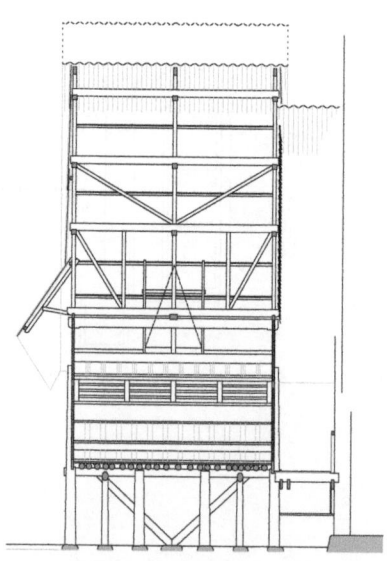

Fig.352 Cross-section 2-2.

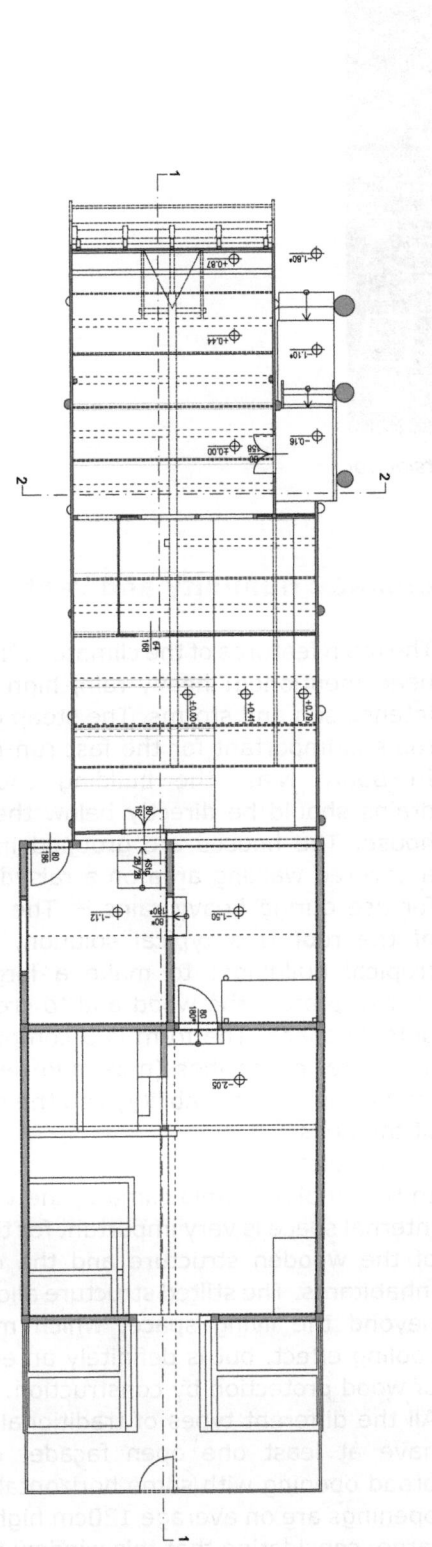

Fig.353 Living floor.

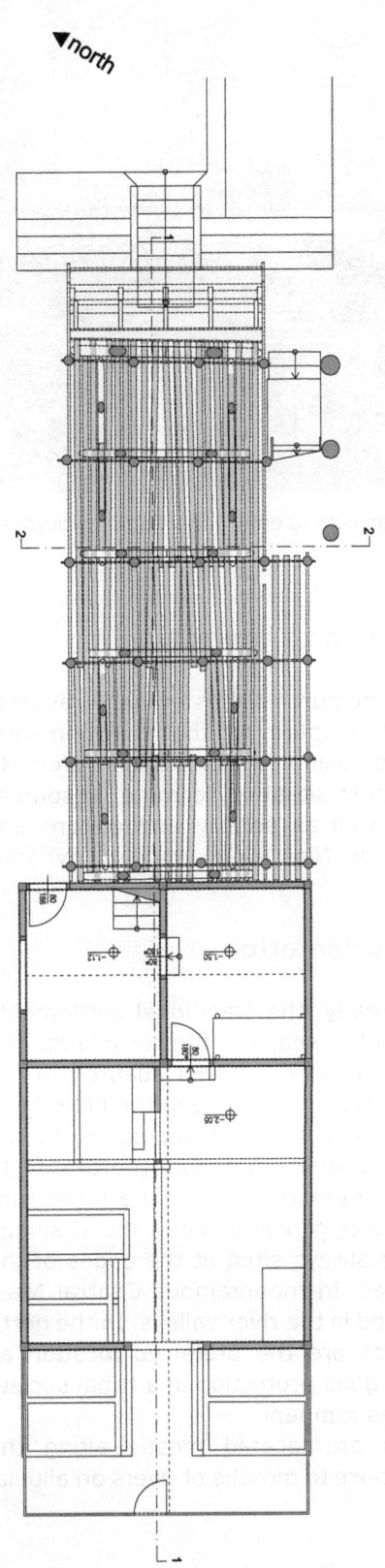

Fig.354 Substructure, long modern extension behind the traditional part.

Fig.355 House in North Nias with open window flap and clothes put putside for drying, 2003.

5.1.5 Adaptation to environment

This "natural architecture" seems in many respects to be very well adapted to the environmental conditions of the island.[428] Adaptation refers to topography, tectonic situation, climate, resources and material as well as society and culture and can be observed on Nias as an ongoing dynamic process.

Topography - orientation

As mentioned already, the traditional settlements are never next to the sea, but further inland. The old settlement structures are well adapted to the topography; they are oriented along the hill ridges. The main direction is often perpendicular to the island's mountains main direction (northeast to southwest). This orientation allows the rapid run-off of huge amounts of water along the drainage ditches which are always sited at the edges of the main village street. In mountainous Central Nias, villages are situated in the river valleys. In the north and south, hilltops are the preferred location as they also provide good protection in a tribal society where warfare was rampant.

New settlements are located mostly along the coastlines, often close to mouths of rivers on alluvial soils.

Climate: humidity and ventilation

The main features of the climate in Nias have already been mentioned: heavy rain, high humidity, heat, intense sun and storms. The steep geometry of the roofs is important for the fast run-off of rainwater. In South Nias, the building code directs that drains should be directly below the eaves of each house. The houses are grouped in rows, forming a covered walking area on a raised stone sidewalk for use during heavy rains.[429] The change in pitch of the roof is a typical solution, found in many tropical buildings: to make a large overhanging roof to protect the wood and to create a sheltered outside space. The form is a compromise between structural necessities (maximum length of beams, load transfer and stability) and the maximum depth of the house.

- Ventilation

In the tropical humid climate, the ventilation of the internal space is very important for the maintenance of the wooden structure and the comfort for the inhabitants. The stilted structure allows open airflow beyond the living space, which may also add a cooling effect, but is definitely an effective method of wood protection by construction.

All the different types of traditional houses in Nias have at least one open façade, consisting of a broad opening with some horizontal louvres. These openings are on average 120cm high, which is quite large, considering that this window primarily serves the big front room, the most public room of the buildings.

---

428 Viaro A. in Feldman, J. A. et al.: Nias, tribal treasures, 1990, pp.60

429 Feldman, J.A.: The Architecture of Nias, 1977, fig.51

Fig.356 Whenever the sun is shining, agricultural products - cacao, patchouli, nutmeg etc. or clothing is spread out to dry, Hilinawalö Mazingö.

The other rooms have only secondary ventilation through connective and roof spaces.

The doors usually stand open most of the time, providing additional ventilation. The location of the "windows" is approximately 250cm above ground: the vegetation around the houses is lower, so the wind becomes stronger with height. There is no way of closing these openings, no curtains or moveable louvres, so a continuous inflow of air is guaranteed.

One of the elements specific to Nias architecture are the big window flaps in the roof, which are open during the day whenever it is possible. They are only closed if rains and during the night. These flaps act like roof-windows and apart from fresh air provide light as well as access to the roof to dry clothing in the sun. Two or more such flaps are placed on the front and rear of the houses.

The construction of these flaps is very simple: they are made of a wooden frame, split bamboo and the usual roof covering, be it traditional palm leaves or corrugated zinc. On rare occasions the flaps consist of more elaborate wooden frames with carved decoration. The flap is opened and held open with a wooden rod, sometimes also carved.

When the large volume of air in the roof section of the house heats up, passive ventilation is initiated: fresh air moves in through the openings in the living floor, and moves out through the roof section above. The roof covering is nowadays often made of zinc. Traditionally, the air would percolate through the thatch of palm leaves. In South Nias, the side walls are closed only on the living floor: above this level, the wooden walls are split into single vertical wooden planks, so that sideways airflow is also possible.

Traditionally, all rooms on the living floor are open to the roof space. In some houses the owners have inserted a second floor, separating the living space from the roof space above. This was done to generate storage room in the attic, but also for reasons of cleanliness, as the palm thatched roofs pollute the living floor below.

But in spite of the changes, the inside atmosphere in all house typologies is much more comfortable than outside.

- Fireplaces

Before the colonisation by the Dutch the houses included kitchens. The fireplaces were part of the core of the houses, like those in the remaining Omo Sebua. Lacking a chimney, the smoke spread in the big roof space, removing insects and protecting the roof materials. The current floor plans stipulate that the fireplaces are situated in annexes behind the traditional houses. According to Viaro, the houses are smoked regularly on purpose to achieve this cleaning and preservation effect.

The semi-permeable covering made of palm leaves is in more and more cases replaced by metal. Its rusty colour may be a matter of personal taste, but the material is bad from the building physics aspect, for humidity, permeability, heating effects and sound. The positive characteristics of metal roofs are price and durability. Considering the fact that the roof is the most important means for protection of wood, metal roofs may be acceptable in spite of their disadvantages. Finding another (cheap) material as good as palm leaf covering would revolutionise building in the tropics. The challenge for future designs is to achieve a good room-climate without active air conditioning systems.

Fig.357 Foundation stone of the façade of the Omo Sebua in Onohondroe.

Fig.358 Palm leaf thatching, laid out to dry. The length of the parts is usually one "drefa", the distance between outstretched arms.

Fig.359 Façade detail, mortise and tenon connections, Hilisimaetanoe.

Fig.360 Paved ground in front of a house in Hilinawalö Fau, the foundation stones being of a different type, and higher than the pavement.

- Light

Direct sunlight in the tropics is merciless, and usually has to be avoided to reduce overheating. So all the different types of houses are very dark inside. The outward slope of the façades, the large front openings and the extensive roof overhangs prevent direct lighting. The only possible way for direct sunlight to come in is through the skylights when the flaps are open. So the public front room with the large front opening is the only one with light conditions that could permit reading. The more modern annexes with kitchens, bathrooms and stables have "normal" windows and usually better natural lighting.

Electricity is still not available in remote areas, so people use kerosene lamps and torches for artificial lighting. Villagers still depend on the cycle of the sun, making the most of the 12 hours of daylight. When it gets dark the big front openings become the only visible elements in the village - if there is electric lighting.

Material and details

Only locally grown plant material was used for the traditional houses: no metal pieces were needed. Elaborate mortise and tenon joints still connect the load bearing parts. Different kinds of wood are used, depending on the location within the construction; the hardwood Manawa Danö in the North Nias houses and huge plates of ebony (imported from Telo) in the South Nias kings' houses being the most impressive ones.

Apart from wood, palm leaves, bamboo and binding is still used for the roofs. Slabs of stone (mostly basalt) provide the foundations, pavement and megaliths.

5.1.6 Adaptation to earthquakes

The hypothesis that these buildings resist earthquakes better than other building types was demonstrated in reality by the big earthquake on 28th of March 2005.

This research project aims at demonstrating the same in theory, through modelling and simulation. The different existing typologies of Nias houses have distinct and common features. The following characteristics of the traditional Nias houses constitute positive factors for their earthquake resistance.

Seismic loads and deformation

The earthquake force, the "shaking" of the ground, is difficult to grasp. The shock in the epicentre initiates a complex wavelike reaction. For engineering purposes such shaking has to be translated into distinct horizontal and vertical forces. The frequency of the waves is of extreme importance, because of resonance effects on the ground and/or the building. The "natural frequency" of the building (natural frequency is the frequency at which the building resonates) determines the probability of heavy damage, because if the frequency of the earthquake waves equals the natural frequency of the building the amplitude of the waves increases until they are so violent that finally, the building may be shaken apart.

The damping factor (trend with which the vibration decreases after the loading) depends on the structure of the house and the material properties. For a house to be earthquake resistant, its natural frequency should be as different as possible from that of an average earthquake to avoid the amplification effects. The natural frequency of the ground must also be taken into account. For static analysis, earthquake frequency is taken from historic examples.

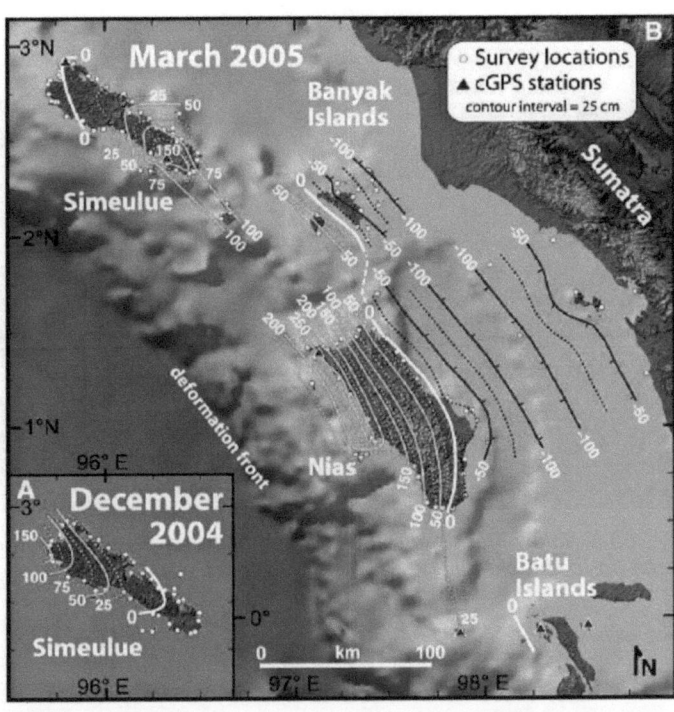

Fig.361 Contour map in cm of vertical deformation during December 2004 (A) and March 2005 (B) earthquakes. Bright contours indicate uplift, dark contours subsidence, solid contours in 50cm intervals, Briggs et al. 2006.

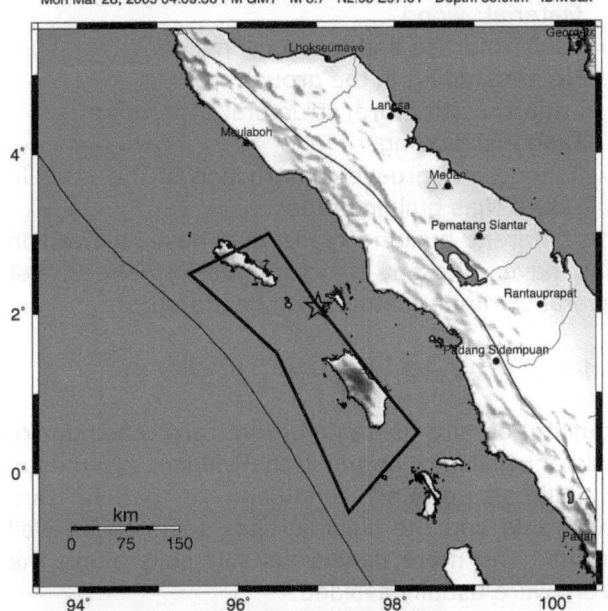

Fig.362 The so-called "shakemap" indicates severe to violent ground acceleration for the Nias region, US Geological Survey March 2005.

Fig.363 Siwalawa landslide 2004, which halted at the foot of the steps up to the old village.

Fig.364 Disruption along the main axis of the settlement, Hilisimaetanoe.

Common characteristics of traditional architecture in terms of earthquake resistance

The following characteristics of the traditional houses provide resistance to damage by earthquake:
- Selection of good sites
- Material: wood
- "Soft" mortise and tenon connections
- No embedding in the ground
- Tripartite structure, structural separation
- Diagonal bracing
- Tongue and groove construction of the walls in South Nias building types.

The different building types that have evolved in Nias have different structural performances, yet several common features can be stated.

Topography and ground condition

Stability of site is required for any earthquake resistant building. As traditional villages are mostly on hilltops, ground conditions in general are favourable and the building sites seem to be well chosen. The more dangerous soft soils along the rivers were usually avoided.

Siwalawa village serves to illustrate the big differences regarding sites: the old settlement cores are on stable ground, whereas a landslide destroyed newer developments in 2004.

We found disruption along the main axis of the villages, a characteristic and often observed damage. This might be a result of general uplift and deformation of the island as a whole.

Influence of material and detailing

The number of victims of the earthquake in March 2005 proved the significance of choice of material - collapsing concrete structures caused most casualties, a result of "pancake"-like collapse due to vehement ground shaking and the poor quality of material and detailing.

The advantageous characteristics of wood, not only in the traditional, but also in the Malayan-style buildings, saved lives. Wood has excellent dampening characteristics and is comparably light. Complete collapse does not occur, no heavy pieces fall down on inhabitants, and the reuse of material is possible.

Apart from the characteristics of materials, skilled workmanship is important: many concrete buildings collapsed because of poor materials and poor workmanship, and step by step progress due to intermittent funding had become an additional handicap. In contrast to that, expertise in building with wood is still high.

All connections and joints of the traditional houses can be opened and the parts disassembled. The mortise and tenon connections allow deformation to a certain extent before failing. Timber joints, secured with wedges, can be fixed again in a slightly different position. They are therefore very "soft" connections, allowing considerable movement before breaking. Dismantling the houses and rebuilding is possible as well.

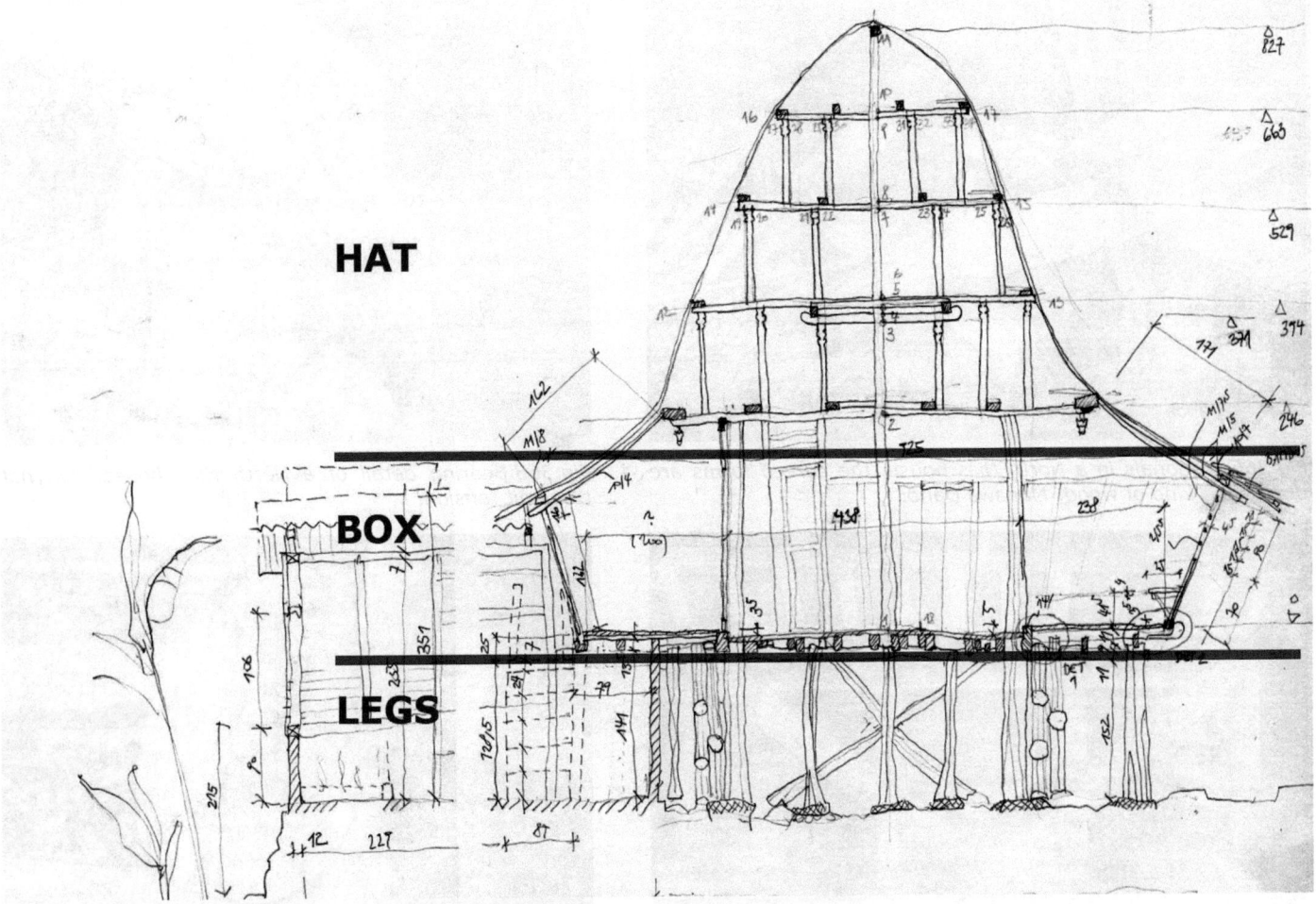

Fig.365 Sketch of a North Nias house, cross-section, indicating the vertical zoning in three main parts.

Foundation

The traditional houses are stilted constructions. The stilts do not settle in the ground but are placed on foundation stones. As far as we could find out, the ground is not treated in any particular way before building. The slabs of stones used as foundation seem to be placed randomly, and the pillars are not really adapted to their shape.

In most cases, the ground below the house is not paved, with some exceptions existing in Central and South Nias. We do not know whether or not the ground below the houses was paved in former times. In any case, the space around the house is more important. In South Nias elaborate drainage systems provide for the runoff of huge amounts of rainwater. This prevents soil erosion, which would render the soil unstable.

The central foundation stones of the kings' houses have elaborate carvings. In case of an external force, movement of the whole building is possible without breaking the stilted substructure. Movement and minimal sliding of the stilts on their foundation stones occurs frequently. The transmission of horizontal forces from ground to the structure is physically impossible.

Tripartite structure

Tripartite structure is not an exclusive feature of Nias houses. This form of building is ubiquitous in the Indonesian archipelago and other regions with tropical climate, for many reasons and advantages. For earthquake resistance, tripartite structures are most advantageous. The "legs" of the substructure provide resistance against the impact of forces in vertical and horizontal directions. Originally the substructure was not used as living space, but as storage space and stable. The living floor constitutes a stable "box" stiffened by the inside walls and open to the roof space. The stiff "box" survives even if the substructure collapses. The roof is a light, open structure, a "hat" that does not endanger the inhabitants. There is considerable structural separation on the different floors.

Altogether, the house resembles a legged box with a light hat. So it combines the two concepts of earthquake resistant design: flexibility to absorb the energy of seismic shocks and stiffness to protect the inhabitants.

Fig.366 Diagonals in a North Nias house, the curved forms are due to the kind of wood: Manawa Danö.

Fig.368 Bearing detail of a North Nias house, can not transmit tension.

Fig.367 V-shaped bracings of South Nias typology, delivering a solution for the bearing.

Fig.369 South Nias house, detail of upper connection: the diagonals transmit forces only to the floor beams, not to the primary structure.

Diagonal bracing

The diagonal bracing in the substructure is the most obvious special feature of Nias architecture and seems to be responsible for earthquake resistant behaviour. According to the respective building typology, the diagonals are placed in rows inside or outside a grid of vertical columns, in both directions. In North Nias the bracings form X shapes behind the vertical posts. In South Nias they are situated as a huge V in the front façade, but also along the length of the houses. With their detailing, the bracings can absorb the compression of the external force but can take tension only to a limited amount, as they are not anchored firmly in the ground.

In all building types, the bracings can transmit compressive forces onto the basement detail, and both compressive and tension forces onto the upper end detail. In North Nias, the triangular space is deliberately filled with heavy stones. The stored weight improves the functioning of the diagonal bracing: the weights' inertia resists the lateral movement of seismic shocks.

Fig.370 The space between the bracings of this North Nias house is filled with posts and stones to reduce tension in the substructure.

Fig.371 Collapsed houses in Hilimondregeraya. Bracing detail: the fixing has moved.

Fig.373 Collapsed houses in Hilimondregeraya.

Fig.372 Broken elements of a highly stressed connection point.

Fig.374 Houses having moved laterally, supposedly collided with each other, Hilisimaetanoe.

The structural integration of the bracings into the primary construction system in the floor between the substructure and the living box is unsatisfactory in all types of building. This indicates that the bracings are a later development, possibly found useful as an adaptation to unstable ground conditions and the frequent earthquakes. Movement of the bracings very often occurred: they seem to take the main force of the seismic shockwaves.

From a structural point of view, the bracing connections are "hard" compared to the overall construction. Furthermore, the connecting parts are not designed to take the transmitted forces, as shown by the typical damage. The overall result of failure in the substructure is a tilting of the upper floors.

Damages stated

We detected few houses that had totally collapsed.[430] But even then the space in the living floor remained unobstructed and free. So in each case the "box" had remained stable and provided good shelter. In the villages that we visited, nobody had been killed in a collapsing house during this earthquake.

Complete collapse of the substructure was observed only in buildings whose upper floors were heavier than normal due to roof extensions, and in those whose state of preservation was so bad that the stilts were already rotten. Damage to the first floor was observed regularly in walls and tension elements. Damage to the roof cover was ubiquitous, but seldom affected the structural parts.

The "usual" form of damage was: movement of roof and first floor in relation to the substructure, and damage to the roof covering.

---

[430] Taking into account that our field trip took place three months after the quake, some of the collapsed houses might have been already torn down. For this reason the empty sites were added in the set of statistics.

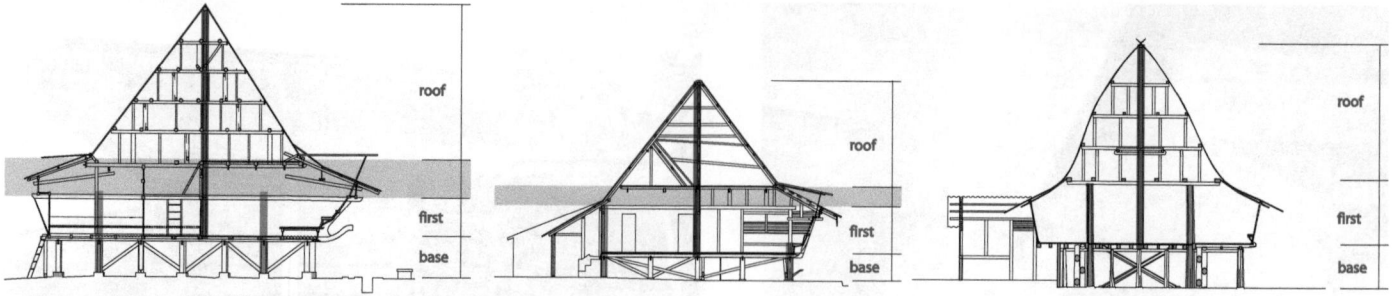

Fig.375 Comparison of the sections of the different house types: South, Middle and North Nias typology from left to right. Grey - vertical elements connecting substructure and living floor, dark grey - vertical elements connecting living floor and roof, light grey - zone of horizontal lateral trusses between living floor and roof, only in South and Middle Nias typology.

Fig.376 Heavily damaged house in Central Nias, the truss zone being visible from outside.

Structural comparison between typologies

A comparison of the cross sections of the different typologies shows the elements that connect the three zones. The most remarkable structural feature is that no single element connects all three zones. Four pillars connect substructure and living floor, two pillars connect living floor and roof. The number of connecting elements seems to be reduced to the absolute minimum.

In Central and South Nias types a structural in-between zone separated living floor and roof. An element similar to a wooden truss spans the entire depth of the building. This again shows the close relationship between the Central and South Nias building types.

Especially in South Nias houses, the roof contains heavy horizontal beams, which are not needed to hold the covering or define the geometry. From a structural point of view, therefore these beams are not necessary. They may be placed there because of their weight, either to keep the centre of gravity at a certain point, to influence the natural frequency, or to increase weight for other loading conditions, e.g. wind. The overall weight and the vertical distribution of weight again refers to the natural frequency of the roof as part of building.

The structural separation of the floors mentioned previously corresponds to the damage symptoms. We observed considerable movement between the zones, especially between the living floor and the substructure. Cases of complete collapse must be attributed to the failure of the substructure.

The movement between the zones stresses the interconnecting elements. Failure of the connection and breaking of the elements show areas of force transmission.

Fig.377 Interior space of the Omo Sebua in Onohondroe in 2005, this part having collapsed already.

Reasons for damage

The violence of forces acting on the houses is not the only parameter responsible for damage. The overall sound condition of the buildings is an important factor for their structural integrity, but is often enough not maintained.

- Bad state of preservation, material integrity

In those South Nias villages accessible by car and thus better off economically it was demonstrated that proper preservation of the houses is needed to maintain their good qualities. In fact, richer settlements were much less affected by damage.

Aid had not reached many remote villages by the time of the field trip in 2005. People obviously had not had the resources for the maintenance of the houses for quite a while, although poor maintenance is the main reason for damage to traditional houses. Two other factors that reduce the durability of traditional houses are inadequate roofing and the humid climate.

- Damage to roof covering

Damage to the roofs was ubiquitous, particularly for the cover (palm leaves on split bamboo), less often the structural parts. The integrity of the roof is crucial for the preservation of the houses. In case of the house being abandoned (or inhabited only by old people after parts of the families moved to Sumatra or Jakarta for work), damage to the roof coverage allows damage below.

- Lack of knowledge

Apart from the renewal of roof covering, there seems to be no tradition of regular repair or renovation measures. Ill-planned adaptations of the houses and bad protection and reconstruction work show a profound misunderstanding of their structural functioning.

Fig.378 House in Bawölato, humid ground and changed proportions because of the roof extension led to considerable movement during the earthquake.

- Endangered structural integrity

Structural integrity was often affected for the worse by changes in traditional construction. Removal of structural parts, esp. of substructure, was carried out to gain unobstructed space in the basement. Roof extensions and second living floors moved the centre of gravity upwards. The collapse of annex buildings led to damage of the traditional parts - e.g. the upper tension ring in the North Nias type.

Fig.379 Samples of Manawa Danö, Afoa and Simalambuo prepared for testing.

Fig.380 Zwick Z250 Universal Materials testing machine, Institute for Building Construction and Technology.

Material tests

The characteristics of the kinds of wood used for the Omo Hada could not be found in literature, and obviously they had never been tested according to European standards. Moreover, it turned out to be difficult to determine the botanical species of the woods. The Nias languages are very individual, and the available sources did not identify the botanical names of the woods. In spite of the obvious differentiation of the material according to the location within the structures, for practical reasons use only one kind of wood was used for calculations.

Wiryomartono describes the following characteristics of woods used for building:

- Afoa: family Laureaceae, genus Litsea Lank, Litsea species, origin Nias; density category 1 (very light), colour light brown or dark brown, used for panels, boards, beams, bars, in the living unit.
- Manawa Danö: family Verbenaceae, genus Vitex L, species Vitex Micrantha, origin Batu islands; density category 4, colour dark red, used for beams, panels in back and front room

Apart from these he describes 19 kinds of wood and their use in building. The substructure is made of timbers with densities of about 4.[431]

Samples of the most commonly used kinds of wood were sent to Vienna by the Pusaka Nias Museum and tested in the facilities of the Institute for Building Construction and Technology at the Faculty of Construction Engineering at the Vienna University of Technology. Due to the remoteness of Nias, only small samples of three different kinds of wood could be tested on the basis of European Norm 408.[432]

The interpretation of tests of small samples is not common, and the results cannot simply be extrapolated to sizes of building parts.[433] The results of the tests were used to find comparable wood which had already been tested in full accordance with the standard defined in EN 408, in order to define the mechanical characteristics for calculation.

A comparison with European woods was carried out by Werkraum Wien[434], the construction engineering company that also did the structural analysis.

[431] Wiryomartono, A.: Cosmological - and spatiotemporal meanings of a traditional dwelling in South Nias, Indonesia, 1989, pp.242, table 20A

[432] The tests were carried out from January to April 2006 by the research assistant Philipp Pongratz at the facilities of the Institute for Building Construction and Technology.

[433] Kollmann, F.: Technologie des Holzes und der Holzwerkstoffe, 1951

[434] Bauer, P. et al.: Statischer Bericht, Projekt Nias, Werkraum Wien Ingenieure, 2007, p.5

Static analysis of South Nias typology

The South Nias typology was selected for analysis for a pragmatic reason: it is the most prevalent type of traditional house, with several hundred houses still existing, and analysis could provide valuable information for reconstruction. A summary of the analysis carried out by Werkraum Wien is given below, followed by discussion of further research and consequences.[435]

Basics

The analysis was carried out on a three-dimensional digital model, including all axes of structural components and dimensions. The investigation was intended to show resistance to earthquake forces that were assumed as quasi-static.
The comparison with a wind load with a velocity of 100km/h reveals that the assumed wind load triggers larger deformation and higher stress than the quasi-static earthquake force.

Interpretation

Because the construction method uses no finishings, the values for deformation of the construction are not important (approximate value in Austria 1:100 of height at maximum). The capacity utilisation is relevant to the failure of the building.
All in all, the results of capacity and overload correspond to the damage that was observed. In many houses, damage occurred in the roof zone. This is easy to repair, but must be carried out without delay. Damage to the joints in the floor zone of living and above are fatal for structural integrity. In this case, the house has to be dismantled and rebuilt with new elements.
In the simulation, the roof and the substructure can transmit the loads without any problems. The deformation and capacity usage due to wind load in cross direction are so serious that the overall stability of single houses cannot be guaranteed. According to the present deformation modelling, the roofs will collide with each other when horizontal forces affect the houses. The results of the calculations put the relevance of the models in question. But as the South Nias houses do not exist as detached buildings but are coupled and in rows, effects of constructive connection or of aerodynamics of the settlement structure may be missing in the model.

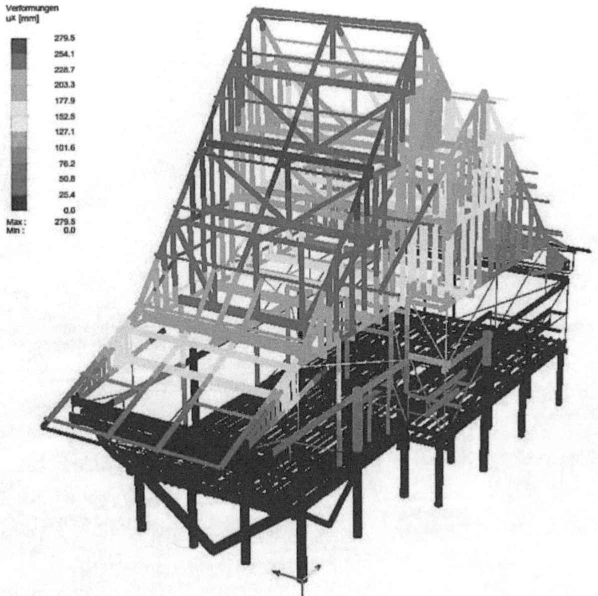

Fig.381 Deformations due to wind, in cross-section, revealing movement in the roof zone of up to 30 cm, Werkraum Wien 2007.

As the analysis shows, the living zone, "the box", is overloaded in the direction parallel to the street.
The substructure works well in the model, the diagonal bracing having a positive effect on the overall construction. The starting points of the diagonal beams in the floor area are overstressed. The difference in deformation between the middle region and the side stringers (Sikhöli) in the living area is remarkable under horizontal force. The brace supports that were added to many houses during the field trip in 2005 may be a reaction to the observed or felt lengthwise movement. (On the other hand this is the easiest method of support.) The lift-off forces occurring in the bearings predicted by the model cannot be withstood by the foundation details in nature. The solution used in North Nias, to put rods and stones onto the slanting diagonals in order to increase pressure and hold the house in position, was not observed in South Nias.
The high stress on the joints in the simulation correspond with the damage observed.

Consequences

Under the precondition that the models are valid, the construction of the South Nias row houses cannot be applied to individual buildings. This should be taken into account in reconstruction. Measures of constructive connection could be introduced. However, well-maintained traditional South Nias structures can withstand earthquake forces with only minor damage in the roof area.

---

435 Bauer, P. et al.: Statischer Bericht, Projekt Nias, Werkraum Wien Ingenieure, 2007

Fig.382 Former palace of the Simalungun in North Sumatra, Pematang Purba. The smaller building to the right represents the typology of a rice storage building, the so-called "sopo".

5.1.7 Differentiation of typologies

The differentiation of the particular architecture in Nias is extraordinary, but has not been given the same attention as the particularities in society and culture, which have been the focus of research for a long time. Documentation of old architecture is scarce.[436] In the 1970s architectural documentation restarted with the works of Feldman, Viaro and Wiryomartono. Viaro documented houses and settlements extensively and thus laid an important foundation for further study. Gaudenz Domenig has done extensive interdisciplinary ethno-architectural research in Indonesia, using a method of typological research based on the investigation of changing construction characteristics, which can deliver information on historical and evolutionary sequences of building typologies.[437] Following this strategy, some particular features of the Nias traditional houses will be discussed here.

Foreign influences and interrelation of typologies

Reimar Schefold stated the similarities of Nias architecture with other Indonesian building typologies[438], referring also to research carried out by Roxana Waterson[439]. He describes the characteristic common features of traditional Indonesian architecture:

- the tripartite house
- the multi-levelled floor
- the outward slanting gable (wall)
- gable finials
- the saddle-backed roof
- differential treatment of root and tip in the use of timber

(In Nias, there are no gable finials and saddle-backed roofs, and instead of the gable the walls are slanting outwards). Schefold supposes that the origin of the common basic traits lies in Neolithic times, and that the further development characterised by a continuous flow of mutual influences is due to long-lasting trade relations between the South-East Asian mainland and the archipelago.[440]

Feldman tried to trace back the extraordinary manifestation of the South Nias façades to foreign influences, especially the Dutch galleons.[441]

Gaudenz Domenig indicates the development of the Nias building types from storage building typologies, existing in different variations in Sumatra. Having investigated construction characteristics, he postulates a common, but meanwhile lost, ancestral typology in Central Nias.[442]

Some of the specific features of Nias architecture are found also in other wood-building traditions. The outward movement of wooden constructions, especially of multi-storey buildings, is found worldwide. This is a converging feature due to constructive issues and protection from water.

The cantilevered benches, for example, can also be found in Chinese architecture.

---

436 The earliest documentations by Schröder 1917 and De Boer 1920

437 Domenig, G.: Typologie als Methode diachronischer Bauforschung, Konstruktionswandel im Hausbau auf Nias (Indonesien), 1993

438 Schefold R. et al.: Indonesian houses Vol 1, 2003, p.23

439 Waterson, R.: The Living House, 1997

440 Schefold R. et al.: Indonesian houses Vol 1, 2003, p.51

441 Feldman, J.A.: Dutch galleons and South Nias palaces, 1984

442 Domenig, G.: Typologie als Methode diachronischer Bauforschung, Konstruktionswandel im Hausbau auf Nias (Indonesien), 1993, p.177

Fig.383 North Nias house in deconstructed state, historical photograph made available by Jerome Feldman 2006. The rectangular construction is extended sideways and one of the rings for the roof structure is visible.

Fig.384 Historical photograph by Schröder, 1917, of a transitional type of building - the right part being an annex to a rectangular type of house, similar to the big houses still found in Central Nias, but the substructure is clearly of North Nias type.

Evolution of typologies in Nias

The ethnological origin of the Niha is unclear and still an issue of investigation and discussion, influencing also hypotheses about the chronological appearance of the architectural typologies.

Research into the development from one typology to the other requires further studies of Central Nias architecture, although this exists in so many variations that a Central Nias typology cannot be clearly defined.

The relationships in construction have been noted, but do not allow a chronological interpretation. Concerning particular characteristics, there may be a correlation between effectiveness and historical development, but the transfer to an evolutionary sequence of the typology as a whole cannot be stated.

Fig.385 Central Nias, Gomo area, house with both bracings in the very front (right) and behind the first row of vertical pillars (left).

Evolution of particular characteristics

The development of additional characteristics may contribute to the discussion. Apart from a formal approach, the following observations are related to the efficiency of the features with respect to their structural performance. Apart from being useful in the discussion of coexisting characteristics, the photographs taken by Schröder and other explorers also allow a historical perspective as well as assumptions about the development.

Bracing in the substructure

The bracings in the substructure are peculiar to Nias, and cannot be found elsewhere in this form. The usual detailing in Nias architecture is highly elaborate and subtle; mostly locked mortise and tenon joints are used.

The unique diagonal beams in all Nias house types are not connected to the other elements in the same elaborate way, but pierced apparently randomly by the floor joists, or hooked onto these. In some houses a smaller hook-like horizontal part underneath the floor joist guarantees the ability of the joint to resist tension. But nonetheless the transmission of forces in this detail is very unsatisfactory: tension cannot be coped with in the ground bearing, and the connection to the upper structure is only at secondary construction elements, as described earlier.

The diagonals are poorly integrated into the overall construction system in all the different types of houses; this indicates the relative novelty of diagonal elements in the construction.

In spite of that, the static analysis showed the excellent resistance of the substructure, but improving the junction of the diagonal beams with the rest of the construction would be one way of improving the structure.

In South Nias, the X-shaped bracings have been transformed into V's. This intelligent variation replaces the bearing detail of North Nias houses with a self-stiffening system of two pieces forming a load bearer for each another. The evolution of X to V-shape is not only related to form, but represents progress regarding transmission of forces.

The V in South Nias is used as the predominant façade element. In Central Nias, two variations, bracings in front of the vertical pillars and bracings behind, exist in neighbouring houses.

Fig.386 North Nias house, with elements that could be transformed sikhölis.

Fig.387 Usual form of sikhöli, South Nias.

Sikhöli

"Sikhöli" is the term for one of the most important parts of the primary structure of the house. In South Nias, the transition between the lateral walls carrying the roof and the stilted substructure is solved with the two big horizontal beams running along the overall length of the house. Their cross-section is narrow but very high. The front end is extended and curved, often carrying elaborate carvings. The two sikhöli are connected by the lower elements of the terraced front façade and by the floor construction.

South Nias houses have very similar forms of sikhöli and show similar decoration. In Hilimondregeraya another variation exists: the sikhöli does not extend from the façade but still ends in a small upward curve.

The sikhöli in Central Nias vary in shape and decoration: some deviant shapes were observed in the Gomo region of Central Nias. The sikhöli of the large adat houses are very large and elaborately decorated. Only the Omo Sebua are decorated with Lasara heads. The sikhöli of ordinary living houses are not decorated and are formed according to the natural growth of the tree. Sometimes the sikhöli is not at all extended or accentuated, and merely carries the lateral façade panel.

In North Nias, some houses show what could have been a sikhöli before the oval floor plan evolved through the addition of semicircles to an initially rectangular floor plan. At the very end of the straight part of the house façade, the cantilevered beams carrying the floor and outside walls emerge outside the lower tension ring which runs around the whole shape. So at these points along the edge of the rectangular part the ring is not only held up as is usual but also held together horizontally. This detail could be a vestige of the supposed common rectangular archetype.

Fig.388 Carved and decorated sikhöli, Central Nias.

Fig.389 Reduced form of sikhöli, common in Hilimondregeraya, South Nias.

Building parts comparable to the sikhöli exist in the houses of the Karo Batak in North Sumatra, which also feature large base beams bordering the living floor. The two beams have carved end parts which are extended through the front and back beam, but not as far as in South Nias. The houses of the Simalungun in the Pematang Purba area and the Toba Batak houses show the same characteristics. The sopos, traditional storage buildings, have the same feature, but one level higher in the roof area.

Fig.390 Elaborately decorated façade in Hilimondregeraya, South Nias.

Façades

The façades of traditional Nias houses have common features. The outward slanting walls enhance the large roof overhang and make the interior more spacious.

The bench is a distinct feature of North Nias architecture as a piece of fixed furniture. Its length corresponds with that of the large window opening. Usually the bench is an added element inside the house and thus not visible in the façade, but exceptions were found in Tumöri.

In Central and South Nias the bench is an integrated part of the cantilevered terraced structure of the front and side façades. It is self-supporting, clamped between sidewalls and held up by vertical elements. The front façade is always partitioned into an even number of sections. The load of the roof is not transmitted to the façade elements, only to the sidewalls.

The system is different in North Nias: the roof hangs from the main pillars like an umbrella. The roof construction pushes the umbrella surface outwards at the upper tension ring, and the outward slanting façade takes the weight down to the floor construction and the substructure. This system is not only self-supporting, but the weight of the roof is partly carried by the façade elements.

The North Nias façade is constructed with a continuous rhythm of vertical posts, fixed to the lower and upper tension rings. The slanting posts can stand parallel or perpendicular to the oval boundary.

The structural system of the houses only holds together with the inclusion of at least the vertical posts of the façades.

Fig.391 Lasara head on the Omo Sebua of Hilinawalö Mazingö.

Fig.392 Carving on the backrest part of a demounted house, Central Nias.

The corner conflict

The terraced front façade which is so elaborate in the South Nias buildings, is a standard feature in Central Nias. In contrast to South Nias, the same cantilevered system is also used on the side facades. The side façade comes in many variations: from plain closed side walls slanting outwards, to those including a bench, and to those with the same lattice openings as the front. Whatever variation may be favoured, the line where two facades meet at a corner is obviously difficult to design. Due to its linearity, the terraced system cannot be bent, and the primary construction of the house has to be integrated. For this geometrical and structural challenge an astonishing variety of solutions has evolved and is found within an area of few kilometres around Gomo.

- The non-corner

Neither façade reaches the corner: they continue as far as the first pillar. The sidewalls are closed, and the site of the corner is outside the house, unused.

- The diagonal wall/façade

The two façades are connected by a diagonal surface. This can be another closed wall, or another bench-window-system. This system also results in a polygonal form, so more space is integrated.
Both elements of the bench, the seat and the backrest, cut through the lateral bordering and bearing panel and cantilever at least a few centimetres to the outside, so the contours of the inside bench are clearly visible from the outside.

- The quarter circle

The connection between the two façades is constructed as a quarter circle.

The underside of the bench and the backrest are often decorated with carvings, sometimes coloured. Animal half-reliefs are preferred, but human figures and abstract patterns may also decorate the façade.

Apart from the non-corner variation, the primary construction is partly inside and partly outside the house. The edge pillar usually reaches up to the height of the floor construction in its full section. Then the remainder may reach the height of the bench, accompanied by another internal pillar, which starts at floor level. Above the height of the bench, only the new internal pillar continues.

Further research on this phenomenon would relate to Domenig's investigation of supports in the South Nias house, and presumably back up his hypothesis of evolution.[443]

[443] Domenig, G.: Typologie als Methode diachronischer Bauforschung, Konstruktionswandel im Hausbau auf Nias (Indonesien), 1993, p.163

Fig.393 The non-corner.

Fig.394 The diagonal façade.

Fig.395 The round corner.

Fig.396 Metal roof coverings in Hilinawalö Fau.

5.1.8 Adaptation of traditional structures

The development of the typologies is an ongoing process, also including the primary structure. Variations mostly occur in size (quantity), number of construction axes, depth of houses and details. Moreover, hybrid constructions between traditional and modern exist in various forms, not limited to the annex buildings, but no typologies could be established.

The ubiquitous annex buildings at the back have their origin in an important functional change that was introduced in colonial times: fireplaces had to be moved outside the house. So a tradition of siting kitchens and bathrooms in an annex has developed, but no building type has yet evolved. Water pipes are provided in the villages, but drainage is an environmental problem and in some cases affects the stability of the soil behind the houses. These shapeless annex buildings are found in all regions of Nias.

The most obvious adaptation of material is the use of corrugated metal for the roof covering. Even if appearance and room conditions suffer, these roofs have saved many houses, including the Omo Sebua in Bawomataluo.

In addition to that, electricity cables and satellite receivers have changed the appearance of the villages. Electric lighting after dark reverses the direction of the day time illumination: the open louvres illuminate the village street.

In North Nias, a change in height has led to the relocation of the entrances from below the living floor to lateral annexes, which are a focus of individual designs, and an obviously recent invention.

Fig.397 Additional floor space in the substructure is provided by lowering the groundlevel and cutting or removing the bracings, Bawomataluo.

Adaptations to the traditional constructions exist in various forms, as the Niha are very inventive. Central Nias in particular offers a great variety of floor plans and constructions.

The need for floor space has sometimes resulted in dangerous changes to the traditional building structure: introduction of a second floor, substructure used as floor space and the like.

The transformation of the characteristic form into another material, e.g. concrete, occurs quite often with peculiar results.

Identity

What has been learnt from tradition is formal reinterpretation. The strong forms delivered by tradition are translated into modern buildings, but they lack their former function of load bearing: we discovered a few extreme examples of imitation of entire traditional South Nias houses in concrete and brick.

What has been introduced most often is the diagonal bracings. They have obviously become a symbol of stability, which could have been provided much more easily by using concrete as the building material.

The diagonal bracing and the steep roof of the Nias house has become a characteristic emblem for Nias, also used for public buildings. In particular entrance gates, annex buildings and porches have been transformed into symbolic Nias houses.

In North Nias, close to Gunung Sitoli, we found a few Rumah Adat, which were either new or renovated according to modern ways. They were obviously owned by rich families who possess an Omo Hada as a status symbol.

Rough models of the Nias houses are ubiquitous, and sold as souvenirs. Very creative models of the Nias houses were carried also around Gunung Sitoli in a national holiday parade in 2005. All this shows that the inhabitants of Nias are well aware of their architecture and will hopefully be able to maintain this unique tradition.

Fig.398 Model of traditional house used for a parade at the National Holiday in 2005, in Gunung Sitoli.

Fig.399 Small Omo Hada as entrance annex, the traditional house already becoming a status symbol in North Nias.

Fig.400 Concrete house in traditional South Nias form for a graveyard, Teluk Dalam.

Fig.401 Tent on a wooden platform, shelter after the earthquake in Hilimondregeraya.

Fig.402 Small hotel in Sorake beach, South Nias, reinterpreting a traditional façade in concrete.

Causes for disappearance

The reasons why building types are "dying" vary with typology and regional situation, but certain major problems can be identified.

The question of identity is no problem for the survival of the traditional architecture in Nias - in contrast to other societies. In our interviews, the majority of the people said they would prefer a traditional house if they could afford it. Many modern houses in Nias show formal features taken from the traditional typology, with no particular function, but showing that Nias' people are still proud of their culture and architecture, and that they associate safe building with the formal style shown by the Omo Hada. Symbolic models and icons are often used in public buildings.

The problems are due to economic conditions, ecological problems and changes in culture and society.

- Economic situation

The economic situation on Nias is appalling, as is the island's infrastructure. There is no building industry, so all material has to be imported. The reconstruction efforts have boosted the economy but they have also boosted inflation.

- Shortage of timber

The important construction parts were made of a durable hardwood which can no longer be easily found in Nias. The general timber shortage has led to the growth of modern settlements constructed from concrete - which resulted in hundreds of the inhabitants of Nias being buried in the devastating earthquake of 2005.

Reconstruction needs worsened the situation, and at the time of our field trip (summer 2005) even cheap softwood was scarce, so that a "timber task group" was set up by aid agencies.

In the forests around the more remote villages in Central and South Nias, inhabitants were forced to cut even the last remaining fruit trees to repair their homes.

- Remoteness

The remoteness of the villages impedes the improvement of economic and social situation. Population density is too high for self-sustaining agriculture and the exchange of goods is difficult due to the inferior infrastructure. Many villages are not even accessible by motorbike but only on foot.

- Social change

The houses in North Nias are too large for today's family sizes. With rare exceptions they are no longer built, and proper maintenance is a real problem for the owners.

In South Nias the houses are smaller and better adapted to the present situation. These houses are still built, but they are very expensive compared to modern ones.

The process of building is ritualised and strongly connected to the traditional law, the Adat. The socially enforced rituals make building even more expensive, so that many people prefer to renounce the ritual and social positioning in favour of improving their economic situation. The integration of traditional architecture into a society which is rapidly changing is dangerous for the survival of these traditions.

The former king's houses have already almost completely disappeared. The main reason for that is the loss of the political function of the king, which was connected with the means to afford such a costly symbol of power.

- Knowledge

A traditional house is built by a local master of building, the tuka. Knowledge about building is still available in Central and South Nias, and tukas still have apprentices.

But knowledge about earthquake resistant design using other building materials, such as concrete and brick, is not available or not applied. Moreover, the material is often of insufficient quality.

Governmental control is difficult in most parts of the island and so is the implementation of new building guidelines.

The perspective for further development has to take into account the social changes and the economic situation. The peculiarity of Nias' culture is one of the foundations on which tourism could prosper as an opportunity to earn money and to save the remaining heritage. Without help from outside the traditional settlements will soon disappear. But the present environmental situation differs from the one in which these building typologies developed, so the qualities that can be found in traditional architecture should be reinterpreted with modern means.

5.1.9 Conclusion

Traditional knowledge and qualities should not be lost but investigated and documented in order to find a new and modern interpretation because:
- Adaptation of traditional architecture to local environmental conditions is usually better than that of modern houses.
- At the moment the environmental conditions stored in traditional architecture may be unknown.
- Passive material- and energy efficient means of adaptation are used.

Documentation is essential as a source of information for the understanding of the many facets of the traditional architecture of Nias. The remaining traditional houses should be evaluated for preservation of a representative selection.

To counter the effect of the frequent earthquakes in Nias, the following future measures are proposed:
- Microzonation maps of Nias with public access are urgently needed
- Heightening the awareness of preservation measures by the owners and government bodies, and implementation of a funding system so that owners of traditional houses are able to preserve them
- Research into and provision of new roof materials

Adaptation will necessarily be carried out, preferably through the following two methods of transformation:
- Adaptation of the traditional buildings to suit modern conditions (security, infrastructure, comfort, material)
- Adaptation of a global building style to local techniques, implementing the qualities of the traditional houses and settlements.

Both methods will lead to a "new tradition", which combines old and new qualities.

Fig.403 High-tech project with a triangular structure, Catrin Huber.

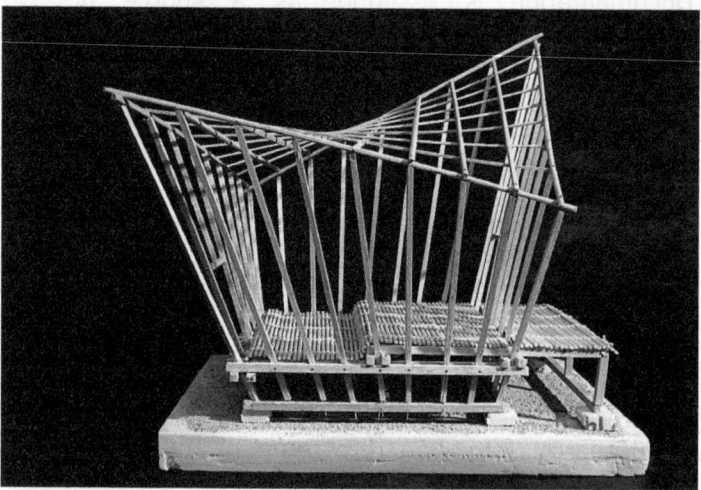

Fig.404 Low-tech project with wood and bamboo, Vera Kumer.

Fig.405 Low-tech project with bamboo as primary structure and a metal roof, Volker Leinich.

5.1.10 Application: design program Surf Resort Nias

The design program "Surf Resort Nias" at the Department for Building Construction was carried out after the field trip in the winter term 2005/06. The program suggested a small hotel development located on Sorake Beach, where a declining economy, the tsunami and the earthquake have devastated a community of small-scale tourism. The preliminary findings were introduced into the students' projects, which are represented here by three successful examples. Triangular bracing, stilted construction, large roof volume, natural ventilation, differentiation in height, conscious selection of materials, careful consideration of building process - these principles used in the traditional architecture were interpreted in the design proposals, which suggest a better adapted, high quality architecture.

5.1.11 Future prospects

The interdisciplinary approach provides information about the development and integration of architectural typologies. The strategy of investigating the inherent qualities of architecture traditions will be continued and used as an information source for new designs. The abundance of matter and energy which we are now accustomed to use is sure to run out soon. Sooner or later, we will be glad to be able to access information on the architectures that we have invented step by step, transformed into artefacts, and lost again, due to complex changes of our environment.

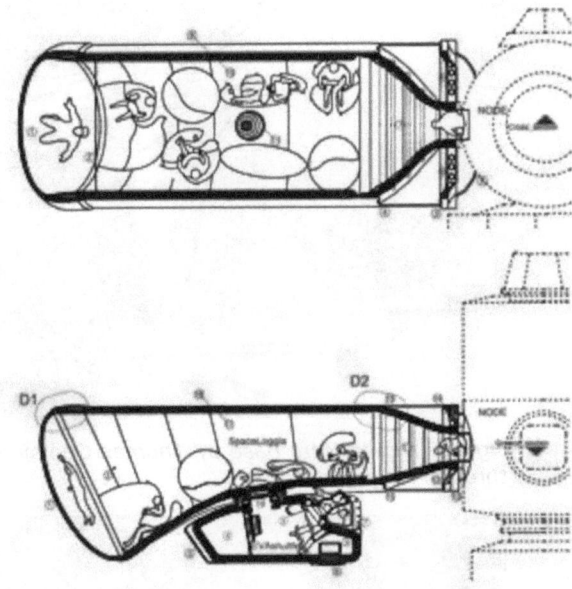

Fig.406 Space Loggia by Stefano Caneppele, docked to ISS, floor plan and section.

5.2 TRANSFORMATION ARCHITECTURE

The architectural design project "Transformation Structure/Space", was carried out at the Department of Building Construction in the summer term of 2004 by space architect Barbara Imhof and the author. It was announced as a set of experimental concepts to investigate the overlapping areas of the two architectural fields of biomimetics and space system design. External experts working in scientific and engineering institutions and organizations supported the project.[444]

Transformation is used for the process of changing ideas and the contexts of abstract ideas in order to apply the findings to design tasks. Transformation can occur as:
- Transformation of concepts from nature, science, art, etc., into architecture
- Transformation of methodologies from one professional field into another
- Transformation of an idea into a built space

Themes of literature research included basic information on space system design, such as living and building in zero-gravity (ISS/MIR), or on Mars/Moon, spin-in and spin-off between technologies used on Earth and technologies used in space and the so-called augmented reality. Parallel to the knowledge about a new environment, promising natural phenomena were investigated: folding techniques and inflatable habitats for space, materials from nature for extreme conditions,

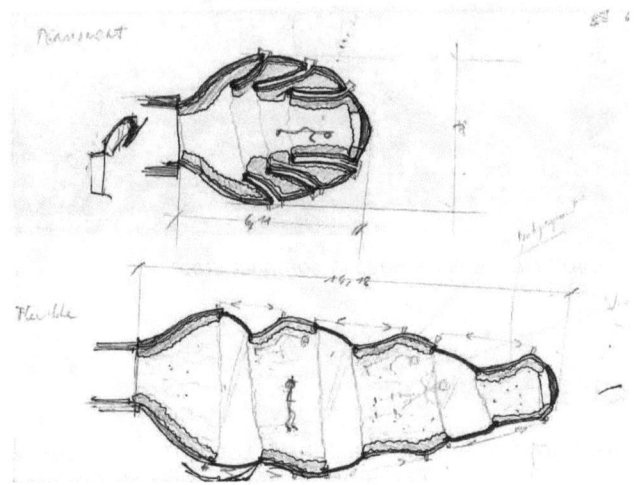

Fig.407 First sketches of "Space Loggia", deployable space.

locomotion in nature, propulsion methods in nature, robotics - functional morphology and sensors, the principle of self-organization and possible technical applications.

The findings of the research were projected onto current topics in space architecture:

inflatable habitats, human factors for long-duration missions and incorporation into the design of a space habitat, human-machine interfaces and spin in/off (technology transfer from space industry to products for earth and vice versa).

The equivalent scenarios were also developed. Four projects were selected as exemplary biomimetic designs.

---

444 Gruber, P.; Imhof, B.: Transformation: Structure/space studies in bionics and space design in Acta Astronautica Volume 60, 2006, p.563

Case studies | Transformation Architecture

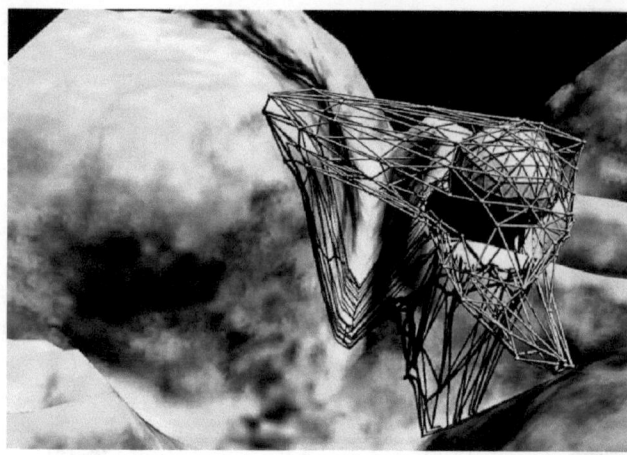

Fig.408 Mobile Mars Base by Thomas Gamsjäger moving through rough terrain.

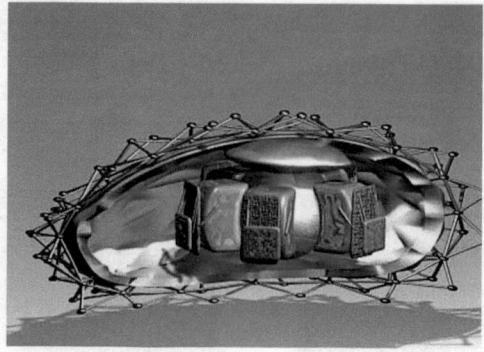

Fig.410 Flexible skin allows deformation.

Fig.409 Mobile Mars Base in compact state.

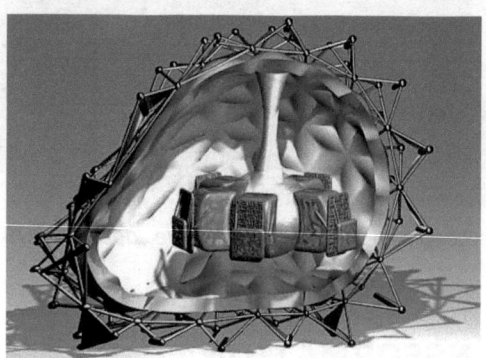

Fig.411 Internal equipment is flexibly attached.

Space Loggia

Stefano Caneppele designed the project "Space Loggia" as an astronaut's lounge where the feeling of being in space is stronger than in an ordinary module of the International Space Station ISS. The starting point for design was the fact that astronauts spend the better part of their free time "earthwatching". The Space Loggia is designed to enhance this activity, and provide space for relaxing.

Biological role models were folding structures and pinecones, which open and close according to humidity. The transparent skin for the window is was designed to be a multilayered system with an integrated liquid or gelatinous material. The challenge is to create a skin that is partly able to withstand the impact of micrometeorites. The specifications of imaginary window materials were included in the design. The scenario assumes that in future such materials will be developed, due to the fact that glass, plexiglass, aerogels and their composites are already undergoing major advances.

The window is interpreted as a structural component, which is life supporting, comparative to the life support system for the astronauts. The initial idea of unfolding is kept for the connection part to the ISS. The system is no longer used to unfold, but to change the direction and position of the large window at the end of the conically shaped space.

Mobile Mars Base

"Mobile Mars base" is a flexible triangular structure designed by Thomas Gamsjäger. Locomotion and structural deformation of cells is transformed into a habitable Mars rover concept. The structural parts evidence analogies with animal skeletons and joints, but as far as construction is concerned the project remains remarkably mechanical. Change of length and flexible nodes provide freedom of movement. The internal space was designed to be flexible enough to allow attachment to the moving structural skeleton, but we still lack materials that could perform such tasks. Adaptation of form and overall space is thought to be interesting for exploratory missions.

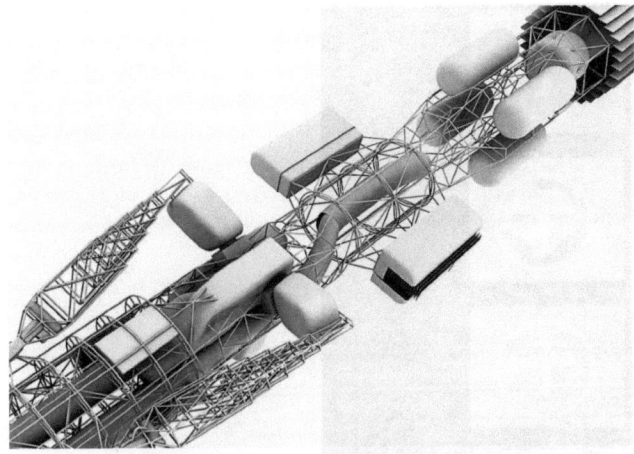

Fig.412 "Marsspaceship+", configuration of modules.

Fig.415 Internal environment of the ship.

Fig.413 Visualisation of Spaceship close to Mars.

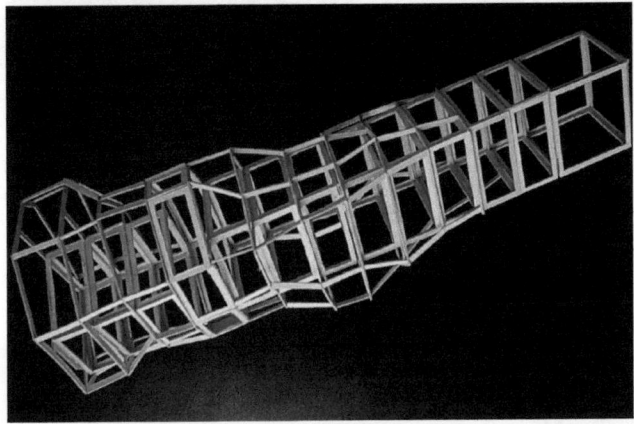

Fig.414 Structural model of the ship.

Marsspaceship+

Melanie Klähn and Thomas Frings designed a spaceship that unites machine and the necessities of human life into a whole organism for the journey to Mars that will last several years. It is a "home on the way", a life-sustaining machine offering safety against the perilous environment, but also serving as a substitute for our natural environment without copying it.

The interior of the ship conforms to the special conditions of zero-gravity, aiming at the architectural utopia where there is no above and below. Moving and orientation is defined anew, stereotypes are dissolved, walls, floors and ceilings do not exist anymore. The biorhythm of the crew plays an important role, being translated into the "behaviour" of the spaceship. This also implies that the skin of the transporters has to be adaptive and reactive. The absence of external influences, such as weather or the change of time, of day and night, is re-interpreted and incorporated into the technology of the spaceship. The layout of the entire spacecraft is subjected to an organism-like interpretation, influencing space, arrangement of functional organs, distribution and circulation systems. The project is an attempt to design a holistic artificial environment similar to a life-supporting organism.

Fig.416 Conceptual visualisation of the Lunarsinglebase by Alessandro Perinelli.

Lunarsinglebase

Alessandro Perinelli's "Lunarsinglebase" project of a lunar habitat for a single person's settlement on the moon's surface also augments reality. Avoiding isolation and depression is very important for future space ventures. The concept for a mobile lunar base combines design research for: habitat architecture, mobility systems, habitability, radiation protection, human factors, living and working environments, communication concepts, autonomous and teleoperated machinery. The most extreme assumption underlying this concept is that the single person has to survive alone on the moon in full autonomy, for 6 months. Augmented reality adds graphics, sounds, haptics and smell to the artificial world. This enhancement can already be useful in teleoperation of modules on lunar ground, superimposing the normal layer of human perception with a layer of computer data and useful information.

A flexible and interactive skin envelope creates a reacting space that behaves according to the use of the unit and in direct answer to the user's activity, providing a better connection to Earth.

The "Transformation Structure / Space" project led to unforeseen and unconventional concepts, although the method of proceeding at first appeared to be purposeless and undirected. The designs developed at the interface of scientific research, experimental architectural approach and the design program.

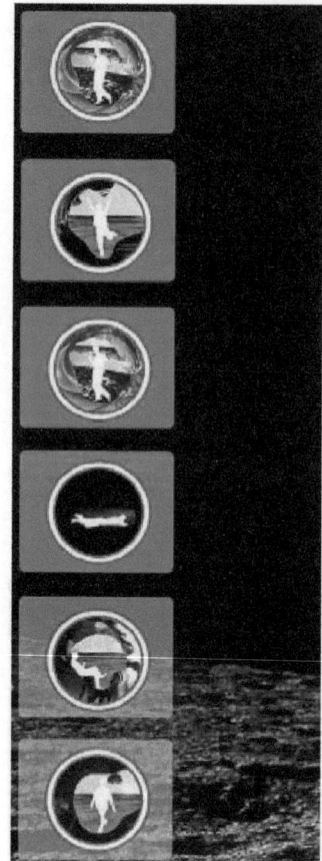

Fig.417 Activities.

Fig.418 Lunex - Mission Scenario Step 1 – First Outpost using Ladybird II.

5.3 LUNAR EXPLORATION ARCHITECTURE

The study "Lunar Exploration Architecture - Deployable Structures for a Lunar Base" was performed within the Alcatel Alenia Space "Lunar Exploration Architecture" study for the European Space Agency in spring 2006 by an interdisciplinary team at the Vienna University of Technology.[445]

Growth of space translated by the deployment of enclosure is especially important in space architecture, as transportation costs are key parameters for space missions. Biomimetic approaches are most interesting for innovative fields, so the European Space Agency financed studies in biomimetics[446] and in 2005 launched a so-called Academia Study to generate novel concepts, mainly in the field of engineering. Lunar Exploration Architecture study was carried out within this framework.

"The purpose of the study was to investigate bionic concepts applicable to deployable structures and to interpret the findings for possible implementation of the concepts.

The study aimed at finding innovative solutions for possible deployments. Translating folding/unfolding principles from nature, candidate geometries were developed and researched using models, drawings and visualisations. The use of materials, joints between structural elements and construction details were investigated for these conceptual approaches.

Reference scenarios were used to identify the technical and environmental conditions, which then served as design drivers. Mechanical issues and the investigation of deployment processes narrowed the selection down to six chosen concepts. Their applicability was evaluated at a conceptual stage in relation to the timescale of the mission."[447]

5.3.1 Methodology

Out of a wide range of role models from nature and a vague mission scenario from ESA a set of differing deployable structures, each with an application scenario, was designed. In contrast to the classic engineering problem, the task for this project was not defined. So the approach to a concept was taken from both sides - role model and application - top-down and bottom-up at the same time.

Literature and web research was the beginning, then the options were gradually narrowed down in collaborative sessions and in a design studio approach. Analogue working models and virtual presentations were important for design.

Steps taken included:
- Identification of relevant role models from nature
- Identification of space application for lunar infrastructure or habitat
- Identification of relevant aspects of evaluation
- Gradual evaluation and selection of role models
- Development of candidate geometries
- Evaluation and selection of the candidate geometries, considering technical and engineering aspects, necessary technology and geometry
- Development of architectural working models
- Mechanical issues and constructive concepts
- Design proposals and development of scenario

---

445 Gruber, P. et al.: Lunar Exploration Architecture, Deployable Structures for a Lunar Base, Study for Alcatel Alenia Spazio, 2006 The study was carried out in collaboration with The Centre for Biomimetics in Reading, George Jeronimidis and Richard Bonser. The team in Vienna consisted of Petra Gruber, space architect Barbara Imhof, Sandra Häuplik, Kürsad Özdemir and René Waclavicek. Maria Antonietta Perino from Alenia Spazio provided support. Structural and mechanical engineers support was done by members of the department. Collaborating students: Pavol Mikolajcak, Phillip Pongratz, Matthias Schwarzgruber, Verena Topaz, Olivia Wimmer, Bernhard Wisser.

446 ESA biomimetics homepage, http://www.esa.int/gsp/ACT/bio/index.htm [06/2007]

447 Gruber, P.et al.; Deployable structures for a human lunar base, in Acta Astronautica Volume 61, 2007, p.484

Fig.419 Tubular Ladybird II structure, folded state.

Fig.420 Tubular Ladybird II structure, unfolding.

Fig.421 Tubular Ladybird II structure, unfolded state.

5.3.2 Selection of role models

A wide range of role models from nature was investigated, and the main criteria for the selection identified:
- Speed of the deployment process
- Reversibility as a possible feature
- Actuation - growth factor, ratio of deployment due to growth, fluid pressure, muscle, Δ-t, Δ-water (osmosis) or stored energy
- Structural performance (mechanical and chemical stabilisation)
- Protection (scale, complexity etc.)
- Material properties for technical applications (non-hybrid design, functionally graded materials, water content)
- Process properties for technical applications (chemical, drying)
- Complexity
- Scalability
- Sensing

The role models were divided into two major categories: role models being close to a structure forming a volume (beech leaf, palm leaf, cactus, earwig, insect wing, morning glory, spine system, giant water lily) and role models that can produce additional features concerning material or actuation etc. (bat, stick insect, earthworm, feather, flea, insect proboscis, lobster, locust, muscle, ovary explosion, scorpion, seedpods, snail shell, snail, snake, spider legs)

As the more interesting one, the first category was further classified into the following groups:
- Fold/deploy: plant leaves (beech leaf, palm leaf, victoria regia), insect wings, flower petals (morning glory), cactus
- Bellows and folded boxes
- Rolled up structures: proboscis, fern
- Spines/backbones: tensegrity systems

Because of the short, three months time frame, many interesting role models that would have needed further basic research had to be dismissed. Several proposals for the application of folding structures which had already been developed in previous studies were looked at.[448] The most promising ladybird wing-folding mechanism is shown as a working model in Figure 422. It was then further developed into a tubular structure. A single translational movement in the structure initiates a complex unfolding process.

448 Kobayashi, H.et al; The geometry of unfolding tree leaves in Proceedings of the Royal Society, 1998 and Haas, F.: Geometry and mechanics of hind-wing folding in Dermaptera and Coleoptera, M.phil.thesis, 1994

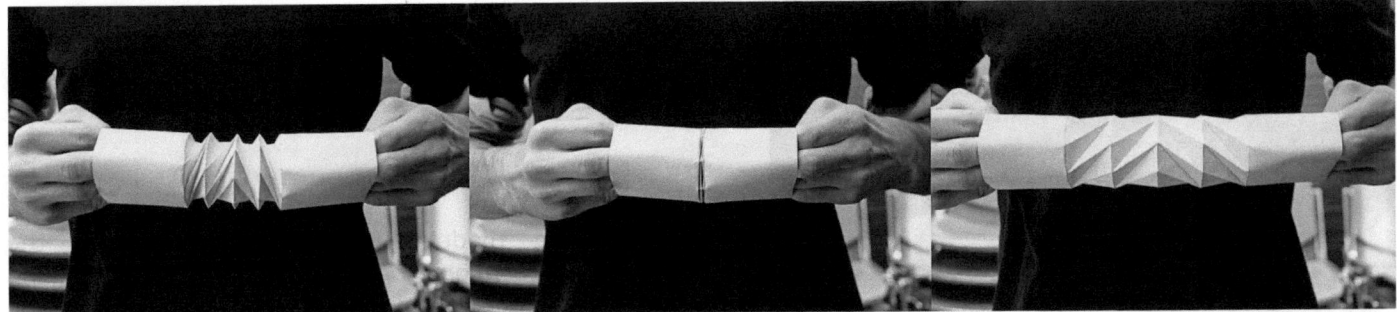

Fig.422 Tubular Ladybird III structure, rotational movement of rings is initiated by linear movement.

The architectural working model "Ladybird II" could be used to construct habitable space providing advanced shielding and as an additional shielding for existing habitats (e.g. radiation shielding, micrometeorite shielding, etc.).

The folded tube delivers different possibilities for applications: vertically positioned and with regolith filling, it provides radiation shielding; horizontally positioned it could be used as additional workspace for short-term use.

In the study a possible evolution of a lunar long-duration mission establishing human presence on the Moon was described in terms of habitation, as can be seen in Mission Scenario Step 1 – First outpost using Ladybird II application to outer shell.

5.3.3 Definition of design task

The possible applications in space were identified by means of a mission scenario from ESA, and were then set into relation with these requirement criteria:
- Test deployable for human and robotic missions
- Test bed for Mars
- Deployable assembly space for construction or repair
- Additional space for crew rotation period
- Habitat modules, e.g.: pre-fabricated membrane structure, unfolded and rigidified, conventional cylinders with additional inflatable modules, inflatable pressurized shell with regolith shielding, long-duration habitat modules, inflatable airlocks, docking ports
- Interiors, e.g.: flexible stowage systems, additional space for crew members, windows
- Logistic modules, e.g.: sunshield, deployable workshop, deployable airlock, deployable solar panels/arrays, tools, antennas, solar reflectors and deflectors, radar

The different uses require different structural parameters and characteristics of strength, stiffness, statics and especially dynamics (e.g. vibration response), mass, resilience, resistance to corrosion and other environmental factors, fatigue, thermal properties, reliability, radiation degradation, manufacturability, availability and cost.

5.3.4 Structural categories according to speed and reversibility

The information about possible tasks was combined with the aspects of possible structures that were already identified. A classification followed, according to issues of speed and reversibility.
- Non-reversible deployable structures
- Habitat and/or associated facilities are packed to the smallest possible volume for transport. The building purpose includes assembly in habitat and deployment outside. Once landed the structure can be reused at a later stage for additional shelter
- Slow reversible deployable structures (process of deployment takes several hours or days)

Possible applications include temporary structures, shelters, roofs, and extendable pieces. The landed structure or parts of it can be reused and transported to another location or used for expansion of the Lunar Base.
- Fast reversible deployable structures (Process of deployment takes several seconds to hours)

The habitat and/or associated facilities can be transported to a different location on the lunar surface. The Lunar Base can be expanded for storage of used structures, etc. Openings, connecting interfaces and moveable parts can be made of fast reversible deployables.[449]

---

449 Gruber, P. et al.; Deployable structures for a human lunar base in Acta Astronautica Volume 61, 2007, pp.488

Fig.423 Ladybird I model in folded state.

Fig.424 Ladybird I model, unfolded.

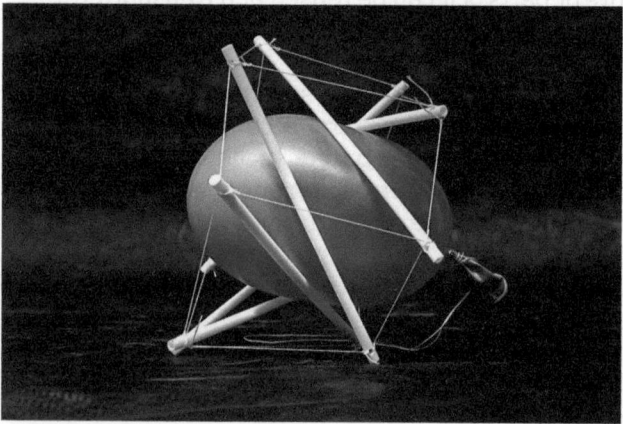

Fig.425 Tensegrity structure built up by pressurising internal volume.

5.3.5 Candidate geometries

Working models were used to develop the candidate geometries. The Ladybird wing folding and the Cactus delivered promising models; some exemplary geometries are shown here.[450]

- Ladybird

The basic structure consists of four infolded surfaces. In flat condition the four surfaces describe one flat angular structure. By pulling apart the end parts of the angular structure along one axis the folded "hinge" deploys along an axis perpendicular to it.
Multiplication and repetition of the basic folding principle leads to tubular structures.

- Cactus

An even number of planar surfaces is multiplied along their longitudinal edges to form a compact starting package. The deployment finally generates tubular volume. The hinges between two panels must allow an opening angle of 180°, and control of the movement.

The other candidate geometries - like mussel shell and tensegrity - were discarded in the later stages. Interesting aspects were displayed by tensegrity and rolled-up structures, but further research would be necessary to make use of these models.

The trade-off matrix is found in the following; the candidate geometries were subjected to evaluation for deployment, connections, suitability for a pressurised volume, structural efficiency and transportation issues. Ladybird III and Cactus III show the most promising characteristics.[451]

Issues of material and construction were considered: diagrammatic solutions were worked out for materials and connections, based on the multilayered system of Transhab and another investigation on carbon-carbon panels. Hinges, sealing and connections were proposed in a schematic way.

---

450 The models were developed together with the students Vera Kumer, Bernhard Wisser, Pavel Mikolaicak, Philipp Pongratz, Matthias Schwarzgruber, Verena Topaz and Olivia Wimmer

451 Gruber, P. et al.; Deployable structures for a human lunar base in Acta Astronautica Volume 61, 2007, p.493

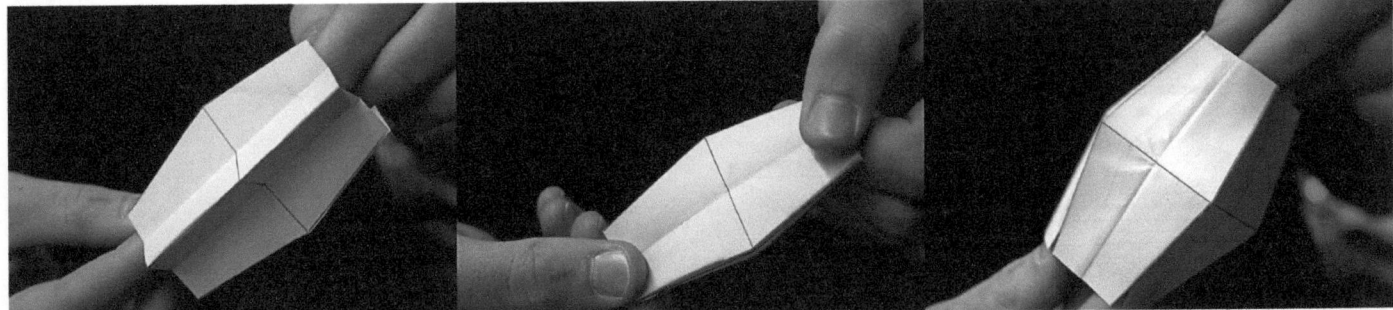

Fig.426 Cactus model, generating tubular space.

Fig.427 Musselshell model, also generating tubular space when unfolding.

	Deployment	Connections hinges/sealing	Suitability for pressurised volume	Structural efficiency	Transportation in conventional rocket
Ladybird I	++	-	-	+	0
Ladybird II	++	0	0	+	++
Ladybird III	++	++	++	++	++
Cactus I	++	0	0	+	+
Cactus II	++	+	+	+	+
Cactus III	++	++	++	++	++
Mussel Shell	+	-	--	0	--
Pineapple folding	+	-	--	-	0
Folded boxes	+	++	-	0	-
Tensegrity structures	-	0	Not applicable	++	+
Rolled-up structures	+	+	--	-	+

Legend: ++ very good, + good, 0 neutral, - poor, -- very poor

Fig.428 Trade-off matrix for evaluation of candidate geometries, best proposals marked in light grey.

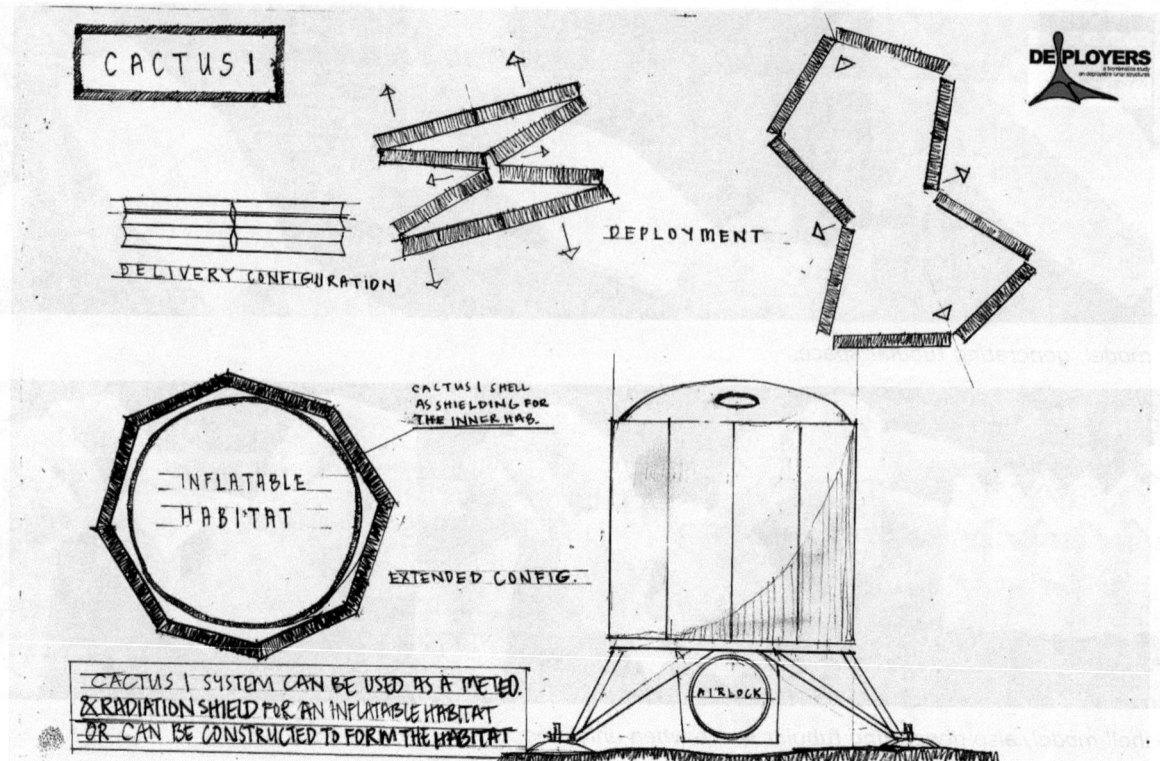

Fig.429 Cactus I used for additional habitat shielding.

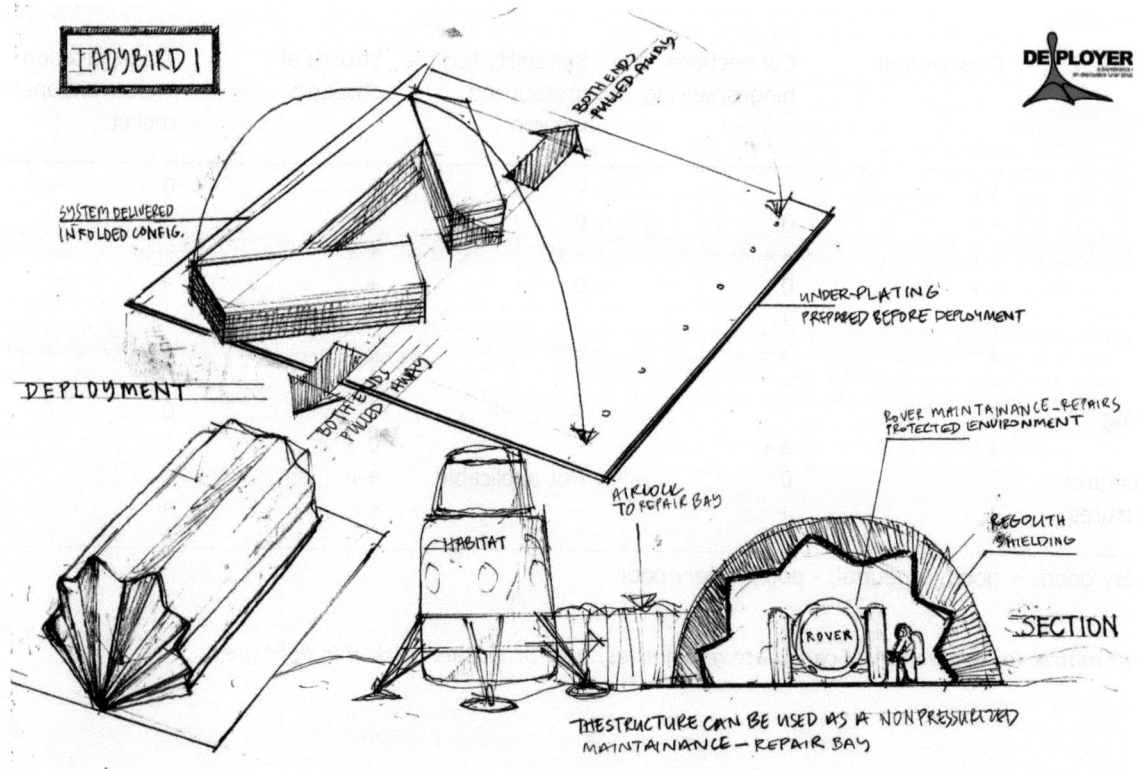

Fig.430 Ladybird I used for additional shielding of maintenance bay.

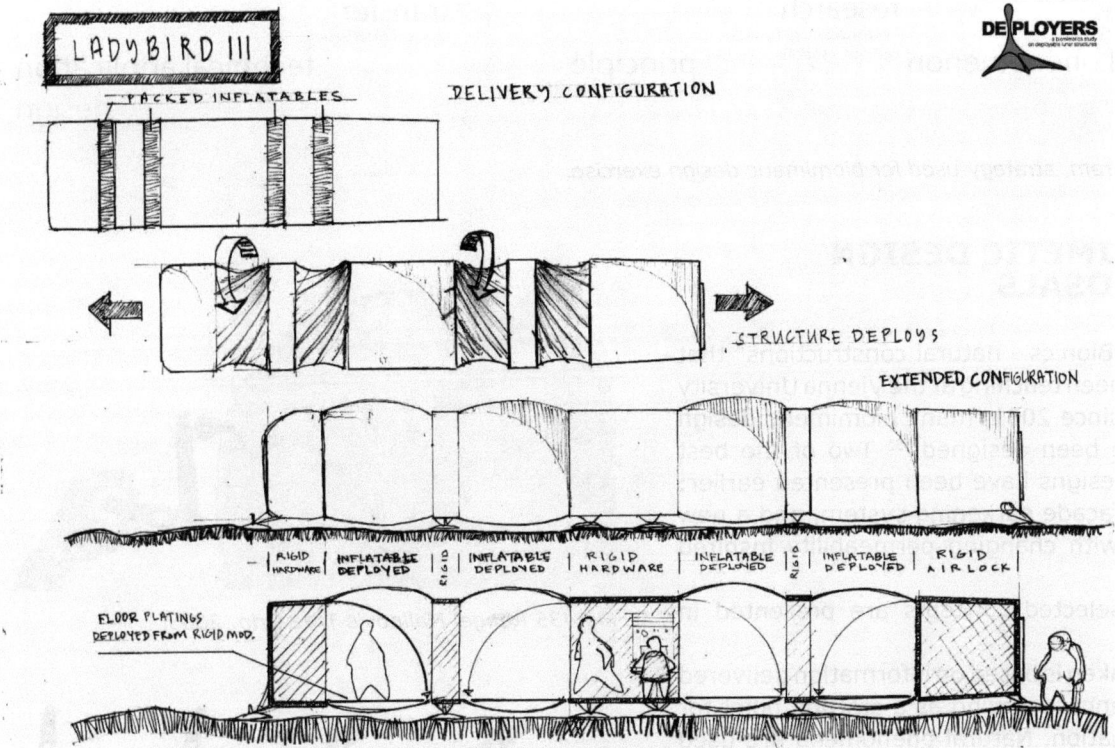

Fig.431 Ladybird III, the most promising proposal for a habitable space.

5.3.6 Scenarios

Ladybird I, Ladybird III and Cactus I were used and set in context with ESA's reference missions for Lunar exploration.
Ladybird I could be implemented as a non-pressurised maintenance - repair bay.
Ladybird III could be used to construct habitable space providing advanced shielding and as an additional shielding for existing habitats.
Cactus I can be used as an additional shielding against micrometeorite impacts and radiation protection for an inflatable habitat. Cactus I can be used to construct a habitat using regolith, a material covering much of the moon's surface.
The structural technologies used for deploying the Lunar base and adjacent infrastructure are derived from an ESA study about a possible evolution of a lunar long-duration mission establishing human presence on the Moon.

5.3.7 Conclusion

The integration of biomimetics into the design process delivers innovative results. The process of gradual advancement by narrowing down options has worked out well.
The detailed search for information on particular role models of the Life Sciences was difficult and time-consuming. Some characteristics although common knowledge are not yet being scientifically investigated. The implementation of mission oriented basic research is recommended.

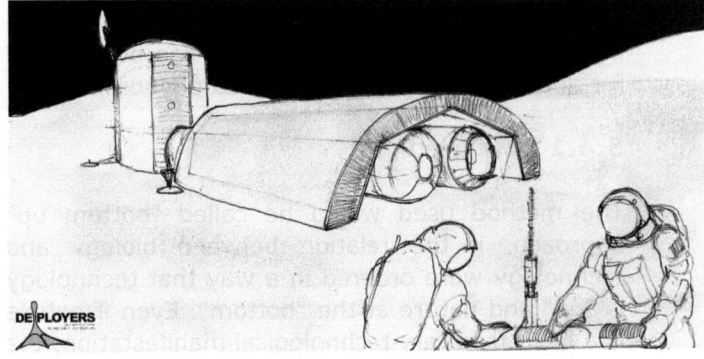

Fig.432 Ladybird I used as a shelter.

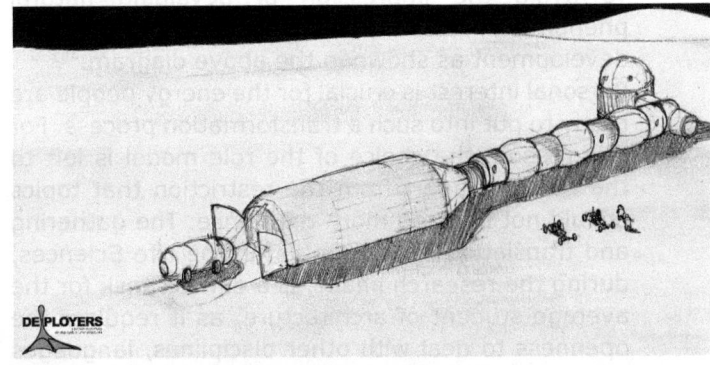

Fig.433 Combination of different designs.

Case studies | Lunar Exploration Architecture

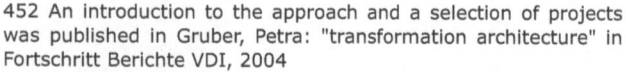

Fig.434 Translation diagram, strategy used for biomimetic design exercise.

5.4 BIOMIMETIC DESIGN PROPOSALS

In the course "Bionics - natural constructions" that the author has been teaching at the Vienna University of Technology since 2001, many biomimetic design proposals have been designed.[452] Two of the best experimental designs have been presented earlier: Aero Dimm, a façade darkening system, and a new kind of fabric with changing permeability inspired by stoma.

Twelve other selected concepts are presented in this chapter.

The approach taken is based on information delivered by the Life Sciences, serving as a starting point for further investigation. Natural phenomena are used as role models for technical, at best architectural application. Phenomena chosen are models of inanimate nature, the plant and animal kingdoms and principles of animate nature. A transfer to some kind of application in architecture has been found for about one third of the selected phenomena.

5.4.1 Method

The method used would be called "bottom up" approach, if the relation between biology and technology were ordered in a way that technology is "up" and nature at the "bottom". Even if nature is a basis for every technological manifestation, the vertical order suggests rating.

To avoid the impression of devaluing natural phenomena, the method is described as a development as shown in the above diagram.[453]

Personal interest is crucial for the energy people are ready to put into such a transformation process. For this reason, the choice of the role model is left to the student, apart from the restriction that topics should not be used more than once. The gathering and translation of information in the Life Sciences, during the research phase, is a difficult task for the average student of architecture, as it requires the openness to deal with other disciplines, languages and methods. The decisive phase is the abstraction of the principle to be applied. If the abstraction is successful, in most cases so is the application.

Fig.435 Rangel Malinov's Tree Grip, 2005.

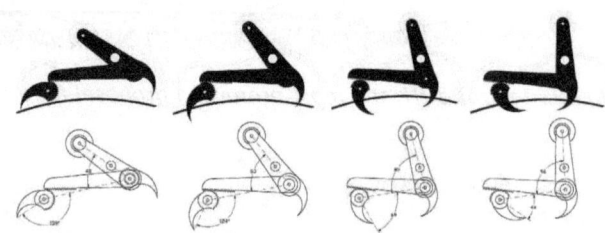

Fig.436 Schematic drawings of the mechanism.

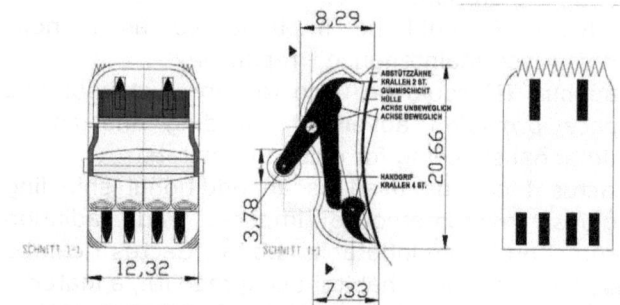

Fig.437 Drawings of the tree grip.

---

452 An introduction to the approach and a selection of projects was published in Gruber, Petra: "transformation architecture" in Fortschritt Berichte VDI, 2004

453 Gruber, Petra: "transformation architecture" in Fortschritt Berichte VDI, 2004, p.16

Fig.438 Caddisfly larvae capture net - Greenhouse construction, Inger Totland, 2005.

5.4.2 Projects

The selection from a wide range of projects reveals the diversity of the biomimetic approach visible, and at the same time represents different aspects, which are issue of investigation and transfer into technological environment.

- Form and function

The Caddisfly Larvae Capture Net was designed by Inger Totland in 2005 based on the capture net of the caddisfly.

"The special feature of the capture net of the Polycentropid larvae is the trumpet shape, which is used to collect as much water as possible, and the ability to filter different materials from the water. The project uses these features in a different setting and for a different purpose; making a collector of rainwater. The water will pass through different filters on the way down the tube. This concept could be used in greenhouses. The normal roof-construction is replaced by several trumpetshapes. The trumpets collect rainwater, and filters inside separate leaves and dirt from the water, add minerals or give the water the right pH-value. Inside the lower end of the trumpet would be a pipe that could reach the various plants that needed watering."[454]

- Function / construction

Rangel Malinov designed The Tree Grip in 2005, a mechanical device the size of a hand with six metal claws following the role model of cat claws: the claws are extended by a change of the grip angle. The more force is used, the better the grip. It could be used to carry wood or to climb. Its passive mechanism is especially interesting.

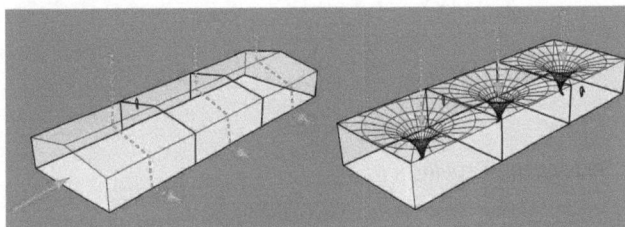

Fig.439 Isometric views of the "net".

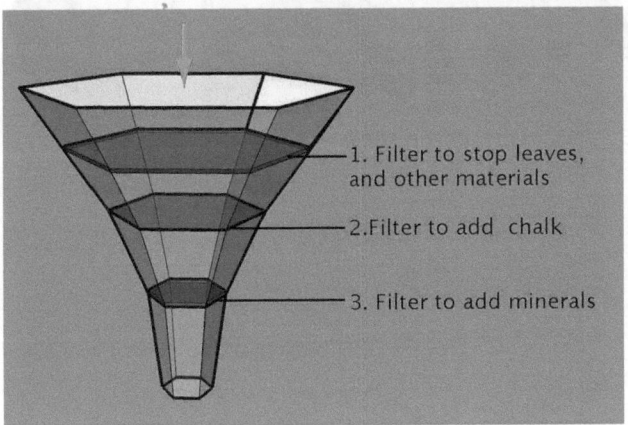

Fig.440 Functional and geometric scheme of the "net".

---

454 Totland, I.: Bionik - natürliche Konstruktionen, 2005

Case studies | Biomimetic Design Proposals

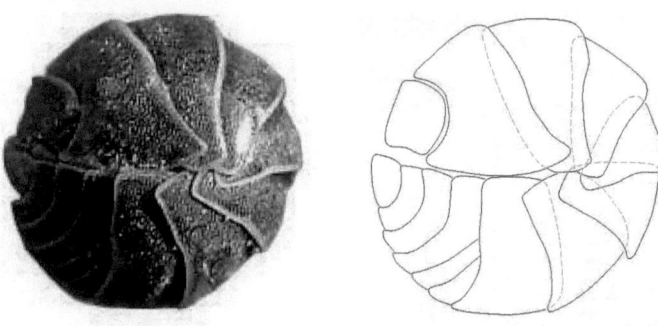

Fig.441 Pillbug shell, Katharina Fuchs, 2005, photograph and drawing.

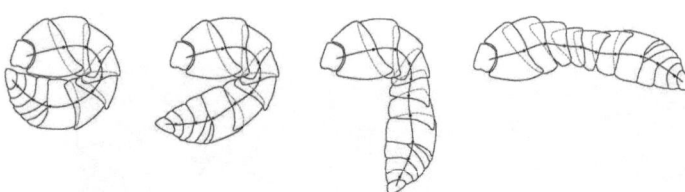

Fig.442 Sequence of rolling up.

Fig.443 Paper model of the shell.

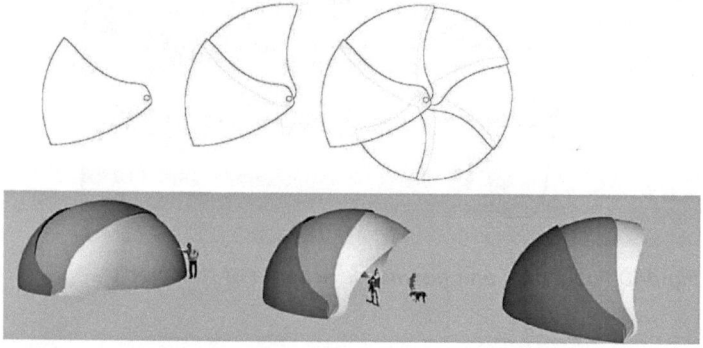

Fig.444 Translation into a rotating roof.

Fig.445 Armadillo flexible roof by Peter Kumhera, 2004, perspective view.

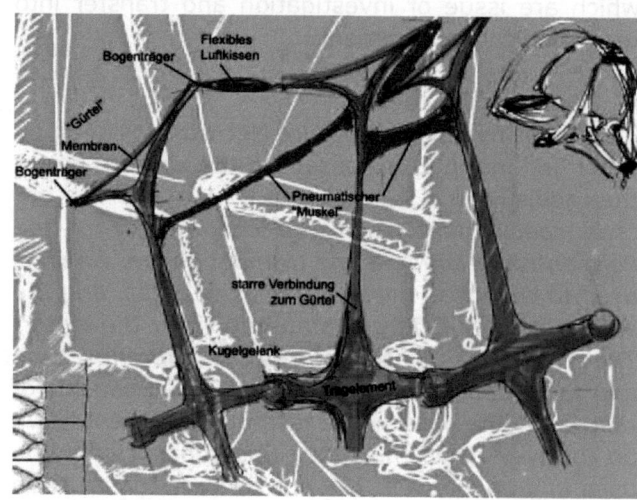

Fig.446 Vertebrae.

The pillbug shell was investigated by Katharina Fuchs in 2005. The pillbug's shell unfolds along one axis. Two adjacent segments twist together and the smaller segment slides under the larger one, ensured by the particular geometry of the tips. A simplified geometry can be used for roof shells or furniture, where movement and extension of space are important qualities. The local geometry of the tips enables the particular movement.

- Construction (Body part)

Constructions modelled on body parts lend themselves as objects of architectural investigation. Many designs were carried out for this kind of role model.

Peter Kumhera investigated the anatomy of the armadillo and designed a flexible roof in 2004. Construction elements inspired by vertebrae are combined flexibly to carry frames covered with membrane. The flexibility between the parts of the shell is provided by pneumatic elements, and the movement and slight change of the internal volume are translated.

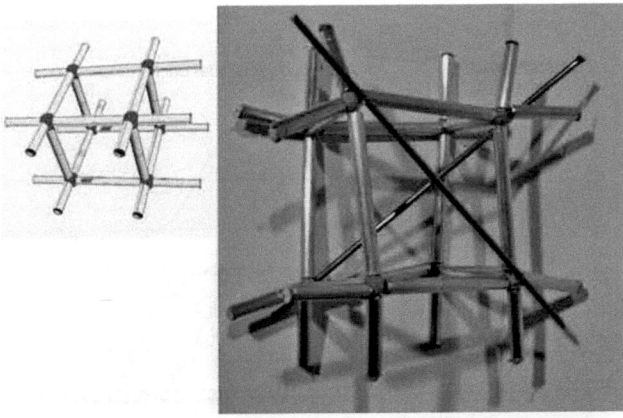

Fig.447 Technical structure similar to the glass sponge, drawing and model, Julia Forster, 2006.

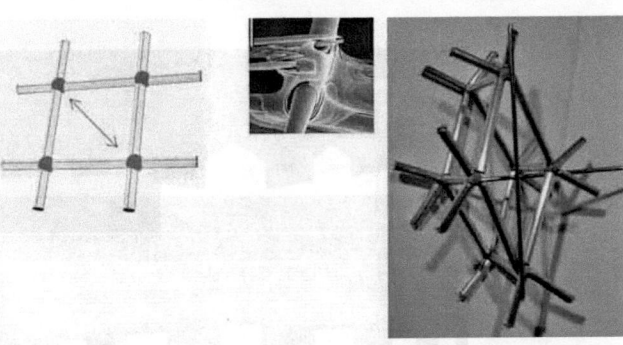

Fig.448 Study of deformation.

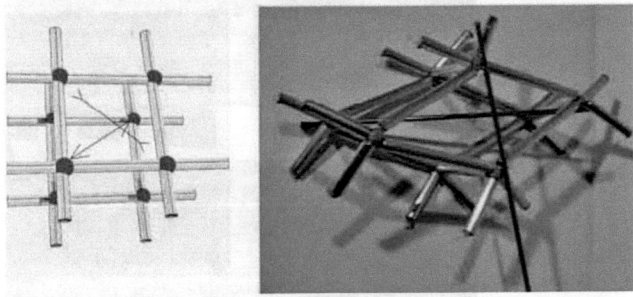

Fig.449 Study of deformation and flexibility of the joints and / or the diagonals.

Julia Forster investigated the glass sponge in 2006. She made a model of the structure and experimented with flexible joints that allow the deformation of the structure. Experiments show the characteristics of movement. The construction could be used for adaptable architecture. The structure was imitated and changed for further development.

- Mechanism / Material

An active transportation system, inspired by the oesophagus, was designed by Christoph Österreicher in 2007. The movement pattern of the so-called P-tube follows the natural role model of the oesophagus. The function of ring muscles and length muscles is translated by electroactive polymers, squeezing pneumatic tubes made of silicon. In contrast to conventional pumps, it can transport both hard pieces and viscous liquids and makes no noise.

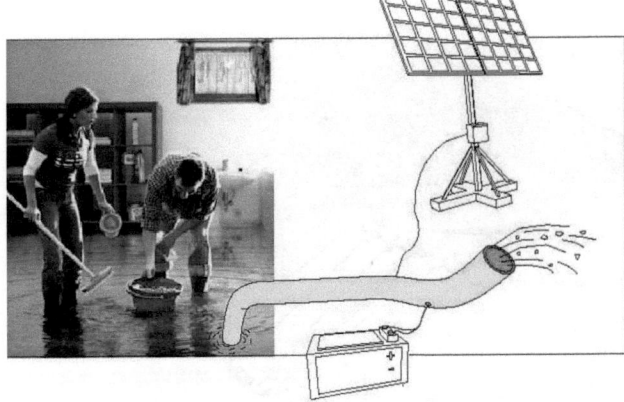

Fig.450 Application scenario of the P-tube.

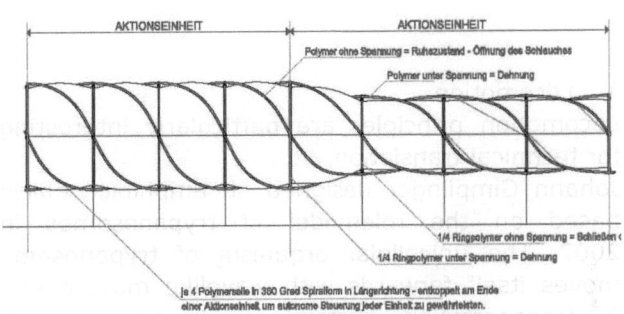

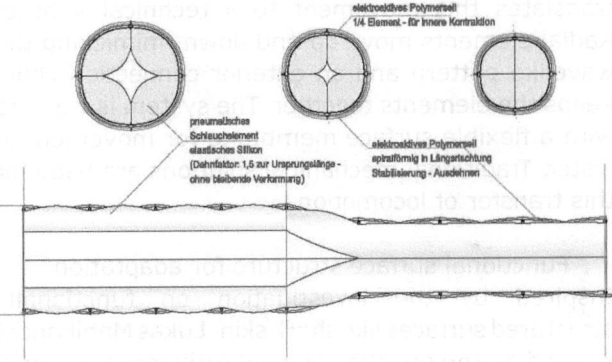

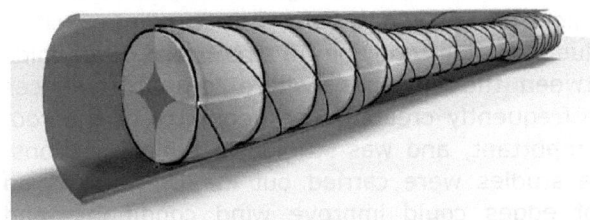

Fig.451 P-tube by Christoph Österreicher, 2007, perspective view and sections.

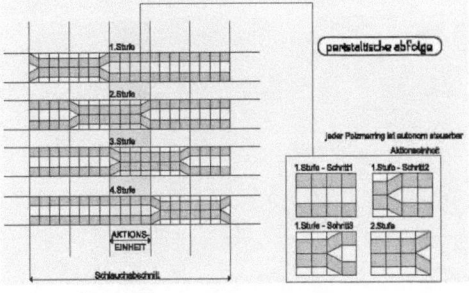

Fig.452 Rhythm of contraction.

Case studies | Biomimetic Design Proposals

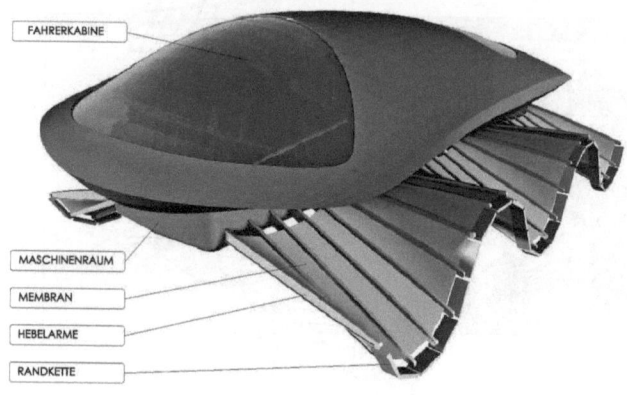

Fig.453 Amphibic vehicle by Johann Gimplinger, 2007.

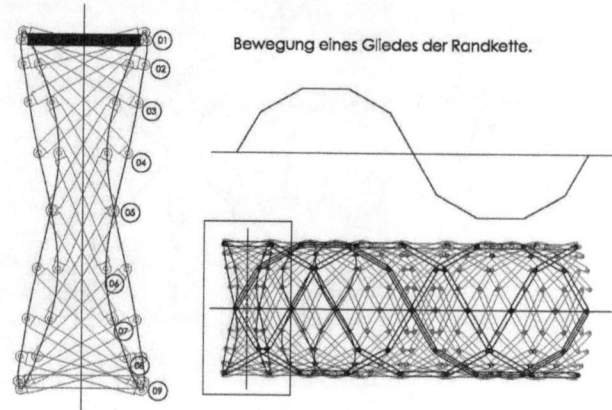

Fig.454 Movement of the exterior chain.

- Locomotion

Locomotion principles are particularly interesting for technical translation.

Johann Gimplinger designed an amphibic vehicle based on the rolemodel of trypanosomes in 2007. The unicellular organism of trypanosome moves itself forwards with wavelike movements. A transportation system is designed, which translates this movement to a technical vehicle. Radial elements move up and down, mimicking the wavelike pattern and an exterior connection chain keeps the elements together. The system is covered with a flexible surface membrane for movement in water. Traditional mechanical solutions are used for this transfer of locomotion.

- Functional surface structure for adaptation

Inspired by the investigation on functionally structured surfaces like shark skin, Lukas Mahlknecht worked on the question of wind turbulence in urban spaces in 2005. He investigated the action of wind spoilers on building roofs. The profile of cities influences wind conditions in streets and other voids between the buildings, where large turbulences are frequently created. The geometry of the roof is important, and was studied in typical sections. The studies were carried out in 2D. Spoilers on roof edges could improve wind conditions, and adaptation to wind impact could save energy and material for other uses.

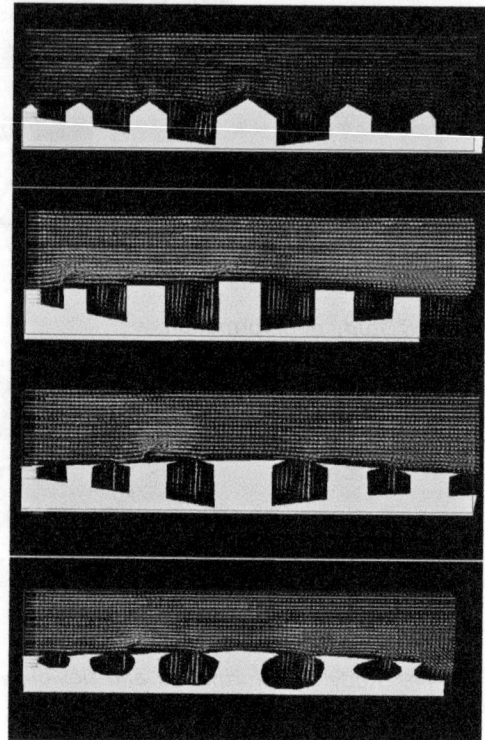

Fig.455 Wind simulation on deformed building shapes, by Lukas Mahlknecht, 2005.

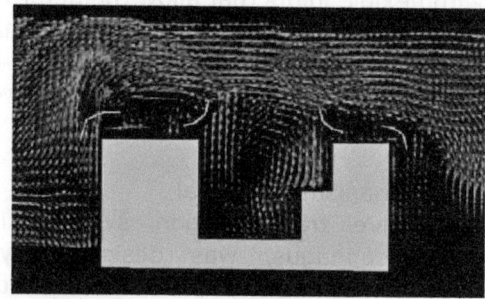

Fig.456 Simulation with spoilers on the roofs edges.

Fig.457 Mimicry building in a natural environment, Gjorgji Mojsov 2005.

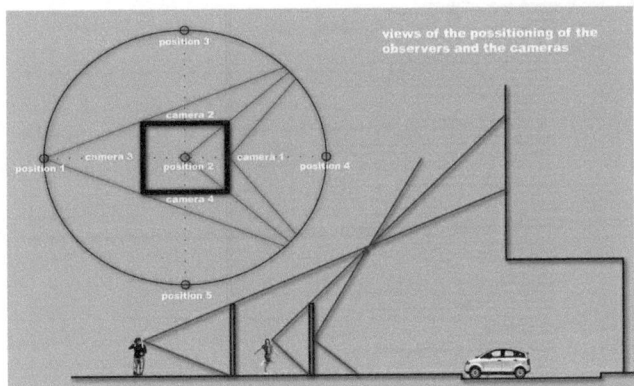

Fig.458 Concept for active mimicry.

Fig.459 Scenario of mimicry building in an urban environment.

Gjorgji Mojsov suggested two concepts for mimicry in architecture in 2005, inspired by the ability of phasmids to blend with their surrounding. The first one is a passive way of reflecting the surroundings, suggested for buildings in a natural environment. The second one is an active method of projecting camera views onto the buildings' surfaces, and this is suggested for an urban environment, to blend buildings into the surroundings. Mimicry is used as an extreme form of visual adaptation.

- Spread

Paul Kweton investigated the spreading of the slime mould myxomycetes in 2003 and on this basis designed a membrane structure. A particular species of myxomycetes spreads in net-like structures. The membrane skin contains the spores, scattering them when bursting due to a drying process. The network structure is translated into a growing membrane structure filled with smaller pneumatic spheres, usable on land or in water. Spatial propagation principles are translated, and adaptation to topography exploited.

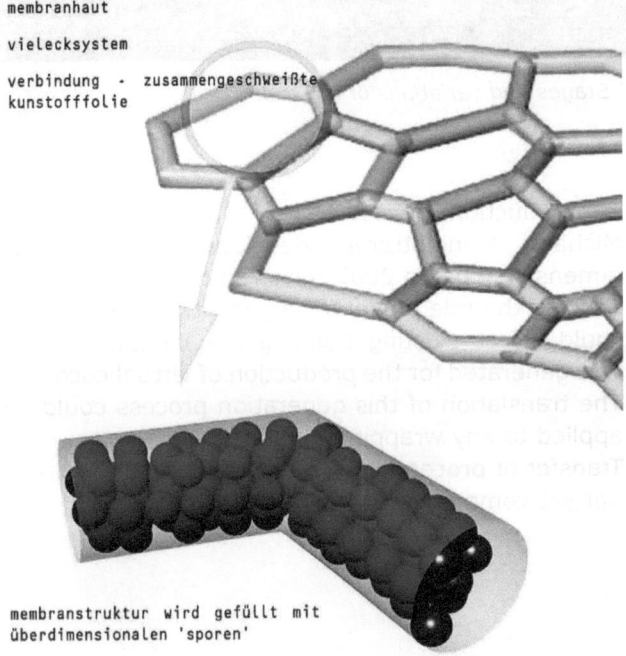

Fig.460 Scheme of the structure by Paul Kweton, 2003.

Case studies | Biomimetic Design Proposals

Fig.461 Parasitic architecture, three dimensional net, Michael Manigatterer 2007.

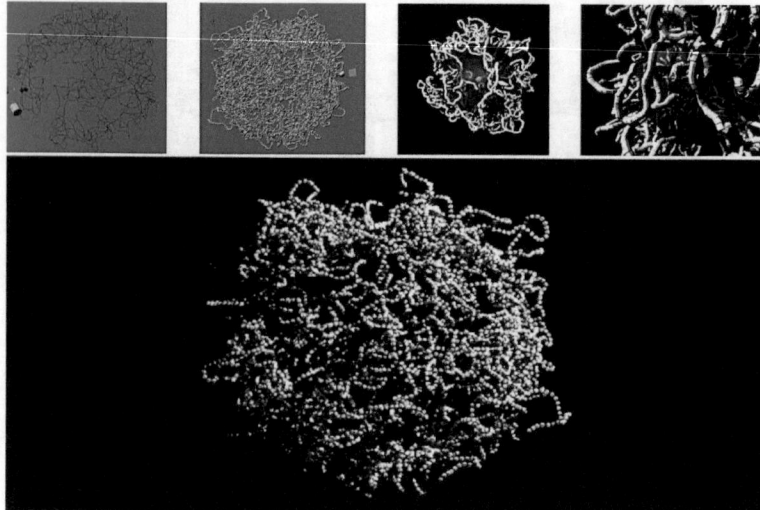

Fig.462 Stages and variations of the code.

- Production process

Michael Manigatterer developed the three-dimensional net in 2007. Wasp spider cocoons were used as the role model for a parasitic cocoon which could affect existing buildings. A computer code was generated for the production of virtual cocoons. The translation of this generation process could be applied to any wrapping application.

Transfer of processes from nature to architecture is not yet common.

XML SCRIPT DES GENERATIONSPROZESSES :

Fig.463 The XML code.

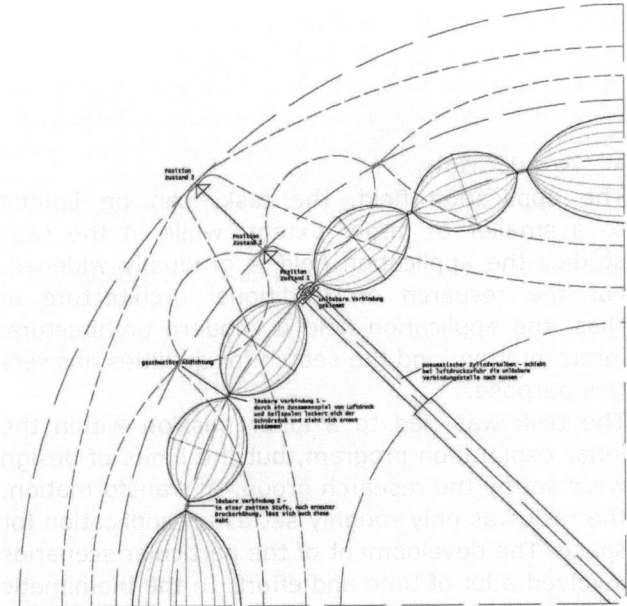

Fig.464 Detailed secion of the Pneumatic Crystal, Solveig Kieser and Julia Oberndorfinger, 2006.

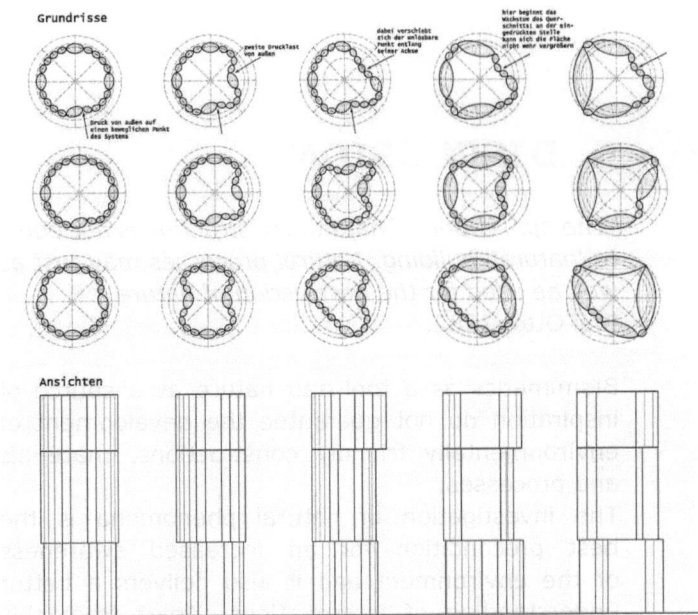

Fig.465 Extreme positions of the Pneumatic Crystal.

Solveig Kieser and Julia Oberndorfinger designed the Pneumatic Crystal in 2006. Crystal growth was investigated and transferred to a pneumatic structure which changes its form according to pressure and external influences. The deformation is analogous to imperfections of crystals in nature, but in contrast to nature, in this project it is dynamic.

- Concluding remarks to the biomimetic design proposals

Although freedom of choice is given, the application of the natural model to a design appears to pose great difficulties. Possibly, individual mental flexibility is more important in the process than the suitability of the model chosen.

Obvious reasons for failure of the transformation are:

- Superficial research because of disinterest, mistaken ideas etc.
- Information from life sciences unavailable or inaccessible
- Unimaginative approach
- Unscaleable phenomenon

All the successful concepts would be worth pursuing in individual research projects.

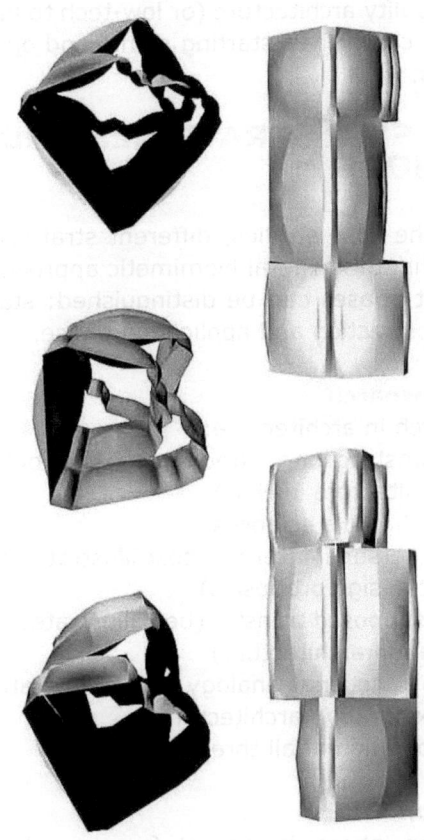

Fig.466 Visualisation of the project.

Case studies | Biomimetic Design Proposals

6 DISCUSSION

"The application of natural construction is not equal to 'natural' building. Natural processes may just as well be used for the destruction of nature."[455]
Frei Otto, 1982.

Biomimetics as a tool and nature as a source of inspiration do not guarantee the development of environmentally friendly constructions, materials and processes.
The investigation of natural phenomena is the best precondition for an increased awareness of the environment and it also delivers a better understanding of interrelations. Apart from this understanding, a firm predelection for ecological design must exist independently. Likewise, a high-tech and low energy approach does not automatically lead to high-quality architecture (or low-tech to bad quality), but it can deliver starting points and open up possibilities.

6.1 TRANSFER STRATEGIES AND METHODS

As shown in the case studies, different strategies were used within the general biomimetic approach. Three different phases can be distinguished: start of research, abstraction and application phase.

- **Start of research**

Start of research in architecture:
- "Natural construction", traditional typologies (traditional architecture in Nias)
Start of research in life sciences:
- According to personal interest (best illustrated by the biomimetic design proposals)
- According to supposed transfer (best illustrated by the transformation architecture)
- According to functional analogy (best illustrated by the lunar exploration architecture)
- Using a combination of all three

- **Abstraction**

The abstraction phase is crucial for the whole transformation process. If the abstraction is unclear it is most likely because the research field is too large, and so no particular knowledge is gained, or the research is carried out at insufficient depth, or no information can be found about the phenomenon. If the abstraction is unclear, application is very difficult.

- **Application**

The application field, the task, can be limited to a smaller or larger extent, while in the case studies the application field is gradually widened. For the research on traditional architecture in Nias, the application field is modern architectural interpretation, and the search for qualities answers this purpose.
The task was tied to a lunar mission within the lunar exploration program, but the limits of design were set by the research group. In transformation, the task was only roughly set as an application for space. The development of the particular scenarios involved a lot of time and effort. In the biomimetic design proposals, an application field is sought according to the abstract principle that had been previously formulated.
The search for the application field is a task in its own right, and it has to be carried out independently of the research on the biological (or otherwise natural) role model.

- **Future strategies**

By using evolutionary adaptation as a role model, architecture could become the object of research and development in the following ways:
- Investigating the definition of architectural criteria to be improved
- Search for projects showing these criteria
- Making slight changes, looking for projects showing the changed criteria, or working with simulation
- Measuring success: use, liking, market price, impact, and biodiversity
- Initiating or simulating the implementation of the more successful solution

In this way a gradual improvement could be reached but creative design of innovation must not be neglected.

Biomimetic design in architecture still lacks an underlying design theory, and the development of transfer strategies and methods is an important issue for the future of the approach.

---

455 Otto, F. et al.: Natürliche Konstruktionen, 1985, p.104

6.2 SUGGESTIONS

The following suggestions for the future result from the work done so far:

- More research on the interaction between architecture and biology would lead to the interaction being better understood

- More specific research in the Life Sciences would increase transfer possibilities

- More interdisciplinary projects would increase the crossover knowledge base

- The structures of decision concerning the built environment should become better adapted to the level of the hierarchy of scale. This means more regional freedom for building design, and stronger centralised or even globalised regulation on the level of environmental impact, land use and urban planning, by using local and regional information.

- Better awareness of natural and built environments is necessary, together with a discussion of environmental quality and the culture of building.

- Better models of the flow of resources and energy would create a better basis for decision-making.

- The creation of a frame for experimental design distanced from existing orders and rules would enhance and facilitate efforts towards innovation.

- Design in nature and in architecture is confronted by limited resources of energy and matter. Pragmatic innovation is characteristic of designs with limited resources, but in technology the limits are exceeded constantly at the expense of both other regions of the world and natural environments. This will be the case until the limitations of energy and resources are felt painfully by the users of modern technology, and it is at this point that creative solutions will finally have to be found.

The superposition of life sciences vocabulary onto the built environment delivers new insights, questions and solutions. As the collection of case studies shows, it enhances innovation in architecture and highlights future fields of design. This biomimetic approach, purposeful research and application, delivers a new method of visionary architectural design.

7 APPENDIX

7.1 LITERATURE

Abel, Chris:
Virtual evolution - a memetic critique of genetic algorithms in design
2006
http://epress.lib.uts.edu.au/ocs/index.php/AASA/2007/paper/viewFile/13/4
[11/2007]

Aldersey-Williams, Hugh:
Zoomorphic, New Animal Architecture
Laurence King Publishing Ltd., London, 2003
ISBN 1-85669-340-6

ARCH+ Verlag GmbH Nikolaus Kuhnert, Sabine Kraft, Günther Uhlig (Ed.):
ARCH+ Zeitschrift für Architektur und Städtebau Nr.157 , Sobeks Sensor oder Wittgensteins Griff
ARCH+ Verlag GmbH, Aachen, 9 2001
ISSN 0587-3452

ARCH+ Verlag GmbH Nikolaus Kuhnert, Sabine Kraft, Günther Uhlig (Ed.):
ARCH+ Zeitschrift für Architektur und Städtebau 142, Architektur natürlich
ARCH+ Verlag GmbH, Aachen, 7 1998
ISSN 0587-3452

Ayre, Mark; European Space Agency (ESA):
Biomimicry - a review, Work Package report
2004

Bar-Cohen, Yoseph:
Biomimetics, Biologically Inspired Technologies
Taylor&Francis Group, Boca Raton, 2006
ISBN 0-8493-3163-3

Bateson, Gregory:
Mind and Nature, A Necessary Unity
Hampton Press, Inc., New Jersey, Cresskill, 2002
ISBN 1-57273-434-5

Bauer, Peter; Eschenbacher, Thomas; Werkraum Wien Ingenieure:
Statischer Bericht, Projekt Nias
Werkraum Wien, Wien, 08/2007

Beukers, Adriaan; van Hinte, Ed; 010 Publishers (Ed.):
Lightness, The inevitable renaissance of minimum energy structures
010 Publishers, Rotterdam, 1998
ISBN 90-6450-334-6

Blaser, Werner:
Renzo Piano: Centre Kanak, Kulturzentrum der Kanak
Birkhäuser Publishers for Architecture, Basel, 2001
ISBN 3-7643-6540-4

Blossfeldt, Karl; Mattenklot, Gert; Kilias, Harald:
Karl Blossfeldt, Urformen der Kunst. Wundergarten der Natur. Das fotografische Werk in einem Band.
Schirmer/Mosel, München, 1994
ISBN 3-88814-718-2

Boeckl, Matthias (Ed.):
Visionäre und Vertriebene: Österreichische Spuren in der modernen amerikanischen Architektur, anlässlich der Ausstellung in der Kunsthalle Wien 24.2. bis 16.4.1995
Ernst & Sohn Verlag für Architektur und technische Wissenschaften GmbH, Berlin, 1995
ISBN 3-433-02445-6

Brebbia, C.A. (Ed.): Design and Nature III, Comparing Design in Nature with Science and Engineering, Wessex Institute of Technology, UK, 2006
ISBN 1-84564-166-3

Brebbia, C.A. (Ed.): Design and Nature IV, Comparing Design in Nature with Science and Engineering, Wessex Institute of Technology, UK, 2008
ISBN 978-1-84564-120-7

Brenner-Felsach, Joachim; Mittersakschmöller, Reinhold (Ed.):
Eine Reise nach Nias, Die Indonesienexpedition 1887
Böhlau Verlag Ges.m.b.H. und Co. KG, Wien, Köln, Weimar, 1998
ISBN 3-205-98959-7

Briggs et.al.; Science:
Deformation and Slip Along the Sunda Megathrust in the Great 2005 Nias-Simeulue Earthquake vol. 311, p. 1897-1901
31. march 2006

Briggs, John:
Chaos, Neue Expeditionen in fraktale Welten
Carl Hanser Verlag, München Wien, 1993
ISBN 3-446-17462-1

Brockhaus und dtv Verlag:
dtv - Lexikon, in 20 Bänden
F.A. Brockhaus GmbH und Deutscher Taschenbuch Verlag GmbH und Co.KG, Mannheim, München, 1997
ISBN 3-423-05998-2

Buchanan, Peter:
Renzo Piano Building Workshop, Sämtliche Bauten Band 1
Verlag Gerd Hatje, Stuttgart, 1994
ISBN 3-7757-0438-8

Bürgin, Toni; Herger, Peter; Künzler, Walter; Vallen, Denis; Naturmuseen Luzern, St. Gallen und Solothurn (Ed.):
HiTechNatur, Drei Museen - Drei Ausstellungen , Begleitbroschüre zu den Sonderausstellungen Wachsen und Bauen, Alles in Bewegung, Von Sinnen
Naturmuseen Luzern, St. Gallen und Solothurn, Luzern, St. Gallen und Solothurn, November 2000
ISBN 1018-2462

Campbell, Neil A.; dt. von Markl, Jürgen:
Biologie
Spektrum Akademie Verlag, Heidelberg, Berlin, Oxford, 2. korr. Nachdruck 2000
ISBN 3-8274-0032-5

Centre for Disaster Mitigation in Bandung; Earthquake Engineering Research Institute; Earthquake Spectra:
Survey of Geotechnical Engineering Aspects of the December 2004 Great Sumatra Earthquake and Indian Ocean Tsunami and the March 2005 Nias–Simeulue Earthquake Volume 22, No. S3, pages S495–S509
June 2006

Cheers, Gordon (Ed.):
Botanica, Das ABC der Pflanzen
Könemann Verlagsgesellschaft mbH, Köln, 3.Auflage 1999
ISBN 3-8290-0868-6

Clements-Croome, Derek, (Ed.):
Intelligent buildings, design, management and operation
Thomas Telford Ltd, London, 2004
ISBN 0-7277-3266-8

Col.legi d´Arquitectes de Catalunya (Ed.):
Quaderns d´arquitectura i urbanisme 225, Las escalas de la sostenibilidad
Editorial Formentera, S.A., Barcelona, 2000
ISSN 1133-8857

Daniels, Klaus:
Low-Tech Light-Tech High-Tech, Bauen in der Informationsgesellschaft
Birkhäuser Verlag, Basel, Boston, Berlin, 1998
ISBN 3-7643-5809-2

Daniels, Klaus:
Technologie des ökologischen Bauens, Grundlagen und Maßnahmen, Beispiele und Ideen
Birkhäuser, Basel, Boston, Berlin, 1999
ISBN 3-7643-6131-X

De Boer, D.W.N.; Meedelingen van het encyclopaedisch bureau betreffende de buitengewesten (Ed.):
Het Niassche Huis
Kolff & Co., Batavia, 1920

Detail Zeitschrift für Architektur+Baudetail:
Detail Zeitschrift für Architektur+Baudetail 8, Mobiles Bauen
Institut für internationale Architektur-Dokumentation GmbH&Co., München, 1998

Domenig, Gaudenz:
Zeitschrift für Ethnologie 117, S. 143-188, Typologie als Methode diachronischer Bauforschung
Konstruktionswandel im Hausbau auf Nias (Indonesien)
Dietrich Reimer Verlag, 1993

Edmaier, Bernhard:
GeoArt: Kunstwerk Erde
BLV Verlagsges.m.b.H., München, Wien, Zürich, 2.Auflage 1999
ISBN 3-405-15327-1

Edwards, Brian; International House, London (Ed.):
AD Architectural Design 71, Green Architecture
John Wiley&Sons Limited, Bognor Regis, West Sussex, 7 2001
ISBN 0-471-49193-4

Edwards, Brian:
Rough Guide to Sustainability
RIBA Enterprises, London, 2005
ISBN-13 978 7 85946 1747

Euler, Manfred:
Biologie in unserer Zeit 30.Jahrgang, Selbstorganisation, Strukturbildung und Wahrnehmung
Wiley-VCH Verlag GmbH, Weinheim, 2000/Nr.1

Faber, Colin:
Candela und seine Schalen
Georg D. Callwey, München, 1965

Feldman, Jerome A. et al.; Gronert, Walther et al. (Ed.) :
Nias, tribal treasures: cosmic reflections in stone, wood and gold
Volkenkundig Museum Nusantara, Delft, 1990
ISBN 90-71423-05-0

Feldman, Jerome Allen:
RES 7&8, Dutch galleons and South Nias palaces
The Peabody Museum of Archaeology and Ethnology, Publications Department, Harvard University, 1984
http://www.hup.harvard.edu/catalog/RES078.html

Feldman, Jerome Allen:
The architecture of Nias, Indonesia, , with special reference to Bawomataluo village, Columbia University Ph.D.
Xerox University Microfilms, Ann Arbor, Michigan, 1977

Feuerstein, Günther:
Biomorphic Architecture, Menschen- und Tiergestalten in der Architektur
Edition Axel Menges, Stuttgart London, 2002
ISBN 3-930698-87-0

Feuerstein, Günther:
Visionäre Architektur, Wien 1958/1988
Ernst&Sohn, Berlin , 1988
ISBN 3-433-02044-2

Francé, Raoul:
Die Pflanze als Erfinder.
Stuttgart, 1920

Frazer, John:
An Evolutionary Architecture
Architectural Association London, London, 1995
ISBN 1-879890-47-7

Fuller, Richard Buckminster; Krausse, Joachim (Ed.):
Bedienungsanleitung für das Raumschiff Erde, und andere Schriften
Rowohlt TB Verlag, Reinbek bei Hamburg, 1973
ISBN 3-499-25013-6

Fuller, Richard Buckminster; Krausse, Joachim; Lichtenstein, Claude (Ed.):
Your private sky Richard Buckminster Fuller, Design als Kunst einer Wissenschaft
Verlag Lars Müller und Museum für Gestaltung, Zürich, 1999
ISBN 3-907044-93-2

Future Systems: Kaplicky, Jan; Levete, Amanda; Field, Marcus:
Future Systems
Springer-Verlag, Wien, 1999
ISBN 3-211-83314-5

Gánti, Tibor:
The principles of life
Oxford University Press, Oxford, 2003
ISBN 0 19 850726 7

Gao, Yang; Ellery Alex; Sweeting Martin, N.; Vincent Julian; Journal of Spacecraft and Rockets:
Bioinspired Drill for Planetary Sampling: Literature Survey, Conceptual Design, and Feasibility Study Vol. 44, No. 3
May–June 2007
DOI 10.2514/1.23025

Goldberg, David, E.:
Genetic Algorithms in Search, Optimization, and Machine Learning
Addison-Wesley, 1989
ISBN 0-201-15767-5

Gordon, J.E.:
Structures, or why Things don't Fall down
Da Capo Press, 1981
ISBN-13: 978-0306801518

Gordon, J.E.(übers. Bewersdorff, Axel):
Strukturen unter Stress, Mechanische Belastbarkeit in Natur und Technik.
Spektrum der Wissenschaften VerlagsgesmbH&Co, Heidelberg, 1989
ISBN 3-922508-94-4

Gruber, Petra; Bannasch, Rudolf; Boblan, Ivo (Ed.):
Fortschritt Berichte VDI, First International Industrial Conference Bionik 2004, Hannover Reihe 15, P. 13 - 21, "transformation architecture"
VDI Verlag GmbH, Düsseldorf, 2004
ISBN 3-18-324915-4

Gruber, Petra; Herbig, Ulrike; Trans Urban (Ed.):
Volume 2 S. 70 - 87, Settlements and Housing on Nias island, Adaptation and Development
Verlag des Instituts für vergleichende Architekturforschung IVA-ICRA, Wien, 2006
ISBN 3-900265-07-0

Gruber, Petra; Herbig, Ulrike (Hg.):
"Traditional Architecture and Art on Nias, Indonesia";
Verlag des Instituts für vergleichende Architekturforschung IVA
Wien, 2009
ISBN 9-783900-265120

Gruber, Petra; Imhof, Barbara; Häuplik, Sandra; Özdemir, Kürsad; Waclaviceka, Rene; Perino; Maria Antoinetta :
Acta Astronautica Volume 61, Issues 1-6, P. 484-495, Deployable structures for a human lunar base, 2007
doi:10.1016/j.actaastro.2007.01.055

Gruber, Petra; Imhof, Barbara; Häuplik, Sandra; Özdemir, Kürsad; Waclaviceka, Rene; Perino; Maria Antoinetta :
Lunar Exploration Architecture, Deployable Structures for a Lunar Base
Study for Alcatel Alenia Spazio, Vienna, 2006

Gruber, Petra; Imhof, Barbara:
Acta Astronautica Volume 60, Issues 4-7, P. 561-570, Transformation: Structure/space studies in bionics and space design, 2006
doi:10.1016/j.actaastro.2006.09.032

Gruber, Petra:
The Signs of Life in Architecture, SEB Glasgow
in: Bioinspiration and Biomimetics, 3, 2008,
doi: 10.1088/1748-3182/3/2/023001

Haase, Walter; Köhnlein, Jochen; Institut für Leichte Flächentragwerke:
Smart Materials, Recherche und Dokumentation
Stuttgart, 1998

Haeckel, Ernst:
Kunstformen der Natur, Neudruck der Farbtafeln aus der Erstausgabe „Kunstformen der Natur" Leipzig und Wien, Bibliograpisches Institut, 1904.
Prestel-Verlag, München New York, 1998
ISBN 3-7913-1978-7

Hämmerle, Johannes Maria; Anthropos Institut e.V. (Ed.):
Nias - eine eigene Welt, Sagen, Mythen, Überlieferungen
Academia Verlag, Sankt Augustin, 1999
ISBN 3-89665-147-1

Hawkes, Dean; McDonald, Jane; Steemers, Koen:
The selective environment, An approach to environmentally responsive architecture
Spon Press, London, New York, 2002
ISBN 0-419-23530-2

Helmcke, Johann-Gerhard; Bach, Klaus; Institut für leichte Flächentragwerke, Frei Otto (Ed.):
IL33 Radiolaria IL33, Schalen in Natur und Technik II
Karl Krämer Verlag, Stuttgart, 9 1990
ISBN 3-7828-2033-9

Helmcke, Johann-Gerhard; Institut für leichte Flächentragwerke, Frei Otto (Ed.):
IL 28 Diatomeen IL 28, Schalen in Natur und Technik I
Karl Krämer Verlag, Stuttgart, 1985

Herbig, Ulrike et al.:
Documentary Film "Architecture and Culture of Northern Sumatra and Nias",
IVA-ICRA, Wien, 2005
ISBN 3-900265-04-6

Hillier, Bill; Harvard Architecture Review:
Specifically architectural theory: a partial account of the ascent from building as cultural transmission to architecture as theoretical concretion.
9. pp.8-27
1993
ISSN 01943650
http://eprints.ucl.ac.uk/archive/00001027/ [12/2007]

Hoffmeyer, Jesper; European Journal for Semiotic Studies:
Biosemiotics: Towards a New Synthesis in Biology Vol. 9 No. 2 pp.355-376
1997

Holland, H. John:
Emergence, From Chaos to Order
Oxford University Press, Oxford, 2002
ISBN 0-19-286211-1

Horden, Richard; Blaser, Werner:
light tech, towards a light architecture
Ausblick auf eine leichte Architektur
Birkhäuser Verlag, Basel, Boston, Berlin, 1995
ISBN 3-7643-5220-5

International House, London:
AD Architectural Design Vol 74 No 3, Emergence Morphogenetic Design Strategies
John Wiley&Sons Ltd., Chichester, West Sussex, 2004
ISSN 0003-8504

International House, London:
AD Architectural Design Vol 76 No 180, Techniques and Technologies in Morphogenetic Design
John Wiley&Sons Ltd., Chichester, West Sussex, 2005
ISSN 0003-8504

International House; Toy, Maggie:
AD Architectural Design Vol 70 No 3, Contemporary Processes in Architecture
John Wiley & Sons Ltd., London, June 2000
ISBN 0-471-49440-2

IOM International Organisation for Migration:
Post Disaster Damage Assessment on Nias and Simeulue Island
June 20, 2005

Isler, Heinz; Ramm, Ekkehard; Schunk, Eberhard; Universität Stuttgart (Ed.):
Heinz Isler, Schalen, Katalog zur Ausstellung
Karl Krämer Verlag, Stuttgart, 1986
ISBN 3-7828-1492-4

Jackson, Martin; Readman, Jo; The Eden Project (Ed.):
Eden: the first book
St.Ives, Roche Ltd, Cornwall, 8 2000

Khoshnevis, Behrokh; Journal of Automation in Construction – Special Issue :
The best of ISARC 2002 Vol 13, Issue 1, , pp 5-19., Automated Construction by Contour Crafting - related Robotics and Information Technologies
January 2004
http://www.contourcrafting.org/ [11/2007]

Kobayashi, H.; Kresling B.; Vincent, J.F.V. ; Proceedings of the Royal Society:
The geometry of unfolding tree leaves 265; 147.154
1998

Kollmann, F.:
Technologie des Holzes und der Holzwerkstoffe Erster Band
Springer Verlag; , Berlin , Zweite Auflage
(1951)

Kresling, Biruta: ; IASS Symposium:
The growing turbinate shell, Model for a deployable technical shell
Montpellier, 2004

Kuhlmann, Dörte:
Lebendige Architektur, Metamorphosen des Organizismus
Universitätsvelag Weimar, Weimar, 1998
ISBN 3-86068-094-3

Kull, Kalevi; Semiotica:
Biosemiotics in the twentieth century: a view from biology vol. 127(1/4), pp. 385-414
1999

Le Corbusier:
Der Modulor, Darstellung eines in Architektur und Technik allgemein anwendbaren harmonischen Maszes im menschlichen Maszstab
Deutsche Verlags-Anstalt, Stuttgart, 4. Auflage 1980
ISBN 3-421-02521-5

Lebedew, Juri.S:
Architektur und Bionik
Verlag MIR, VEB Verlag für Bauwesen , Moskau, Berlin, 1.Auflage 1983

Lehner, Erich:
Wege der architektonischen Evolution, Die Polygenese von Pyramiden und Stufenbauten, Aspekte zu einer vergleichenden Architekturgeschichte
Phoibos Verlag, Wien, 1998

Lynn, Greg:
animate form
Princeton Architectural Press, New York, 1999
ISBN 1-56898-083-3

Mahdavi, A.; M.-L. Chiu (Ed.):
A. CAAD Talks 5. Insight of Smart Environments p.45-66, Sentient buildings, from concept to implementation
Archidata Co., Ltd., 2005
ISBN 957-0454-66-0

Margulies, Lynn; Sagan, Doris:
Leben, vom Ursprung zur Vielfalt
Spektrum Akademischer Verlag GmbH, Heidelberg, Berlin, Oxford, 1997
ISBN 3-8274-0100-3

Margulis, Lynn; Sagan, Dorion:
What is Life?
University of California Press, Berkeley and Los Angeles, California, 2000
ISBN 0-520-22021-8

Mattheck, Claus:
Design in der Natur, Der Baum als Lehrmeister
Rombach GmbH Druck- und Verlagshaus, Freiburg, 1997
ISBN 3-7930-9150-3

Maturana, Humberto R.; Varela, Francisco J.:
Autopoiesis and Cognition, The Realization of the Living
Springer, 1991
ISBN-10: 9027710161

Millennium Ecosystem Assessment Series:
ECOSYSTEMS AND HUMAN WELL-BEING: CURRENT STATE AND TRENDS, Findings of the Condition and Trends Working Group
Island Press, 2005
ISBN: 1-55963-228-3

Mollerup, Per:
Collapsibles, Ein Album platzsparender Objekte
Stiebner Verlag GmbH, München, 2001
ISBN 3-8307-1268-5

Müller, Christpoh:
Architekturtradition, Traditioneller Wohn- und Siedlungsbau in der Provinz Nordsumatra
dissertation, Technische Universität Wien, 2005

Munteanu, Andrea; Solé, Richard V.; Santa Fe Institute:
Chaos in chemoton dynamics
05.02.2007
http://www.santafe.edu/research/publications/wpabstract/200505017
[10/2007]

Museen der Stadt Wien (Ed.):
Friedrich Kiesler 1890-1965 Inside the Endless House, anlässlich der Ausstellung im Historischen Museum der Stadt Wien, 12.12.1997-1.3.1998
Holzhausen, Wien, 1997
ISBN 3-205-98838-8

Nachtigall, Werner; Blüchel, Kurt G. (Ed.):
Das grosse Buch der Bionik., Neue Technologien nach dem Vorbild der Natur.
Deutsche Verlags-Anstalt, Stuttgart/München, 2000
ISBN 3-421-05379-0

Nachtigall, Werner:
Bau-Bionik, Natur, Analogien, Technik
Springer, Berlin, New York, Heidelberg, 2005
ISBN 3-540-44336-3

Nachtigall, Werner:
Biomechanik, Grundlagen Beispiele Übungen
Friedr. Vieweg&Sohn Verlagsges. mbH, Braunschweig, Wiesbaden, 2000
ISBN 3-528-03926-4

Nachtigall, Werner:
Bionik, Grundlagen und Beispiele für Ingenieure und Naturwissenschaftler
Springer-Verlag, Berlin Heidelberg, 1998
ISBN 3-540-63403-7

Nachtigall, Werner:
Bionik, Grundlagen und Beispiele für Ingenieure und Naturwissenschaftler
Springer, Berlin, New York, Heidelberg, 2. Auflage 2002
ISBN 3-540-43660-X

Nachtigall, Werner:
Vorbild Natur, Bionik-Design für funktionelles Gestalten.
Springer-Verlag, Berlin Heidelberg, 1997
ISBN 3-540-63245-x

Novak, Marcos:
transarchitecture, learning from aliens
http://www.mat.ucsb.edu/~marcos/Centrifuge_Site/MainFrameSet.html
[03/2007]

Novak, Marcos:
Transmitting Architecture: The Transphysical City
http://www.ctheory.net/articles.aspx?id=76 [11/2007]

Oliver, Paul:
Dwellings, The Vaernacular House World Wide
Phaidon Press Inc, London, 2003
ISBN 0 7148 4202 8

Oosterhuis, Kas et al.:
Sculpture City
010 publishers, Rotterdam, 1994
ISBN 90 6450 229 3
http://www.oosterhuis.nl/quickstart/index.php?id=269 [11/07]

Oosterhuis, Kas:
Hyperbodies, Towards an E-motive architecture
Birkhäuser, Basel, Boston, Berlin, 2003
ISBN 3-7643- 6736-9

Oosterhuis, Kas:
The Genes of Architecture, Are buildings organisms?
1995
ISBN 9064502293
http://www.oosterhuis.nl/quickstart/index.php?id=269 [03/2007]

Otto, Frei (Rainer Barthel, Berthold Burkhardt, Heide Drüsedau, Rainer Graefe, Jürgen Hennicke, Georgina Krause-Valdovinos, Eda Schaur, Ilse Schmall, Cornelius Thywissen):
Natürliche Konstruktionen, Formen und Strukturen in Natur und Technik und Prozesse ihrer Entstehunng.
Deutsche Verlags-Anstalt GmbH, Stuttgart, 2. Auflage 1985
ISBN 3-421-02591-6

Otto, Frei et al.; Institut für leichte Flächentragwerke, Frei Otto (Ed.):
IL18 Seifenblasen IL18, Forming Bubbles
Karl Krämer Verlag, Stuttgart, 1987

Otto, Frei et al.; Institut für leichte Flächentragwerke, Frei Otto (Ed.):
IL6 SFB64 IL6, Biologie und Bauen 3
Karl Krämer Verlag, Stuttgart, 1973
ISBN 3-7828-2006-1

Otto, Frei; Burkhardt, Berthold; Institut für leichte Flächentragwerke:
Il 14 Anpassungsfähig bauen IL 14
Institut für leichte Flächentragwerke, Stuttgart, 1975
ISBN 3-1828-2014-2

Pasquire, Christine; Soar, Rupert; Gibb, Alistair; Proceedings IGLC-14:
Beyond pre-fabrication, The potential of next generation technologies to make a step change in construction manufacturing
Santiago, Chile, July 2006

Patzelt, Otto:
Wachsen und Bauen, Konstruktionen in Natur und Technik.
VEB Verlag für Bauwesen, Berlin, 2.Auflage 1974

Pawley, Martin
(übers. von Haase, Axel und Stein, Ria):
Future Systems, Die Architektur von Jan Kaplicky und Amanda Levete
Birkhäuser Verlag, Basel, Berlin, Boston, 1993
ISBN 3-7643-2853-3

Pawley, Martin
Norman Foster, A Global Architecture
Thames&Hudson Ltd, London 1999
ISBN 0-500-28123-8

Pevsner, Nikolaus; Honour, Hugh; Fleming, John:
Lexikon der Weltarchitektur
Prestel Verlag, München, 3.Auflage
1992
ISBN 3-7913-2095-5

Polano, Sergio:
Santiago Calatrava, Gesamtwerk
Deutsche Verlags-Anstalt, Stuttgart, 1997
ISBN 3-421-03138-X

Portoghesi, Paolo:
Nature and Architecture
Skira editore, Milan, 2000
ISBN 88-8118-658-6

Powers, Alan:
natur und design, Inspirationen für Architektur, Mode und angewandte Kunst.
Verlag Paul Haupt, Berlin, Stuttgart, Wien, 1999
ISBN 3-258-06183-1

Rechenberg, Ingo:
Evolutionsstrategie '94.
Frommann Verlag, Stuttgart, 1994
ISBN 3-7728-1642-8

Rice, Peter::
Peter Rice, An Architect Imagines
ellipsis, London, 1994
ISBN 1-899858-11-3

Riley, Terence; The Museum od Modern Art (Ed.):
LightConstruction, anlässlich der Ausstellung im Museum of Modern Art in New York, 21.9.1995-2.1.1996
The Museum of Modern Art, New York, 1995
ISBN 0-07070-129-0
ISBN 0-8109-6154-7

Robbin, Tony:
Engineering a New Architecture
Yale University Press, New Haven and London, 1996
ISBN 0-300-06116-1

Rudofsky, Bernard:
Architektur ohne Architekten, Eine Einführung in die anonyme Architektur
Residenz Verlag, Salzburg und Wien, 2.Auflage 1993
ISBN 3-7017-0565-8

Sato, Atsushi (Ed.):
Eisaku Ushida + Kathryn Findlay, Parallel Landscapes
Gallery MA Books 02, Tokyo, 1996
ISBN 4-88706-135-8

Sato, Atsushi (Ed.):
The Architectural Map of Tokyo
Gallery MA, Tokyo, 1994
ISBN 4-88706-098-X

Schaur Eda:
Il 41, Intelligent Bauen
Institut für leichte Flächentragwerke, Stuttgart, 1995
ISBN 3-7828-2041-X

Schefold, Reimar; Nas, Peter J.M.; Domenig, Gaudenz et al.; Schefold, Reimar; Nas, Peter J.M. (Ed.):
Indonesian houses Vol 1, Tradition and transformation in vernacular architecture
KTILV Press, Leiden, 2003
ISBN 90-6718-205-2

Schittich, Christian (Ed.):
im Detail: Gebäudehüllen, Konzepte, Schichten, Material
Birkhäuser Verlag, Basel, Boston, Berlin, 2001
ISBN 3-7643-6464-5

Schock, Hans-Joachim:
Segel, Folien und Membranen, Innovative Konstruktionen in der textilen Architektur
Birkhäuser Verlag, Basel, Berlin, Boston, 1997
ISBN 3-7643-5449-6

Schröder, E.E.W.Gs.:
Nias., Ethnographische, geographische en historische aanteekeningen en studien.
Teil I: Tekst, Teil II: Platen en kaarten
Brill E.J., Leiden, 1917

Schrödiger, Erwin:
What is Life?
Canto, Cambridge University Press, Cambridge, 1967
ISBN 0 521 42708 8

Schweitzer, Frank; Zimmermann, Jörg:; Fischer, M.M.; Fröhlich, J. (Eds.):
Knowledge, Complexity and Innovation Systems, Advances in Spatial Sciences pp. 275-296, Communication and Self-Organisation in Complex Systems: A Basic Approach
Springer, Berlin, 2001

Scientific American (Ed.):
Spektrum der Wissenschaft Digest 3, Moderne Werkstoffe
Spektrum der Wissenschaft Verlagsgesellschaft mbH, Heidelberg, März 1996
ISSN 0945-9537

Scientific American (Ed.):
Spektrum der Wissenschaft Spezial 4, Schlüsseltechnologien im 21. Jahrhundert
Spektrum der Wissenschaft Verlagsgesellschaft mbH, Heidelberg, Oktober 1995
ISSN 0943-7096

Scully, Vincent Jr.:
Grosse Meister der Architektur Bd.4, frank lloyd wright
Otto Maier Verlag, Ravensburg, 1960

Senosiain, Javier:
Bio-Architecture
Elsevier, Oxford, 2003
ISBN 0-7506-5604-2

Siemens Forum, Landesmuseum für Technik und Arbeit Mannheim (Ed.):
Bionik, Zukunfts-Technik lernt von der Natur
Landesmuseum für Technik und Arbeit in Mannheim, Mannheim, 1999
ISBN 3-9804930-5-9

Smithson Robert ; Schmidt, Eva; Vöckler, Kai (Ed.):
Robert Smithson: Gesammelte Schriften, anlässlich der Ausstellung in der Kunsthalle Wien 23.11.2000 bis 25.2.2001
Verlag der Buchhandlung Walther König, Köln, 2000
ISBN 3-88375-388-2

Speck, Thomas et al; Proceedings of the Fifth Plant Biomechanics Conference:
Self-repairing membranes for pneumatic structures: transferring nature's solutions into technical applications Volume 1
Stockholm, Sweden, 2006

Stattmann, Nicola; Rat für Formgebung (Ed.):
Handbuch Material-Technologie
avedition GmbH Verlag für Architektur und Design, Ludwigsburg, 2000
ISBN 3-929638-44-4

Stevens, Peter S. (übers. Weichert, Uta und Martin):
Formen in der Natur (Patterns in Nature).
R. Oldenbourg, München, Wien, 2. Auflage 1988
ISBN 3-486-20926-4

Stewart, Ian:
Die Zahlen der Natur, Mathematik als Fenster zur Welt
Spektrum Verlag, 2001
ISBN 978-3827411235

Strasburger, E.; Neubearbeitung von Sitte, Peter; Ziegler, Hubert; Ehrendorfer, Friedrich; Bresinsky, Andreas:
Strasburger, Lehrbuch der Botanik
Gustav Fischer Verlag, Stuttgart, Jena, Lübeck, Ulm, 34. Auflage 1998
ISBN 3-437-25500-2

Tarassow, Lev V.:
Symmetrie, Symmetrie!, Strukturprinzipien in Natur und Technik.
Spektrum, Akademischer Verlag, Heidelberg, 1982
ISBN 3-86025-300-x

Teichmann, Klaus und Wilke, Joachim (Ed.):
Prozess und Form „Natürlicher Konstruktionen", Der Sonderforschungsbereich 230
Verlag für Architektur und technische Wissenschaften GmbH, Berlin, 1996
ISBN 3-433-02883-4

Teuffel, Patrick; Institut für Leichtbau Entwerfen und Konstruieren Universität Stuttgart:
Entwerfen adaptiver Strukturen, Lastpfadmanagement zur Optimierung tragender Leichtbaustrukturen Dissertation
Universität Stuttgart, Stuttgart, 2004

Thompson , D´Arcy W.:
On Growth and Form, The Complete Revised Edition.
Dover Publications Inc., New York, 1992
ISBN 0-486-67135-6

Trask, Richard; Bond, Ian; European Space Agency:
Enabling Self-Healing Capabilities – A Small Step to Bio-Mimetic Materials,
ESA CONTRACT No 18131/04/NL/PA
University of Bristol, UK, 2005

Tsui, Eugene:
Evolutionary Architecture, Nature as a basis for design
John Wiley & Sons, Inc, 1999
ISBN 0-471-11726-9

Turner, Will, R.; Nakamura, Toshihiko; Dinetti, Marco:
BioScience Vol. 54, No. 6, Global Urbanization and the Separation of Humans from Nature
June 2004

Turnovsky, Jan; Bauwelt Fundamente:
Die Poetik eines Mauervorsprungs Bd. 77
Vieweg, 1987
ISBN 3-528-08777-3

Venturi, Robert; Scott Brown, Denise; Izenour Steven:
Learning from Las Vegas
MIT Press, Cambridge, 1977

Viaro, Alain M.; UNESCO (Ed.):
Urbanisme et architecture traditionels du sud de Iîle de Nias 21
Etablissements humains et environnement socio-culturel UNESCO, Geneve, 1980

Viaro, Alain M.:
Nias: Habitat et Megalithisme
Archipel 27/1984, Paris, 1984

Vincent, Julian F.V.; Bogatyreva, Olga A.; Bogatyrev, Nikolaj R.; Bowyer, Adrian; Pahl, Anja-Karina; Journal of the Royal Society:
Biomimetics - its practice and theory Volume 3, Number 9 pp.471-482
August 22, 2006
DOI 10.1098/rsif.2006.0127

Vogel, Steven (übers. Filk, Thomas):
Von Grashalmen und Hochhäusern, Mechanische Schöpfungen in Natur und Technik.
Wiley-VCH Verlag GmbH, Weinheim, BRD, 2000
ISBN 3-527-40303-5

Vogel, Steven:
Cats' Paws and Catapults, Mechanical Worlds of Nature and People
W.W.Norton & Company, New York, London, 1998
ISBN 0-939-31990-3

Wassermann, Jack (Ed.):
Leonardo da Vinci
DuMont Buchverlag, Köln, 1977
ISBN 3-7701-0960-0

Waterson, Roxana:
The Living House, An Anthropology of Architecture in South-East Asia
Thames and Hudson, 1997
ISBN-13: 9780500280300

Williams, Chris .J .K.:
The analytic and numerical definition of the geometry of the British Museum Great Court roof
http://staff.bath.ac.uk/abscjkw/BritishMuseum/ ChrisDeakin2001.pdf
[11/2007]

Williams, Chris .J .K.:
The definition of curved geometry for widespan enclosures
http://staff.bath.ac.uk/abscjkw/OrganicForms/WideSpan.pdf [11/2007]

Wiryomortono, Achmad Bagoes Poerwono:
Cosmological - and spatiotemporal meanings of a traditional dwelling in South Nias, Indonesia, Dissertation, Technische Hochschule Aachen
1989

Yeang, Ken:
The Green Skyscraper, The Basis for Designing Sustainable Intensive Buildings
Prestel Verlag, Munich, London, New York, 1999
ISBN 3-7913-1993-0

Zöhrer, Günter:
Moaro - Die kanakische Architektur Neukaledoniens und ihre Stellung in Ozeanien , Diplomarbeit an der TU Wien, CD
IVA Institut für vergleichende Architekturforschung, Wien, 2005
ISBN-13: 978-3900265038

7.2 FIGURES AND PHOTOGRAPHY

If not stated otherwise in the list, the copyrights of the images used in this book are held by the authors of the cited publication.

Fig.1. Diagram by the author, 2007
Fig.2. Photograph by the author, 2002
Fig.3. Nachtigall, W.: Vorbild Natur, 1997, fig.1, translation by the author
Fig.4. Nachtigall, W.: Bionik, Grundlagen und Beispiele für Ingenieure und Naturwissenschaftler, 1998, chapter 1, fig.6
Fig.5. Wassermann, J. (Ed.): Leonardo da Vinci, 1977, fig.19
Fig.6. Nachtigall, W.: Bionik, Grundlagen und Beispiele für Ingenieure und Naturwissenschaftler, 2002, chapter 1, fig.1-3
Fig.7. Wassermann, J. (Ed.): Leonardo da Vinci, 1977, fig.18
Fig.8. Wassermann, J. (Ed.): Leonardo da Vinci, 1977, fig.56
Fig.9. Nachtigall, W.: Bionik, Grundlagen und Beispiele für Ingenieure und Naturwissenschaftler, chapter 3, fig.3
Fig.10. Nachtigall, W.: Bionik, Grundlagen und Beispiele für Ingenieure und Naturwissenschaftler, chapter 3, fig.4
Fig.11. Bürgin,T. et al.: HiTechNatur, 2000, p.91, fig.3 from Invention of the Aeroplane, 1799-1909, Charles Harvard Gibbs-Smith, 1966
Fig.12. Bürgin,T. et al.: HiTechNatur, 2000, p.81
Fig.13. Nachtigall, W.: Bionik, Grundlagen und Beispiele für Ingenieure und Naturwissenschaftler, 2002, p.234, fig.10-64
Fig.14. Photograph by the author, 2004
Fig.15. Nachtigall, W.: Bionik, Grundlagen und Beispiele für Ingenieure und Naturwissenschaftler, 1998, chapter 6, fig.1
Fig.16. Nachtigall, W.: Bionik, Grundlagen und Beispiele für Ingenieure und Naturwissenschaftler, 2002, p.41, fig.4-5
Fig.17. Nachtigall, W. et al.: Das große Buch der Bionik, 2000, p.173
Fig.18. Nachtigall, W. et al.: Das große Buch der Bionik, 2000, p.184
Fig.19. Haeckel, E.: Kunstformen der Natur, 1998, p.27
Fig.20. Haeckel, E.: Kunstformen der Natur, 1998, p.19
Fig.21. Haeckel, E.: Kunstformen der Natur, 1998, p.9
Fig.22. Blossfeldt, K. et al.: Karl Blossfeldt, 1994, table 13
Fig.23. Blossfeldt, K. et al.: Karl Blossfeldt, 1994, table 2
Fig.24. Blossfeldt, K. et al.: Karl Blossfeldt, 1994, table 53
Fig.25. Blossfeldt, K. et al.: Karl Blossfeldt, 1994, table 107 abc
Fig.26. Blossfeldt, K. et al.: Karl Blossfeldt, 1994, table 11
Fig.27. Nachtigall, W.: Bionik Grundlagen und Beispiele für Ingenieure und Naturwissenschaftler, 2002, fig.5-5, p.41
Fig.28. Helmuth Goldammer, Cell Imaging, Ultrastructure Research, University of Vienna
Fig.29. Nachtigall, W.: Bionik, Grundlagen und Beispiele für Ingenieure und Naturwissenschaftler, 2002, fig.5-12, p.46
Fig.30. Ingo Rechenberg, http://www.bionik.tu-berlin.de/institut/s2anima.html [10/2007]
Fig.31. Ingo Rechenberg, http://www.bionik.tu-berlin.de/institut/s2foshow/show.php?show=Polyox [10/2010]
Fig.32. C. Mattheck : Design in nature – learning from trees Springer Verlag, Heidelberg, 1998
Fig.33. Siemens Forum, et al. (Ed.): Bionik, Zukunfts-Technik lernt von der Natur, 1999, fig.32, © TECHNOSEUM Mannheim
Fig.34. Siemens Forum, et al. (Ed.): Bionik, Zukunfts-Technik lernt von der Natur, 1999, fig.123, © W. Barthlott, Universität Bonn
Fig.35. © W. Barthlott, Universität Bonn
Fig.36. © W. Barthlott, Universität Bonn
Fig.37. Nachtigall, W.: Bionik, Grundlagen und Beispiele für Ingenieure und Naturwissenschaftler, 1998, chapter 7, fig.15, © W. Barthlott, Universität Bonn
Fig.38. © Alfred Wisser
Fig.39. Nachtigall, W. et al.: Das große Buch der Bionik, 2000, p.92, fig.1
Fig.40. Siemens Forum, et al. (Ed.): Bionik, Zukunfts-Technik lernt von der Natur, 1999, fig.78, © Karlsruher Institut für Technologie (KIT)
Fig.41. © Knut Braun
Fig.42. Photograph by the author, 2010
Fig.43. Nachtigall, W. et al.: Das große Buch der Bionik, 2000, p.243
Fig.44. Otto, F. (Ed.): IL6, SFB64 Biologie und Bauen 3, 1973, p.15, fig.6
Fig.45. C. Mattheck: Design in nature – learning from trees Springer Verlag, Heidelberg, 1998
Fig.46. © Knut Braun
Fig.47. Siemens Forum exhibition, press photographs, fig.10, © Karlsruher Institut für Technologie (KIT)
Fig.48. Siemens Forum exhibition, press photographs, fig.10, © Karlsruher Institut für Technologie (KIT)
Fig.49. Bürgin,T. et al.: HiTechNatur, 2000, p.70, fig.2, Urs Hochuli © Naturmuseum St.Gallen
Fig.50. © Ingo Rechenberg
Fig.51. Photograph by the author 2010
Fig.52. Drawing by the author 2010
Fig.53. © Ingo Rechenberg
Fig.54. © Ingo Rechenberg
Fig.55. © Ingo Rechenberg
Fig.56. Nachtigall, W.: Vorbild Natur, 1997, fig.56, © Ingo Rechenberg
Fig.57. Photograph by the author, 2007
Fig.58. FlickR Bob Gutowski [08/2010]
Fig.59. Otto, F. et al. (Ed.): IL18 Seifenblasen, 1987, p.56, fig.1, © ILEK Institut für Leichtbau Entwerfen und Konstruieren, Universität Stuttgart
Fig.60. Otto, F. et al. (Ed.): IL18 Seifenblasen, 1987, p.56, fig.2, © ILEK Institut für Leichtbau Entwerfen und Konstruieren, Universität Stuttgart
Fig.61. Nachtigall, W.: Bionik, Grundlagen und Beispiele für Ingenieure und Naturwissenschaftler, 1998 chapter 6 fig.16
Fig.62. Nachtigall, W. et al.: Das große Buch der Bionik, 2000, p.220, fig.2
Fig.63. Nachtigall, W. et al.: Das große Buch der Bionik, 2000, p.220, fig.3
Fig.64. Nachtigall, W. et al.: Das große Buch der Bionik, 2000, p.221, fig.5
Fig.65. Nachtigall, W. et al.: Das große Buch der Bionik, 2000, p.222, fig.4, translated by the author
Fig.66. Nachtigall, W.: Bionik, Grundlagen und Beispiele für Ingenieure und Naturwissenschaftler, 1998, chapter 1, fig.10
Fig.67. Helmuth Goldammer, Cell Imaging, Ultrastructure Research, University of Vienna
Fig.68. Fuller, R.B. et al.: Bedienungsanleitung für das Raumschiff Erde, 1973, fig.61

Fig.69. Krausse, J. et al. (Eds.): Your private sky
Richard Buckminster Fuller, 1999, p.423
Fig.70. Erich Lehner, private archive
Fig.71. Photograph by the author, 2006
Fig.72. © COOP HIMMELB(L)AU / Erwin Reichmann
Fig.73. Feuerstein, G.: Visionäre Architektur
1988, p.91, fig.105, © Haus-Rucker-Co
Fig.74. Feuerstein, G.: Visionäre Architektur
1988, p.92, fig.104, © Haus-Rucker-Co
Fig.75. The estate of R. Buckminster Fuller, © R.B. Fuller
Fig.76. Photograph by the author, 2010
Fig.77. Photograph by the author, 2009
Fig.78. Photograph by Erich Lehner, 2003
Fig.79. © 2010 Austrian Frederick and Lillian Kiesler Private Foundation
Fig.80. Aldersey-Williams H.: Zoomorphic,
2003, p.67, © Michael Sorkin
Fig.81. © Michael Sorkin
Fig.82. http://gallica.bnf.fr/ark:/12148/btv1b7703120m [10/2007]
Fig.83. http://www.bl.uk/learning/images/bodies/leonardo2.jpg [08/2007]
Fig.84. Le Corbusier: Der Modulor, 1980, © FLC/VBK, Wien 2010
Fig.85. Photograph by Robert Gruber, 1995
Fig.86. Sato, A. (Ed.): Eisaku Ushida + Kathryn
Findlay, 1996, p.49, © Katsuhisa Kida
Fig.87. Sato, A. (Ed.): Eisaku Ushida + Kathryn Findlay,
1996, p.63, © Eisaku Ushida + Kathryn Findlay
Fig.88. © Katsuhisa Kida
Fig.89. Sato, A. (Ed.): Eisaku Ushida + Kathryn Findlay,
1996, p.38, © Eisaku Ushida + Kathryn Findlay
Fig.90. Photograph by the author, 2005
Fig.91. Photograph by the author, 2005
Fig.92. Photograph by the author, 2005
Fig.93. Photograph by the author, 2005
Fig.94. Rice, P.: Peter Rice, 1994, p.61, © Arup
Fig.95. Robbin, T.: Engineering a New Architecture,
1996, p.28, © Kenneth Snelson
Fig.96. wikimedia commons [08/2010]
Fig.97. FlickR Evan Chakroff [08/2010]
Fig.98. Polano, S.: Santiago Calatrava,
1997, p.89, © Santiago Calatrava
Fig.99. Polano, S.: Santiago Calatrava,
1997, p.88, © Santiago Calatrava
Fig.100. www.eerosaarinen.net [08/2010] © Eero Saarinen
Collection. Manuscripts and Archives, Yale University Library
Fig.101. Otto, F. et al.: Natürliche Konstruktionen,
1985, p.51, fig.4, © ILEK Institut für Leichtbau
Entwerfen und Konstruieren, Universität Stuttgart
Fig.102. Photograph by the author, 2008
Fig.103. FlickR Gisela Schmoll [07/2010]
Fig.104. FlickR Jorge Ayala/ Ay_A Studio [07/2010]
Fig.105. FlickR Jorge Ayala/ Ay_A Studio [07/2010]
Fig.106. FlickR Jean-Pierre Dalbéra [08/2010]
Fig.107. FlickR Jean-Pierre Dalbéra [08/2010]
Fig.108. Photograph by the author 2007

Fig.109. Otto, F. et al.: Natürliche Konstruktionen,
1985, p.64, fig.1, © ILEK Institut für Leichtbau
Entwerfen und Konstruieren, Universität Stuttgart
Fig.110. Photograph by the author, 2009
Fig.111. Beukers, A. et al. (Ed.): Lightness, 1998,
p.22, © Adriaan Beukers, Ed van Hinte
Fig.112. Beukers, A. et al. (Ed.): Lightness,
1998, p.14, © Mike Ashby
Fig.113. © Waltraut Hoheneder, 2010
Fig.114. Photograph by the author, 2010
Fig.115. © Mirtsch GmbH
Fig.116. © Mirtsch GmbH
Fig.117. Photograph by the author, 2010
Fig.118. Photograph by the author, 2010
Fig.119. Photograph by the author, 2010
Fig.121. Photograph by the author, 2007.
Fig.122. Mayser GmbH & Co. KG archive
Fig.123. © Fraunhofer ISE
Fig.124. Bürgin,T. et al.: HiTechNatur, 2000, p.36 fig.a
Fig.125. Bürgin,T. et al.: HiTechNatur, 2000, p.37
Fig.126. Photograph by the author, 2010
Fig.127. Photograph by the author, 2010
Fig.128. Yeang, K.: The Green Skyscraper, 1999, fig.37
Fig.129. Photograph by the author, 2008
Fig.130. Photograph by the author, 2008
Fig.131. Photograph by the author, 2008
Fig.132. Otto, F. et al.: Natürliche Konstruktionen,
1985, p.42, fig.1, © ILEK Institut für Leichtbau
Entwerfen und Konstruieren, Universität Stuttgart
Fig.133. Otto, F. et al.: Natürliche Konstruktionen,
1985, p.41, fig.2, © ILEK Institut für Leichtbau
Entwerfen und Konstruieren, Universität Stuttgart
Fig.134. Hawkes, D. et al.: The selective environment,
2002, fig.1.3, amended by the author
Fig.135. Yeang, K.: The Green Skyscraper, 1999, fig.17
Fig.136. Yeang, K.: The Green Skyscraper, 1999, fig.1
Fig.137. Edwards, B. (Ed.): AD Architectural Design
Nr.71, Green Architecture, 2001, p.108, fig.1
Fig.138. Daniels, K.: Low-Tech Light-Tech High-Tech, 2000, p.105
Fig.139. Yeang, K.: The Green Skyscraper, 1999, fig.52
Fig.140. Yeang, K.: The Green Skyscraper, 1999, fig.54
Fig.141. Photograph by the author, 2002
Fig.142. © Future Systems
Fig.143. Future Systems: Future Systems, 1999, p.116 fig.2
Fig.144. Future Systems: Future Systems,
1999, p.118, © Future Systems
Fig.145. Future Systems: Future Systems,
1999, p.114 fig.2, © Future Systems
Fig.146. Col.legi d´Arquitectes de Catalunya
(Ed.): Quaderns d´arquitectura i urbanisme
No.225, 2000, p.91 fig.2, © Ken Yeang
Fig.147. Edwards, B. (Ed.): AD Architectural Design 71,
Green Architecture, 2001, p.15, © Ken Yeang

Fig.149. International house (Ed.): AD Architectural Design Vol 70 No 3, Contemporary Processes in Architecture, 2000, p.88 © Reiser Umemoto

Fig.150. Future Systems: Future Systems, 1999, p.105

Fig.151. Photograph by the author, 2006

Fig.152. Smithson, R. et al. (Ed.): Robert Smithson: Gesammelte Schriften, 2000, p.176, Robert Smithson : Spiral Jetty, 1970/© VBK, Wien 2010

Fig.153. © MVRDV Hans Werlemann

Fig.154. Photograph by the author, 2005

Fig.155. Photograph by the author, 2005

Fig.156. Jackson, M. et al. (Ed.): Eden: the first book, 2000, © Eden Project

Fig.157. © Grimshaw

Fig.158. Photograph by the author, 2005

Fig.159. © Eden Project

Fig.160. © Grimshaw

Fig.161. © Eden Project

Fig.162. Image provided and copyright by Dennis Kunkel Microscopy, Inc.

Fig.163. Kull, U. et al in Teichmann, K. et al. (Ed): Prozess und Form natürlicher Konstruktionen 1996, p.34

Fig.164. Helmuth Goldammer, Cell Imaging, Ultrastructure Research, University of Vienna

Fig.165. FlickR Arenamontanus [07/2010], © Anders Sandberg

Fig.166. FlickR Arenamontanus [07/2010], © Anders Sandberg

Fig.167. Briggs, J.: Chaos, 1993, pp.22

Fig.168. Teichmann, K. et al. (Ed.): Prozess und Form Natürlicher Konstruktionen, 1996, table 63, fig.1, © ILEK Institut für Leichtbau Entwerfen und Konstruieren, Universität Stuttgart

Fig.169. Teichmann, K. et al. (Ed.): Prozess und Form Natürlicher Konstruktionen, 1996, table 63, fig.2, © ILEK Institut für Leichtbau Entwerfen und Konstruieren, Universität Stuttgart

Fig.170. Photograph by the author, 2010

Fig.171. Photographs by the author, 2010

Fig.172. Teichmann, K. et al. (Ed.): Prozess und Form Natürlicher Konstruktionen, 1996 table 65 fig.2, © ILEK Institut für Leichtbau Entwerfen und Konstruieren, Universität Stuttgart

Fig.175. Photograph by the author, 2007

Fig.173. © Knut Braun

Fig.174. Vogel, S.: Cats' Paws and Catapults, 1998, p.23, fig.2.1

Fig.176. Photograph by the author, 2009

Fig.177. Teichmann, K. et al. (Ed.): Prozess und Form Natürlicher Konstruktionen, 1996, table 65 fig.3, © ILEK Institut für Leichtbau Entwerfen und Konstruieren, Universität Stuttgart

Fig.178. Teichmann, K. et al. (Ed.): Prozess und Form Natürlicher Konstruktionen, 1996, table 65 fig.1, © ILEK Institut für Leichtbau Entwerfen und Konstruieren, Universität Stuttgart

Fig.179. Photograph by the author, 2010

Fig.180. Photograph by the author, 2010

Fig.181. Vogel, S.: On Cats' Paws and Catapults, 1998, pp.289

Fig.182. Image provided and copyright by Dennis Kunkel Microscopy, Inc.

Fig.183. Photograph by the author, 2008

Fig.184. Helmuth Goldammer, Cell Imaging, Ultrastructure Research, University of Vienna

Fig.185. Helmuth Goldammer, Cell Imaging, Ultrastructure Research, University of Vienna

Fig.186. © Andrew R Kirby, Imaging Partnership, Institute of food research

Fig.187. Image provided and copyright by Dennis Kunkel Microscopy, Inc.

Fig.188. Helmuth Goldammer, Cell Imaging, Ultrastructure Research, University of Vienna

Fig.189. Photograph by the author, 2010

Fig.190. Helmuth Goldammer, Cell Imaging, Ultrastructure Research, University of Vienna

Fig.191. Helmuth Goldammer, Cell Imaging, Ultrastructure Research, University of Vienna

Fig.192. Photograph by the author, 2010

Fig.193. Photograph by the author, 2009

Fig.194. Photograph by the author, 2008

Fig.195. Photograph by the author, 2008

Fig.196. Photograph by the author, 2008

Fig.197. Photographs by the author, 2007

Fig.198. Photograph by the author, 2010

Fig.199. Photograph by the author, 2009

Fig.200. Photograph by the author, 2010

Fig.201. Photograph by the author, 2006

Fig.202. Diagram by the author, 2007

Fig.203. Photograph by the author, 2006

Fig.204. Flickr Inti [07/2010], CC2.0 © Inti

Fig.205. http://www.wernersobek.com [12/2007], © Werner Sobek, Stuttgart

Fig.206. © Werner Sobek, Stuttgart

Fig.207. FlickR Jessica Spenger [07/2010]

Fig.208. http://www.coop-himmelblau.at [10/2007], © Hélène Binet

Fig.210. Photograph by the author, 2008

Fig.211. Photograph by the author, 2008

Fig.209. Photograph by the author, 2008

Fig.212. S.R. White, N.R. Sottos, J. Moore, P. Geubelle, M. Kessler, E. Brown, S. Suresh, and S. Viswanathan, Autonomic healing of polymer composites, Nature, 409, 2001, pp. 794-797, © Scott White

Fig.213. Trask, R. et al.: Enabling Self-Healing Capabilities – A Small Step to Bio-Mimetic Materials, 2005, fig.1.3, © Richard Trask

Fig.214. Speck, T. et al: Self-repairing membranes for pneumatic structures: transferring nature's solutions into technical applications, 2006, fig.2, p.117

Fig.215. © Rpbw, Renzo Piano Building Workshop - Via Rubens, 29 - 16158 Genova Tel. 010-61711 Fax 010-6171350

Fig.216. Future Systems: Future Systems, 1999, p.128

Fig.217. International House: AD 76 No 180, Techniques and Technologies in Morphogenetic Design, 2005, p.37, © FOA Foreign office architects

Fig.218. Photograph by Gruber/Herbig, 08/2005

Fig.219. Horden, R. et al (Ed.): light tech, 1995, p.31

Fig.220. Horden, R. et al (Ed.): light tech, 1995, p.30

Fig.221. Horden, R. et al (Ed.): light tech, 1995, p.28

Fig.222. Future Systems: Future Systems, 1999, p.128, fig.2

Fig.223. Photograph by the author, 08/2005

Fig.224. Schittich, C. (Ed.): im Detail: Gebäudehüllen, 2001, p.62 fig.1, © Fink
Fig.225. Schittich, C. (Ed.): im Detail: Gebäudehüllen, 2001, p.63 fig.1, © Fink
Fig.226. Vogel, S.: Cats' Paws and Catapults, 1998, p.40, fig.3.1.
Fig.227. FlickR Joi Ito [07/2010]
Fig.228. FlickR Jerome Rigaud [07/2010]
Fig.229. Photograph by the author, 2004
Fig.230. Photograph by the author, 2007
Fig.231. Photograph by the author, 2006
Fig.232. The Hundertwasser Non Profit Foundation, www.hundertwasser.at © Hundertwasser Archive, Vienna, Photograph by Gerhard Deutsch
Fig.233. © Tomio Ohashi
Fig.234. © Kisho Kurokawa architect & associates
Fig.235. © Kisho Kurokawa architect & associates
Fig.236. http://www.fosterandpartners.com/Projects/0828/Default.aspx [12/2007], © Foster + Partners
Fig.237. © Nigel Young/Foster + Partners
Fig.239. Williams, C.J.K.: The analytic and numerical definition of the geometry of the British Museum Great Court roof, http://staff.bath.ac.uk/abscjkw/BritishMuseum/ChrisDeakin2001.pdf [11/2007], fig.6, © Foster + Partners
Fig.238. © Foster + Partners
Fig.240. © Behrokh Khoshnevis
Fig.241. © Behrokh Khoshnevis
Fig.242. © Behrokh Khoshnevis
Fig.243. © Behrokh Khoshnevis
Fig.244. © NAI Nederlands Architectuurinstituut
Fig.245. Detail 8/1998: Mobiles Bauen, p.1422-25, © Böthlingk, Photograph by Roos Aldershoff, Amsterdam
Fig.246. http://www.werkraumwien.at [12/2007], © Hans Kupelwieser
Fig.247. http://www.werkraumwien.at [12/2007] © Hans Kupelwieser
Fig.248. http://www.baubotanik.de/ [11/2007], © Ferdinand Ludwig
Fig.249. http://www.baubotanik.de/ [11/2007], © Ferdinand Ludwig
Fig.250. Photograph by the author, 2006
Fig.251. International House, (Ed.): AD 70, Contemporary Processes in Architecture, 2000, p.29, © Greg Lynn FORM, http://www.glform.com/embryonic/embryonic.htm
Fig.252. Photograph by the author, 2006
Fig.253. Photograph by the author, 2006
Fig.254. Rice, P.: Peter Rice, 1994, p.91, fig.1, © Berengo Gardin Gianni, Via San Michele del Carso 21, 20144 Milano Italia Tel/Fax 024692877
Fig.255. Photograph by Richard Bryant © Rpbw, Renzo Piano Building Workshop - Via Rubens, 29 - 16158 Genova Tel. 010-61711 Fax 010-6171350
Fig.256. Rice, P.: Peter Rice, 1994, p.92, fig.2, © Rpbw, Renzo Piano Building Workshop - Via Rubens, 29 - 16158 Genova Tel. 010-61711 Fax 010-6171350

Fig.257. Rice, P.: Peter Rice, 1994, p.92, fig.2, © Rpbw, Renzo Piano Building Workshop - Via Rubens, 29 - 16158 Genova Tel. 010-61711 Fax 010-6171350
Fig.258. Rice, P.: Peter Rice, 1994, p.86, Photograph by Ben Smusz © Rpbw, Renzo Piano Building Workshop - Via Rubens, 29 - 16158 Genova Tel. 010-61711 Fax 010-6171350
Fig.259. Buchanan, P.: Renzo Piano Building Workshop, Sämtliche Bauten Band 1, 1994, p.151, Photograph by Hester Paul © Rpbw, Renzo Piano Building Workshop - Via Rubens, 29 - 16158 Genova Tel. 010-61711 Fax 010-6171350
Fig.260. © Werner Sobek, Stuttgart
Fig.261. © Werner Sobek, Stuttgart
Fig.262. © Werner Sobek, Stuttgart
Fig.263. Waterson, R.: The Living House, 1997, title, © Walter Imber, walterimberphoto@bluemail.ch
Fig.264. Photograph by the author, 2007
Fig.265. Photograph by the author, 2004
Fig.266. Häuplik, S.: Diploma Thesis, 2004
Fig.267. http://www.rolfdisch.de [11/2007], © Rolf Disch SolarArchitektur
Fig.268. Vincent, J.F.V. presentation in Hannover, 2004, © Arup Associates
Fig.269. Vincent, J.F.V. presentation in Hannover, 2004, © Arup Associates
Fig.270. Vincent, J.F.V. presentation in Hannover, 2004, © Arup Associates
Fig.271. © Rpbw, Renzo Piano Building Workshop - Via Rubens, 29 - 16158 Genova Tel. 010-61711 Fax 010-6171353
Fig.272. © Rpbw, Renzo Piano Building Workshop - Via Rubens, 29 - 16158 Genova Tel. 010-61711 Fax 010-6171353
Fig.273. Photograph by the author, 2005
Fig.274. © Rpbw, Renzo Piano Building Workshop - Via Rubens, 29 - 16158 Genova Tel. 010-61711 Fax 010-6171353
Fig.275. Murauer, M.: Bionik - natürliche Konstruktionen, 2005
Fig.276. Pfaffstaller, S.: Bionik - natürliche Konstruktionen, 2004
Fig.277. FlickR Colb [08/2010], © Colb
Fig.278. Teuffel, P.: Entwerfen adaptiver Strukturen, diss., 2004, fig.6.1, p.71, © ILEK Institut für Leichtbau Entwerfen und Konstruieren, Universität Stuttgart
Fig.279. © FESTO
Fig.280. © FESTO
Fig.281. http://www.nitinol.com/4applications.htm [12/2007], © NDC, 30 Glenbarr Ct, Pinehurst, NC 28374
Fig.282. Photograph by the author, 2007
Fig.283. http://www.sial.rmit.edu.au/Projects/Aegis_Hyposurface.php [11/2007]
Fig.284. http://www.sial.rmit.edu.au/Projects/Aegis_Hyposurface.php [11/2007]
Fig.285. © ONL [Oosterhuis_Lénárd], Rotterdam
Fig.286. © ONL [Oosterhuis_Lénárd], Rotterdam
Fig.287. © Kisho Kurokawa architect & associates
Fig.288. © Kisho Kurokawa architect & associates
Fig.289. http://grin.hq.nasa.gov/images/medium/gpn-2000-001046.jpg [12/2007]
Fig.290. Photograph by the author, 2004

Fig.291. © Foreign Office Architects, London
Fig.292. FlickR Colb [08/2010], © Colb
Fig.293. © ONL [Oosterhuis_Lénárd], Rotterdam
Fig.294. http://www.mat.ucsb.edu/~marcos/Centrifuge_Site/MainFrameSet.html [11/2007], © Marcos Novak
Fig.295. Diagram by the author, 2007
Fig.296. http://www.thecityreview.com/arcnowt.jpg [11/2007], © Courtesy of Diller Scofidio + Renfro
Fig.297. Diagram by the author, 2007
Fig.298. Photograph by Gruber/Herbig, 08/2005
Fig.299. Photograph by the author, 08/2005
Fig.300. Viaro, A.M.: Nias: Habitat et Megalithisme, 1984, © Alain Viaro
Fig.301. Photograph by Gruber/Herbig, 08/2005
Fig.302. Photograph by Gruber/Herbig, 08/2005
Fig.303. Photograph by Gruber/Herbig, 08/2005
Fig.304. Schröder, E.E.W.Gs.: Nias, 1917, fig.208
Fig.305. Schröder, E.E.W.Gs.: Nias, 1917, fig.16
Fig.306. Photograph by Gruber/Herbig, 08/2005
Fig.307. Photograph by Gruber/Herbig, 08/2005
Fig.308. Photograph by Gruber/Herbig, 08/2005
Fig.309. Photograph by Gruber/Herbig, 08/2005
Fig.310. Photograph by Gruber/Herbig, 08/2005
Fig.311. Diagram by the author, 2006
Fig.312. Original map in Museum Pusaka Nias, changed by the author
Fig.313. Müller, C.: Architekturtradition, diss., 2005, p.262, fig.284
Fig.314. Photograph by Erich Lehner, 2003
Fig.315. Photograph by Gruber/Herbig, 08/2005
Fig.316. Photograph by Gruber/Herbig, 08/2005
Fig.317. Photograph by Gruber/Herbig, 08/2005
Fig.318. Photograph by Gruber/Herbig, 08/2005
Fig.319. Müller, C.: Architekturtradition, diss., 2005, p.247, fig.251
Fig.320. Photograph by Gruber/Herbig, 08/2005
Fig.321. Photograph by Gruber/Herbig, 08/2005
Fig.322. Photograph by Gruber/Herbig, 08/2005
Fig.323. Photograph by Gruber/Herbig, 08/2005
Fig.324. Photograph by Gruber/Herbig, 08/2005
Fig.325. Photograph by Gruber/Herbig, 08/2005
Fig.326. Photograph by Gruber/Herbig, 08/2005
Fig.327. Photograph by Gruber/Herbig, 08/2005
Fig.329. Photograph by Gruber/Herbig, 08/2005
Fig.330. Rendering by Marc Lorenz, 2006.
Fig.331. Rendering by Marc Lorenz, 2006.
Fig.332. Rendering by Marc Lorenz, 2006.
Fig.333. Rendering by Marc Lorenz, 2006.
Fig.334. Rendering by Marc Lorenz, 2006.
Fig.335. Rendering by Marc Lorenz, 2006.
Fig.336. Drawing by the author and Thomas Schmidle, 2006
Fig.337. Drawing by the author and Thomas Schmidle, 2006
Fig.338. Photograph by Gruber/Herbig, 08/2005
Fig.339. Drawing by the author and Thomas Schmidle, 2006
Fig.340. Drawing by the author and Thomas Schmidle, 2006
Fig.341. Drawing by the author, 2006
Fig.342. Drawing by the author, 2006
Fig.343. Photograph by Gruber/Herbig, 08/2005
Fig.344. Drawing by the author, 2006
Fig.345. Photograph by Gruber/Herbig, 08/2005
Fig.346. Photograph by Gruber/Herbig, 08/2005
Fig.347. Drawing by the author, 2006
Fig.351. Photograph by Gruber/Herbig, 08/2005
Fig.349. Drawing by the author and Thomas Schmidle, 2006
Fig.350. Photograph by Gruber/Herbig, 08/2005
Fig.352. Drawing by the author and Thomas Schmidle, 2006
Fig.353. Drawing by the author and Thomas Schmidle, 2006
Fig.354. Drawing by the author and Thomas Schmidle, 2006
Fig.355. Photograph by the author, 2003
Fig.356. Photograph by Gruber/Herbig, 08/2005
Fig.357. Photograph by Gruber/Herbig, 08/2005
Fig.358. Photograph by Gruber/Herbig, 08/2005
Fig.359. Photograph by Gruber/Herbig, 08/2005
Fig.360. Photograph by Gruber/Herbig, 08/2005
Fig.361. Briggs et.al.: Deformation and Slip Along the Sunda Megathrust in the Great 2005 Nias-Simeulue Earthquake, 2006, fig.4, © Richard Briggs
Fig.362. USGS US Geological Survey: shake map 28th of March 2005, http://earthquake.usgs.gov/eqcenter/shakemap/global/shake/weax/#download [06/2007], © USGS
Fig.363. Photograph by Gruber/Herbig, 08/2005
Fig.364. Photograph by Gruber/Herbig, 08/2005
Fig.365. Drawing by the author, 08/2005
Fig.366. Photograph by Gruber/Herbig, 08/2005
Fig.367. Photograph by Gruber/Herbig, 08/2005
Fig.368. Photograph by Gruber/Herbig, 08/2005
Fig.369. Photograph by Gruber/Herbig, 08/2005
Fig.370. Photograph by Gruber/Herbig, 08/2005
Fig.371. Photograph by Gruber/Herbig, 08/2005
Fig.372. Photograph by Gruber/Herbig, 08/2005
Fig.373. Photograph by Gruber/Herbig, 08/2005
Fig.374. Photograph by Gruber/Herbig, 08/2005
Fig.375. Drawing by the author, 08/2005
Fig.376. Photograph by Gruber/Herbig, 08/2005
Fig.377. Photograph by Gruber/Herbig, 08/2005
Fig.378. Photograph by Gruber/Herbig, 08/2005
Fig.379. Photograph by Phillip Pongratz, 2006
Fig.380. Photograph by Phillip Pongratz, 2006
Fig.381. Bauer, P. et al.: Statischer Bericht, Projekt Nias, Werkraum Wien Ingenieure, 2007
Fig.382. Photograph by Angela Lehner Wieternik, 2003
Fig.383. Photograph made available by Jerome Feldman in 2006, source unknown.
Fig.384. Schröder, E.E.W.Gs.: Nias, 1917, fig.135
Fig.385. Photograph by Gruber/Herbig, 08/2005
Fig.386. Photograph by Gruber/Herbig, 08/2005
Fig.387. Photograph by Gruber/Herbig, 08/2005
Fig.388. Photograph by Gruber/Herbig, 08/2005
Fig.389. Photograph by Gruber/Herbig, 08/2005
Fig.390. Photograph by Gruber/Herbig, 08/2005
Fig.391. Photograph by Gruber/Herbig, 08/2005
Fig.392. Photograph by Gruber/Herbig, 08/2005
Fig.393. Photograph by Gruber/Herbig, 08/2005

Fig.394. Photograph by Gruber/Herbig, 08/2005
Fig.395. Photograph by Gruber/Herbig, 08/2005
Fig.396. Photograph by Gruber/Herbig, 08/2005
Fig.397. Photograph by Gruber/Herbig, 08/2005
Fig.398. Photograph by Gruber/Herbig, 08/2005
Fig.399. Photograph by Gruber/Herbig, 08/2005
Fig.400. Photograph by Gruber/Herbig, 08/2005
Fig.401. Photograph by Gruber/Herbig, 08/2005
Fig.402. Photograph by Gruber/Herbig, 08/2005
Fig.403. Photograph by Irmgard Derschmid, 2006
Fig.404. Photograph by Irmgard Derschmid, 2006
Fig.405. Photograph by Irmgard Derschmid, 2006
Fig.406. Drawings and rendering by Stefano Caneppele, 2005
Fig.407. Drawings and rendering by Stefano Caneppele, 2005
Fig.408. Rendering by Thomas Gamsjäger, 2004
Fig.409. Rendering by Thomas Gamsjäger, 2004
Fig.410. Rendering by Thomas Gamsjäger, 2004
Fig.411. Rendering by Thomas Gamsjäger, 2004
Fig.412. Rendering by Melanie Klähn and Thomas Frings, 2004
Fig.413. Rendering by Melanie Klähn and Thomas Frings, 2004
Fig.414. Rendering by Melanie Klähn and Thomas Frings, 2004
Fig.415. Rendering by Melanie Klähn and Thomas Frings, 2004
Fig.416. Rendering by Alessandro Perinelli, 2004
Fig.417. Rendering by Alessandro Perinelli, 2004
Fig.418. Gruber, P. et al.: Lunar Exploration Architecture, Deployable Structures for a Lunar Base, Study for Alcatel Alenia Spazio, 2006, p.90, fig.61
Fig.419. Gruber, P. et al.: Lunar Exploration Architecture, Deployable Structures for a Lunar Base, Study for Alcatel Alenia Spazio, 2006, p.48 fig.15
Fig.420. Gruber, P. et al.: Lunar Exploration Architecture, Deployable Structures for a Lunar Base, Study for Alcatel Alenia Spazio, 2006, p.48 fig.15
Fig.421. Gruber, P. et al.: Lunar Exploration Architecture, Deployable Structures for a Lunar Base, Study for Alcatel Alenia Spazio, 2006, p.48 fig.15
Fig.422. Gruber, P. et al.: Lunar Exploration Architecture, Deployable Structures for a Lunar Base, Study for Alcatel Alenia Spazio, 2006, p.50 fig.17
Fig.423. Gruber, P. et al.: Lunar Exploration Architecture, Deployable Structures for a Lunar Base, Study for Alcatel Alenia Spazio, 2006, p.47 fig.14
Fig.424. Gruber, P. et al.: Lunar Exploration Architecture, Deployable Structures for a Lunar Base, Study for Alcatel Alenia Spazio, 2006, p.47 fig.14
Fig.425. Gruber, P. et al.: Lunar Exploration Architecture, Deployable Structures for a Lunar Base, Study for Alcatel Alenia Spazio, 2006, p.51 fig.18
Fig.426. Gruber, P. et al.: Lunar Exploration Architecture, Deployable Structures for a Lunar Base, Study for Alcatel Alenia Spazio, 2006, p.54 fig.21
Fig.427. Gruber, P. et al.: Lunar Exploration Architecture, Deployable Structures for a Lunar Base, Study for Alcatel Alenia Spazio, 2006, p.57 fig.24
Fig.428. Gruber, P. et al.: Lunar Exploration Architecture, Deployable Structures for a Lunar Base, Study for Alcatel Alenia Spazio, 2006, p.59
Fig.429. Gruber, P. et al.: Lunar Exploration Architecture, Deployable Structures for a Lunar Base, Study for Alcatel Alenia Spazio, 2006, p.85, fig.56
Fig.430. Gruber, P. et al.: Lunar Exploration Architecture, Deployable Structures for a Lunar Base, Study for Alcatel Alenia Spazio, 2006, p.87, fig.54
Fig.431. Gruber, P. et al.: Lunar Exploration Architecture, Deployable Structures for a Lunar Base, Study for Alcatel Alenia Spazio, 2006, p.88, fig.53
Fig.432. Gruber, P. et al.: Lunar Exploration Architecture, Deployable Structures for a Lunar Base, Study for Alcatel Alenia Spazio, 2006, p.91, fig.62
Fig.433. Gruber, P. et al.: Lunar Exploration Architecture, Deployable Structures for a Lunar Base, Study for Alcatel Alenia Spazio, 2006, p.91, fig.63
Fig.437. Gruber, P.: "transformation architecture" in Fortschritt Berichte VDI 2004, p.16, fig.5
Fig.434. Malinov, R.: Bionik - natürliche Konstruktionen, 2005
Fig.435. Malinov, R.: Bionik - natürliche Konstruktionen, 2005
Fig.436. Malinov, R.: Bionik - natürliche Konstruktionen, 2005
Fig.438. Totland, I.: Bionik - natürliche Konstruktionen, 2005
Fig.439. Totland, I.: Bionik - natürliche Konstruktionen, 2005
Fig.440. Totland, I.: Bionik - natürliche Konstruktionen, 2005
Fig.441. Fuchs, K.: Bionik - natürliche Konstruktionen, 2005
Fig.442. Fuchs, K.: Bionik - natürliche Konstruktionen, 2005
Fig.443. Fuchs, K.: Bionik - natürliche Konstruktionen, 2005
Fig.444. Fuchs, K.: Bionik - natürliche Konstruktionen, 2005
Fig.445. Kumhera, P.: Bionik - natürliche Konstruktionen, 2004
Fig.446. Kumhera, P.: Bionik - natürliche Konstruktionen, 2004
Fig.447. Foster, J.: Bionik - natürliche Konstruktionen, 2006
Fig.448. Foster, J.: Bionik - natürliche Konstruktionen, 2006
Fig.449. Foster, J.: Bionik - natürliche Konstruktionen, 2006
Fig.450. Österreicher, C.: Bionik - natürliche Konstruktionen, 2007
Fig.451. Österreicher, C.: Bionik - natürliche Konstruktionen, 2007
Fig.452. Österreicher, C.: Bionik - natürliche Konstruktionen, 2007
Fig.453. Gimplinger, J.: Bionik - natürliche Konstruktionen, 2007
Fig.454. Gimplinger, J.: Bionik - natürliche Konstruktionen, 2007
Fig.455. Mahlknecht, L.: Bionik - natürliche Konstruktionen, 2005
Fig.456. Mahlknecht, L.: Bionik - natürliche Konstruktionen, 2005
Fig.457. Mojsov, G.: Bionik - natürliche Konstruktionen, 2005
Fig.458. Mojsov, G.: Bionik - natürliche Konstruktionen, 2005
Fig.459. Mojsov, G.: Bionik - natürliche Konstruktionen, 2005
Fig.460. Kweton, P.: Bionik - natürliche Konstruktionen, 2003
Fig.461. Manigatterer, M.: Bionik - natürliche Konstruktionen, 2007
Fig.462. Manigatterer, M.: Bionik - natürliche Konstruktionen, 2007
Fig.463. Manigatterer, M.: Bionik - natürliche Konstruktionen, 2007
Fig.464. Kieser, S.; Oberndorfinger, J.: Bionik - natürliche Konstruktionen, 2006
Fig.465. Kieser, S.; Oberndorfinger, J.: Bionik - natürliche Konstruktionen, 2006
Fig.466. Kieser, S.; Oberndorfinger, J.: Bionik - natürliche Konstruktionen, 2006

Appendix | Figures and Photography

Special thanks for fantastic collaboration,
patient support and inspiring discussion to:

Erich Lehner
George Jeronimidis
Helmut Richter
Ulrike Herbig
Barbara Imhof
Peter Bauer
Phillip Pongratz
Alfred Brandhofer
Regine Muskens
Roman Bönsch
Jo Lakeland
Robert Gruber
my students for their committed work and vision

Department for Design and Building Construction, Institute for Building Construction and Technology
and Institute for History of Architecture and Art, Building Research and Preservation,
Vienna University of Technology
Centre for Biomimetics, The University of Reading

Petra Gruber, Vienna 2010

Bei Fragen zur Produktsicherheit wenden Sie sich bitte an:
If you have any questions regarding product safety,
please contact:

Birkhäuser Verlag GmbH
Im Westfeld 8
4055 Basel, Schweiz
productsafety@degruyterbrill.com